MEMS Mechanical Sensors

For a listing of recent titles in the *Artech House Microelectromechanical Systems (MEMS) Series*, turn to the back of this book.

MEMS Mechanical Sensors

Stephen Beeby
Graham Ensell
Michael Kraft
Neil White

Artech House, Inc.
Boston • London
www.artechhouse.com

Library of Congress Cataloging-in-Publication Data
A catalog record for this book is available from the U.S. Library of Congress.

British Library Cataloguing in Publication Data
Beeby, Stephen.
    MEMS mechanical sensors.— (Artech House MEMS library)
1. Microelectricalmechanical systems—Design and construction   2. Transducers
I. Beeby, Stephen
621.3'81

        ISBN 1-58053-536-4

Cover design by Igor Valdman

International Standard Book Number: 1-58053-536-4

10 9 8 7 6 5 4 3 2 1

Contents

Preface   *ix*

CHAPTER 1
Introduction   1

1.1   Motivation for the Book   1
1.2   What Are MEMS?   2
1.3   Mechanical Transducers   3
1.4   Why Silicon?   4
1.5   For Whom Is This Book Intended?   5
      References   5

CHAPTER 2
Materials and Fabrication Techniques   7

2.1   Introduction   7
2.2   Materials   7
     2.2.1   Substrates   7
     2.2.2   Additive Materials   11
2.3   Fabrication Techniques   11
     2.3.1   Deposition   12
     2.3.2   Lithography   17
     2.3.3   Etching   21
     2.3.4   Surface Micromachining   28
     2.3.5   Wafer Bonding   29
     2.3.6   Thick-Film Screen Printing   32
     2.3.7   Electroplating   33
     2.3.8   LIGA   34
     2.3.9   Porous Silicon   35
     2.3.10   Electrochemical Etch Stop   35
     2.3.11   Focused Ion Beam Etching and Deposition   36
      References   36

CHAPTER 3
MEMS Simulation and Design Tools   39

3.1   Introduction   39
3.2   Simulation and Design Tools   40
     3.2.1   Behavioral Modeling Simulation Tools   40
     3.2.2   Finite Element Simulation Tools   43
      References   56

CHAPTER 4
Mechanical Sensor Packaging   57

4.1   Introduction   57
4.2   Standard IC Packages   58
    4.2.1   Ceramic Packages   58
    4.2.2   Plastic Packages   59
    4.2.3   Metal Packages   59
4.3   Packaging Processes   59
    4.3.1   Electrical Interconnects   60
    4.3.2   Methods of Die Attachment   63
    4.3.3   Sealing Techniques   65
4.4   MEMS Mechanical Sensor Packaging   66
    4.4.1   Protection of the Sensor from Environmental Effects   67
    4.4.2   Protecting the Environment from the Sensor   71
    4.4.3   Mechanical Isolation of Sensor Chips   71
4.5   Conclusions   80
    References   81

CHAPTER 5
Mechanical Transduction Techniques   85

5.1   Piezoresistivity   85
5.2   Piezoelectricity   89
5.3   Capacitive Techniques   92
5.4   Optical Techniques   94
    5.4.1   Intensity   94
    5.4.2   Phase   95
    5.4.3   Wavelength   96
    5.4.4   Spatial Position   96
    5.4.5   Frequency   96
    5.4.6   Polarization   97
5.5   Resonant Techniques   97
    5.5.1   Vibration Excitation and Detection Mechanisms   98
    5.5.2   Resonator Design Characteristics   99
5.6   Actuation Techniques   104
    5.6.1   Electrostatic   104
    5.6.2   Piezoelectric   107
    5.6.3   Thermal   107
    5.6.4   Magnetic   109
5.7   Smart Sensors   109
    References   112

CHAPTER 6
Pressure Sensors   113

6.1   Introduction   113
6.2   Physics of Pressure Sensing   114
    6.2.1   Pressure Sensor Specifications   117
    6.2.2   Dynamic Pressure Sensing   120

	6.2.3	Pressure Sensor Types	121
6.3	Traditional Pressure Sensors		121
	6.3.1	Manometer	121
	6.3.2	Aneroid Barometers	122
	6.3.3	Bourdon Tube	122
	6.3.4	Vacuum Sensors	123
6.4	Diaphragm-Based Pressure Sensors		123
	6.4.1	Analysis of Small Deflection Diaphragm	125
	6.4.2	Medium Deflection Diaphragm Analysis	127
	6.4.3	Membrane Analysis	127
	6.4.4	Bossed Diaphragm Analysis	128
	6.4.5	Corrugated Diaphragms	129
	6.4.6	Traditional Diaphragm Transduction Mechanisms	129
6.5	MEMS Technology Pressure Sensors		130
	6.5.1	Micromachined Silicon Diaphragms	130
	6.5.2	Piezoresistive Pressure Sensors	132
	6.5.3	Capacitive Pressure Sensors	137
	6.5.4	Resonant Pressure Sensors	139
	6.5.5	Other MEMS Pressure Sensing Techniques	142
6.6	Microphones		143
6.7	Conclusions		145
	References		145

CHAPTER 7

Force and Torque Sensors   153

7.1	Introduction	153
7.2	Silicon-Based Devices	154
7.3	Resonant and SAW Devices	157
7.4	Optical Devices	159
7.5	Capacitive Devices	160
7.6	Magnetic Devices	162
7.7	Atomic Force Microscope and Scanning Probes	164
7.8	Tactile Sensors	166
7.9	Future Devices	168
	References	168

CHAPTER 8

Inertial Sensors   173

8.1	Introduction		173
8.2	Micromachined Accelerometer		175
	8.2.1	Principle of Operation	175
	8.2.2	Research Prototype Micromachined Accelerometers	180
	8.2.3	Commercial Micromachined Accelerometer	192
8.3	Micromachined Gyroscopes		195
	8.3.1	Principle of Operation	195
	8.3.2	Research Prototypes	199
	8.3.3	Commercial Micromachined Gyroscopes	204

8.4 Future Inertial Micromachined Sensors 206
References 207

CHAPTER 9
Flow Sensors 213
9.1 Introduction to Microfluidics and Applications for
Micro Flow Sensors 214
9.2 Thermal Flow Sensors 217
9.2.1 Research Devices 219
9.2.2 Commercial Devices 225
9.3 Pressure Difference Flow Sensors 229
9.4 Force Transfer Flow Sensors 232
9.4.1 Drag Force 232
9.4.2 Lift Force 235
9.4.3 Coriolis Force 236
9.4.4 Static Turbine Flow Meter 238
9.5 Nonthermal Time of Flight Flow Sensors 239
9.5.1 Electrohydrodynamic 239
9.5.2 Electrochemical 240
9.6 Flow Sensor Based on the Faraday Principle 241
9.7 Flow Sensor Based on the Periodic Flapping Motion 242
9.8 Flow Imaging 243
9.9 Optical Flow Measurement 245
9.9.1 Fluid Velocity Measurement 245
9.9.2 Particle Detection and Counting 246
9.9.3 Multiphase Flow Detection 246
9.10 Turbulent Flow Studies 247
9.11 Conclusion 248
References 250

About the Authors 257
Index 259

Preface

The field of microelectromechanical systems (MEMS), particularly micromachined mechanical transducers, has been expanding over recent years, and the production costs of these devices continue to fall. Using materials, fabrication processes, and design tools originally developed for the microelectronic circuits industry, new types of microengineered device are evolving all the time—many offering numerous advantages over their traditional counterparts. The electrical properties of silicon have been well understood for many years, but it is the mechanical properties that have been exploited in many examples of MEMS. This book may seem slightly unusual in that it has four editors. However, since we all work together in this field within the School of Electronics and Computer Science at the University of Southampton, it seemed natural to work together on a project like this. MEMS are now appearing as part of the syllabus for both undergraduate and postgraduate courses at many universities, and we hope that this book will complement the teaching that is taking place in this area.

The prime objective of this book is to give an overview of MEMS mechanical transducers. In order to achieve this, we provide some background information on the various fabrication techniques and materials that can be used to make such devices. The costs associated with the fabrication of MEMS can be very expensive, and it is therefore essential to ensure a successful outcome from any specific production or development run. Of course, this cannot be guaranteed, but through the use of appropriate design tools and commercial simulation packages, the chances of failure can be minimized. Packaging is an area that is sometimes overlooked in textbooks on MEMS, and we therefore chose to provide coverage of some of the methods used to provide the interface between the device and the outside world. The book also provides a background to some of the basic principles associated with micromachined mechanical transducers. The majority of the text, however, is dedicated to specific examples of commercial and research devices, in addition to discussing future possibilities.

Chapter 1 provides an introduction to MEMS and defines some of the commonly used terms. It also discusses why silicon has become one of the key materials for use in miniature mechanical transducers. Chapter 2 commences with a brief discussion of silicon and other materials that are commonly used in MEMS. It then goes on to describe many of the fabrication techniques and processes that are employed to realize microengineered devices. Chapter 3 reviews some of the commercial design tools and simulation packages that are widely used by us and other researchers/designers in this field. Please note that it is not our intention to provide critical review here, but merely to indicate the various features and functionality offered by a selection of packages. Chapter 4 describes some of the techniques and structures that can be used to package micromachined mechanical sensors. It also discusses ways to minimize unwanted interactions between the device and its packaging. Chapter 5 presents some of the fundamental principles of mechanical transduction. This chapter is largely intended for readers who might not have a background in mechanical engineering. The remaining four chapters of the book are dedicated to describing specific mechanical microengineered devices including pressure sensors (Chapter 6), force and torque sensors (Chapter 7), inertial sensors (Chapter 8), and flow sensors (Chapter 9). These devices use many of the principles and techniques described in the earlier stages of the book.

Acknowledgments

We authors express our thanks to all the contributing authors of this book. They are all either present or former colleagues with whom we have worked on a variety of MEMS projects over the past decade or so.

Steve Beeby
Graham Ensell
Michael Kraft
Neil White
Southampton, United Kingdom
April 2004

CHAPTER 1

Introduction

1.1 Motivation for the Book

As we move into the third millennium, the number of microsensors evident in everyday life continues to increase. From automotive manifold pressure and air bag sensors to biomedical analysis, the range and variety are vast. It is interesting to note that pressure sensors and ink-jet nozzles currently account for more than two-thirds of the overall microtransducer market share. Future predications indicate that the mechanical microsensor market will continue to expand [1]. One of the main reasons for the growth of microsensors is that the enabling technologies are based on those used within the integrated circuit (IC) industry. The production cost of a commercial pressure sensor, for example, is around 1 Euro, and this is largely because the cost of producing ICs is inversely proportional to the volume produced. The trend in IC technology since the 1960s has been for the number of transistors on a chip to double every 18 months; this is referred to as Moore's law. This has profound implications for the electronic systems associated with microsensors. In addition to the reduction of size there is added functionality and also the possibility of producing arrays of individual sensor elements on the same chip.

Another feature that has influenced the popularity trend of microsensors is that many (but certainly not all) are based on silicon (Si). The electrical properties of silicon have been studied for many years and are well understood and thoroughly documented. Silicon also possesses many desirable mechanical properties that make it an excellent choice for many types of mechanical sensor.

Today there are many companies working in the field of microelectromechanical systems (MEMS). A quick search on the Internet in July 2003 revealed several hundred in the United States, Europe, and the Far East, including multinational corporations such as TRW Novasensor, Analog Devices, Motorola, Honeywell, Senso-Nor, Melexis, Infineon, and Mitsubishi, as well as small start-up companies. There are also many conferences dedicated to the subject. A selection of examples (but by no means an exhaustive list) is given here:

- *Transducers—International Conference on Solid-State Sensors and Actuators* (held biennially and rotating location between Asia, North America, and Europe);
- *Eurosensors* (held annually in Europe);
- *IEEE Sensors Conference* (first held in 2002, annually United States and Canada);
- *Micro Mechanics Europe—MME* (held annually in Europe);

- *IEEE International MEMS Conference* (rotates annually between the United States, Asia, and Europe);
- *Micro and Nano Engineering—MNE* (held annually in Europe);
- *Japanese Sensor Symposium* (held annually in Japan);
- *Micro Total Analysis Systems—μTAS* (held annually in the United States, Asia, Europe, and Canada);
- *SPIE* hold many symposia on MEMS at worldwide locations.

In addition, there are several journals that cover the field of microsensors and sensor technologies, including:

- *Sensors and Actuators* (A-Physical, B-Chemical);
- *IEEE/ASME Journal of Microelectromechanical Systems (JMEMS)*;
- *Journal of Micromechanics and Microengineering*;
- *Measurement, Science and Technology*;
- *Nanotechnology*;
- *Microelectronic Engineering*;
- *Journal of Micromechatronics*;
- *Smart Materials and Structures*;
- *Journal of Microlithography, Microfabrication, and Microsystems*;
- *IEEE Sensors Journal*;
- *Sensors and Materials*.

The major advancements in the field of microsensors have undoubtedly taken place within the past 20 years, and there is good reason to consider these as a modern technology. From an historical point of view, the interested reader might wish to refer to a paper titled "There's Plenty of Room at the Bottom" [2]. This is based on a seminar given in 1959 by the famous physicist Richard Feynman where he considered issues such as the manipulation of matter on an atomic scale and the feasibility of fabricating denser electronic circuits for computers. He also considered the issues of building smaller and smaller tools that could make even smaller tools so that eventually the individual atoms could be manipulated. The effects of gravity become negligible while those of surface tension and Van der Waals forces do not. Feynman even offered a prize (subsequently claimed in 1960) to the first person who could make an electric motor 1/64 in^3 (about 0.4 mm^3). These size limits turned out to be slightly too large and the motor was actually made using conventional mechanical engineering methods that did not require any new technological developments.

1.2   What Are MEMS?

MEMS means different things to different people. The acronym MEMS stands for microelectromechanical systems and was coined in the United States in the late 1980s. Around the same time the Europeans were using the phrase *microsystems technology* (MST). It could be argued that the former term refers to a physical entity, while the latter is a methodology. The word "system" is common to both, implying that there is some form of interconnection and combination of components. As an example, a microsystem might comprise the following:

- A sensor that inputs information into the system;
- An electronic circuit that conditions the sensor signal;
- An actuator that responds to the electrical signals generated within the circuit.

Both the sensor and the actuator could be MEMS devices in their own right. For the purpose of this book, MEMS is an appropriate term as it specifically relates to mechanical (micro) devices and also includes wider areas such as chemical sensors, microoptical systems, and microanalysis systems.

There is also a wide variety of usage of terms such as transducer, sensor, actuator, and detector. For the purpose of this text, we choose to adopt the definition proposed by Brignell and White [3], where *sensors* and *actuators* are two subsets of *transducers*. Sensors input information into the system from the outside world, and actuators output actions into the external world. Detectors are merely binary sensors. While these definitions do not specifically relate to energy conversion devices, they are simple, unambiguous, and will suffice for this volume.

As we will see in the following, micromachined transducers are generally (but not exclusively) those that have been designed and fabricated using tools and techniques originating from the IC industry. In general, there are two methods for silicon micromachining: bulk and surface. The former is a subtractive process whereby regions of the substrate are removed; while with the latter technique layers are built up on the surface of the substrate in an additive manner.

1.3 Mechanical Transducers

The market for micromachined mechanical transducers has, in the past, had the largest slice of the pie of the overall MEMS market. This is likely to be the case in the immediate future as well. The main emphasis of this text is on mechanical sensors, including pressure, force, acceleration, torque, inertial, and flow sensors. Various types of actuation mechanism, relevant to MEMS, will also be addressed together with examples of the fundamental techniques used for mechanical sensors. The main methods of sensing mechanical measurands have been around for many years and are therefore directly applicable to microsensors. There is, however, a significant effect that must be accounted for when considering mesoscale devices (i.e., those that fit into the palm of your hand) and microscale devices. This is, of course, scaling. Some physical effects favor the typical dimensions of micromachined devices while others do not. For example, as the linear dimensions of an object are reduced, other parameters do not shrink in the same manner. Consider a simple cube of material of a given density. If the length l is reduced by a factor of 10, the volume (and hence mass) will be reduced by a factor of 1,000 (l^3). There are many other consequences of scaling that need to be considered for fluidic, chemical, magnetic, electrostatic, and thermal systems [4]. For example, an interesting effect, significant for microelectrostatic actuators operating in air, is Paschen's law. This states that the voltage at which sparking occurs (the breakdown voltage) is dependent on the product of air pressure and the separation between the electrodes. As the gap between two electrodes is reduced, a plot of breakdown voltage against the gap separation and gas pressure product (Paschen curve) reveals a minimum in the characteristic, as shown in Figure 1.1. The consequence is that for air gaps of less than several microns, the breakdown voltage *increases*.

1.4   Why Silicon?

Micromachining has been demonstrated in a variety of materials including glasses, ceramics, polymers, metals, and various other alloys. Why, then, is silicon so strongly associated with MEMS? The main reasons are given here:

- Its wide use within the microelectronic integrated circuit industry;
- Well understood and controllable electrical properties;
- Availability of existing design tools;
- Economical to produce single crystal substrates;
- Vast knowledge of the material exists;
- Its desirable mechanical properties.

The final point is, of course, particularly desirable for mechanical microsensors. Single crystal silicon is elastic (up to its fracture point), is lighter than aluminum, and has a modulus of elasticity similar to stainless steel. Its mechanical properties are *anisotropic* and hence are dependent on the orientation to the crystal axis. Table 1.1 illustrates some of the main properties of silicon in relation to other materials. Typical values are given and variations in these figures may be found in the literature as some of the listed properties are dependent upon the measurement conditions used to determine the values. Stainless steel is used as a convenient reference as it is widely used in the manufacture of traditional mechanical transducers. It must be noted, however, that there are many different types of stainless steel exhibiting a broad variation to those values listed here.

Silicon itself exists in three forms: crystalline, amorphous, and polycrystalline (polysilicon). High purity, crystalline silicon substrates are readily available as

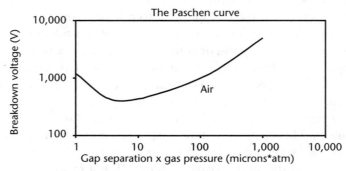

Figure 1.1   A plot of breakdown voltage against electrode separation (in air at 1 atmosphere of pressure).

Table 1.1 Properties of Silicon and Selected Other Materials

Property	Si {111}	Stainless Steel	Al	Al_2O_3 (96%)	SiO_2	Quartz
Young's modulus (GPa)	190	200	70	303	73	107
Poisson's ratio	0.22	0.3	0.33	0.21	0.17	0.16
Density (g/cm^3)	2.3	8	2.7	3.8	2.3	2.6
Yield strength (GPa)	7	3.0	0.17	9	8.4	9
Thermal coefficient of expansion (10/K)	2.3	16	24	6	0.55	0.55
Thermal conductivity at 300K (W/cm·K)	1.48	0.2	2.37	0.25	0.014	0.015
Melting temperature (°C)	1,414	1,500	660	2,000	1,700	1,600

circular wafers with typical diameters of 100 mm (4 inches), 150 mm (6 inches), 200 mm (8 inches), or 300 mm (12 inches) in a variety of thicknesses. Amorphous silicon does not have a regular crystalline form and contains many defects. Its main use has been in solar cells, photo-sensors, and liquid crystal displays. Both amorphous and polysilicon can be deposited as thin-films, usually less than about 5 μm thickness. Other materials that are often used within the MEMS fabrication process include glasses, quartz, ceramics, silicon nitride and carbide, alloys of various metals, and a variety of specialist materials that are used for very specific purposes.

1.5 For Whom Is This Book Intended?

This book is intended for graduate researchers who have taken a first degree in electronics, electrical engineering, or the physical sciences. It is also aimed at senior undergraduate students (years three or four) who are studying one of these courses. The main subject area of the text is that of mechanical microsensors, and in order to assist the reader in this respect, we have covered some of the fundamental principles of applied mechanics that might not have been covered in detail during some of these courses. Those who have a background in mechanical engineering will find that this book provides an overview of some of the main transducer microfabrication techniques that can be used to make a variety of transducer systems. Overall, it should become clear that there is a synergy between the electrical and mechanical engineering disciplines, and those who work in the field of sensors and actuators will have the joy of participating in one of the truly interdisciplinary fields in the whole of science.

References

[1]   Nexus MST market analysis, http://www.nexus-mems.com.

[2]   Feynman, R. P., "There's Plenty of Room at the Bottom," *Journal of Microelectromechanical Systems*, Vol. 1, No. 1, 1992, pp. 60–66.

[3]   Brignell, J. E., and N. M. White, *Intelligent Sensor Systems*, Bristol, England: IOP Publishing, 1994.

[4]   Judy, J. W., "Microelectromechanical Systems (MEMS): Fabrication, Design and Applications," *Smart Materials and Structures*, Vol. 10, 2001, pp. 1115–1134.

CHAPTER 2

Materials and Fabrication Techniques

2.1 Introduction

MEMS devices and structures are fabricated using conventional integrated circuit process techniques, such as lithography, deposition, and etching, together with a broad range of specially developed micromachining techniques. Those techniques borrowed from the integrated circuit processing industry are essentially two dimensional, and control over parameters in the third dimension is only achieved by stacking a series of two-dimensional layers on the workpiece, which is usually a silicon wafer. There are practical and economic limits, however, to the number of layers that can be managed in such a serial process, and therefore, the expansion of devices into the third dimension is restricted. Micromachining techniques enable structures to be extended further into the third dimension; however, it has to be understood that these structures are simply either extruded two-dimensional shapes or are governed by the crystalline properties of the material. True three-dimensional processing would allow any arbitrary curved surface to be formed, and this is clearly not possible with the current equipment and techniques. An important aspect of MEMS is to understand the limitations of the micromachining techniques currently available. Although the range of these techniques is continually being expanded, there are some core techniques that have been part of the MEMS toolkit for many years. This chapter deals mainly with these core techniques, but also with those process techniques borrowed from integrated circuit manufacturing.

2.2 Materials

2.2.1 Substrates

2.2.1.1 Silicon

Just as silicon has dominated the integrated circuit industry, so too is it predominant in MEMS. There are a number of reasons for this: (1) pure, cheap, and well-characterized material readily available; (2) a large number and variety of mature, easily accessible processing techniques; and (3) the potential for integration with control and signal processing circuitry. In addition to these reasons, the mechanical and physical properties of silicon give it a powerful advantage for its use in mechanical sensors, and therefore, this book deals mainly with devices fabricated in bulk silicon and silicon on insulator (SOI).

Crystalline silicon has a diamond structure. This is a face-centered cubic lattice with two atoms (one at the lattice point and one at the coordinates ¼, ¼, ¼ normalized to the unit cell) associated with each lattice point. The crystal structure is shown in Figure 2.1. The crystal planes and directions are designated by Miller indices, as shown in Figure 2.2. Any of the major coordinate axes of the cube can be designated as a <100> direction, and planes perpendicular to these are designated as {100} planes. The {111} planes are planes perpendicular to the <111> directions, which are parallel to the diagonals of the cube. Bulk silicon from material manufacturers is usually either {100} or {111} orientation, although other orientations can be obtained from specialist suppliers. This orientation identifies the plane of the top surface of the wafer. The wafers are cut at one edge to form a primary flat in a {110} plane. A secondary flat is also cut on another edge to identify the wafer orientation and doping type, which is either n- or p-type. The doping is done with impurities to give a resistivity of between 0.001 and 10,000 Ωcm. For mainstream integrated circuit processing wafers are typically of the order of 10 to 30 Ωcm corresponding to an impurity level of $\sim 3 \times 10^{14}$ cm^{-3} for n-type and $\sim 9 \times 10^{14}$ cm^{-3} for p-type. Table 2.1 shows some of the properties of crystalline silicon. It should be remembered that some of the properties are anisotropic, and therefore, the orientation of the silicon needs to be taken into account in the design of any mechanical sensor. For example, the piezoresistance coefficient of single crystal silicon depends on the orientation of the resistor with respect to the crystal orientation; Young's modulus is orientation dependent; cracks initiated through mechanical loading will tend to propagate along certain crystal planes.

In the last few years, SOI wafers have become available and are now being employed in MEMS applications. As shown in Figure 2.3, there are a number of distinct types of SOI wafer, each of which has its own particular features. Separation by ion implantation of oxygen (Simox) wafers are fabricated by implanting bulk silicon wafers with high-energy oxygen ions, followed by anneal at 1,300°C. This process forms a buried oxide (BOX) layer at a fixed depth below the surface, leaving a single-crystalline silicon layer (SOI layer) on the top surface. Although the SOI layer

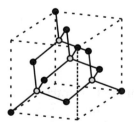

Figure 2.1   Unit cell of silicon. The crystalline structure is face-centered cubic with two silicon atoms associated with each lattice point. The dark atoms are on the lattice points and the gray atoms are at (¼ ¼ ¼), (¼ ¾ ¾), (¾ ¼ ¾), and (¾ ¾ ¼).

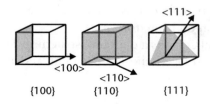

Figure 2.2   Diagram illustrating the important planes and directions in crystalline silicon.

Table 2.1 Selected Properties of Crystalline Silicon

Yield strength (10^9 Nm^{-2})	7
Knoop hardness (kgmm^{-2})	850
Young's modulus (GPa), (100) orientation	160
Poisson's ratio, (100) orientation	0.28
Density (gcm^{-3})	2.33
Lattice constant (Å)	5.435
Thermal expansion coefficient (10^{-6}K^{-1})	2.6
Thermal conductivity (Wm^{-1}K^{-1})	157
Specific heat (Jg^{-1}K^{-1})	0.7
Melting point (°C)	1,410
Energy gap (eV)	1.12
Dielectric constant	11.9
Dielectric strength (10^7 Vm^{-1})	3
Electron mobility (cm^{2}V^{-1}s^{-1})	1,450
Hole mobility (cm^{2}V^{-1}s^{-1})	505

can be thickened by epitaxy, the thicknesses of the SOI and BOX layers are limited due to the range and distribution of the implanted ions. Typically, these are ~0.2 and ~0.1 μm, respectively. Wafer bonding is an alternative technique for producing thick layers of silicon on a buried oxide. Two wafers, at least one of which is covered with a thick oxide layer, are bonded together by van der Waals forces, and subsequent annealing at ~1,100°C causes a chemical reaction that strengthens the bonded interface. One of the wafers is then thinned down by mechanical grinding, and a final polish can produce SOI films 1 μm thick with a uniformity of 10% to 30%. The BOX layer can be between 0.5 and 4 μm thick. These wafers are sometimes referred to as bonded and etched SOI (BESOI) wafers. Both ion implantation and wafer bonding are used in the production of UNIBOND SOI wafers. Starting with two wafers, the silicon surface of one wafer is first oxidized to form what will become the buried oxide layer of the SOI structure. An ion implantation step, using

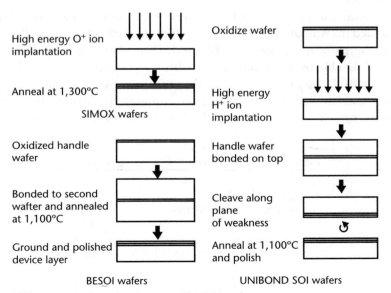

Figure 2.3 Different manufacturing processes for SOI wafers.

hydrogen ions, is then executed through the oxide layer by a standard high-current ion implanter to form the Smart Cut layer. The implanted hydrogen ions alter the crystallinity of the silicon, creating a plane of weakness in the wafer. After the wafers are bonded together, the implanted wafer can be cleaved along this plane to leave a thin layer of silicon on top of the oxide layer. The wafer is then annealed at 1,100°C to strengthen the bond, and the surface of the silicon is polished to reduce the defect level to a level approaching that of bulk silicon. The buried oxide layer is pinhole free. SOI layers in the range from 0.1 to 1.5 μm and BOX layers from 200 nm to 3 μm can be fabricated by this method.

Other substrates, however, should not be ignored. Among those that have been used in micromachining are glasses, quartz, ceramics, plastics, polymers, and metals. Quartz and glass are often used in MEMS mechanical sensors; therefore, a short description of these materials is given here.

2.2.1.2   Quartz and Glasses

Quartz is mined naturally but is more commonly produced synthetically in large, long faceted crystals. It has a trigonal trapezohedral crystal structure and is similar to silicon in that it can be etched anisotropically by selectively etching some of the crystal planes in etchants such as ammonium bifluoride or hydrofluoric acid. Unlike silicon, however, this has not been extensively used as an advantage but has been identified more as a disadvantage due to the development of unwanted facets and poor edge definition after etching. Since the fastest etch rate is along the z-axis [1], most crystalline quartz is cut with the z-axis perpendicular to the plane of the wafer. The property of quartz that makes it useful in MEMS mechanical sensors is that it is piezoelectrical. Quartz has been used to fabricate resonators, gyroscopes, and accelerometers. Another form of quartz is fused quartz, but be careful not to confuse this material with crystalline quartz, as fused quartz is used to denote the glassy noncrystalline, and, therefore, isotropic form better known as silica. It is tough and hard and has a very low expansion coefficient.

Glass can be etched in hydrofluoric acid solutions and is often electrostatically bonded to silicon to make more complicated structures. Both phosphosilicate and borosilicate glasses can be used. One of the more favored glasses is Pyrex, which is a borosilicate glass composition with a coefficient of thermal expansion of $3.25 \times 10^{-6}/°C$, which is close to that of silicon, an essential property for structures to be used in thermally unstable environments. Some of the properties of quartz and Pyrex are shown in Table 2.2. The substrate is sometimes used purely as a

Table 2.2   Selected Properties of Quartz and Pyrex

Property	Quartz	Pyrex
Young's modulus (GPa)	107	64
Poisson's ratio, (100) orientation	0.16	0.20
Density (gcm^{-3})	2.65	2.33
Dielectric constant	3.75	4.6
Thermal expansion coefficient (10^{-6}K^{-1})	0.55	3.25
Thermal conductivity (Wm^{-1}K^{-1})	1.38	1.13
Specific heat (Jg^{-1}K^{-1})	0.787	0.726
Refractive index	1.54	1.474

foundation on which a micromachined device is built, in which case the substrate material may be unimportant and need only be compatible with the processing equipment used. Both quartz and Pyrex can be obtained in forms suitable for processing using standard silicon processing equipment. Sometimes, however, the device is formed in the substrate itself, in which case the material properties become important.

2.2.2 Additive Materials

The materials deposited on the substrates include all those associated with integrated circuit processing. These are either epitaxial, polycrystalline, or amorphous silicon, silicon nitride, silicon dioxide, silicon oxynitride, or a variety of metals and metallic compounds, such as Cu, W, Al, Ti, and TiN, deposited by chemical (CVD) or physical vapor deposition (PVD) processes. Organic polymer resists with thicknesses up to the order of a few micrometers are deposited by optical or electron beam lithography.

Additional materials used in MEMS mechanical sensors are: ceramics (e.g., alumina, which can be sputtered or deposited by a sol-gel process); polymers, such as polyimides and thick X-ray resists and photoresists; a host of other metals and metallic compounds (e.g., Au, Ni, ZnO) deposited either by PVD, electroplating, or CVD; and alloys (e.g., SnPb) deposited by cosputtering or electroplating. Some alloys, such as TiNi, have a shape memory effect that causes the material to return to a predetermined shape when heated. This is caused by atomic shuffling within the material during phase transition. At low temperatures the phase is martensite, which is ductile and can be easily deformed. By simply heating, the phase of the deformed material changes to austenite and the deformation induced at low temperature can be fully recovered. The transition temperature depends on the impurity concentration, which can be controlled to give values between −100°C and 100°C. Therefore, by repeated deformation and heating the shape memory alloy (SMA) can be incorporated in a useful mechanical device. For micromechanical devices the high power-to-weight ratio, large achievable strain, low voltage required for heating, and large mean time between failure suggest that SMAs have the potential for superior actuators. The maximum frequency of operation, however, is only of the order of 100 Hz [2]. Diamond and silicon carbide deposited by CVD have some potentially useful mechanical and thermal properties. Each has high wear resistance and hardness, is chemically inert, and has excellent heat resistance. Neither has been extensively explored for their use in MEMS sensors.

It is safe to say that, unless there is an issue of contamination or the sensors are integrated with circuitry, it is possible to deposit almost any material on the substrate. The issues that are likely to need addressing, however, are how well does it adhere to the substrate, are there any stresses in the deposited layer that may cause it to deform, and can it be patterned and etched using lithographic techniques?

2.3 Fabrication Techniques

The fabrication techniques used in MEMS consist of the conventional techniques developed for integrated circuit processing and a variety of techniques developed specifically for MEMS. The three essential elements in conventional silicon processing are deposition, lithography, and etching. These are illustrated in Figure 2.4. The common deposition processes, which include growth processes, are oxidation, chemical vapor deposition, epitaxy, physical vapor deposition, diffusion, and ion implantation. The types of lithography used are either optical or electron beam, and etching is done using either a wet or dry chemical etch process. Many of these conventional techniques have been modified for MEMS purposes, for example, the use of thick photoresists, grayscale lithography, or deep reactive ion etching. Other processes and techniques not used in conventional integrated circuit fabrication have been developed specifically for MEMS, and these include surface micromachining, wafer bonding, thick-film screen printing, electroplating, porous silicon, LIGA (the German acronym for Lithographie, Galvansformung, Abformung), and focused ion beam etching and deposition. For a more general reference covering MEMS fabrication techniques, see the book by Kovaks [3].

2.3.1 Deposition

2.3.1.1 Thermal Growth

Silicon dioxide is grown on silicon wafers in wet or dry oxygen ambient. This is done in a furnace at temperatures in the range from 750°C to 1,200°C. For oxides grown at atmospheric pressure the thickness of the oxide can be as small as 1.5 nm or as large as 2 μm. For each micron of silicon dioxide grown, 0.45 μm of silicon is consumed and this generates an appreciable compressive stress at the interface. Furthermore, there is a large difference between the thermal expansion coefficients of silicon and silicon dioxide, which leaves the oxide in compression after cooling from the growth temperature, adding to the intrinsic stress arising during growth. Stress is, of course, an important issue for MEMS mechanical devices and

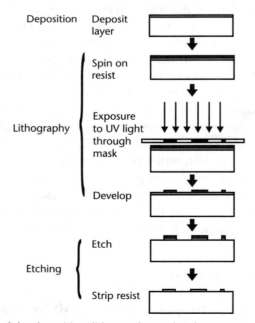

Figure 2.4 Illustration of the deposition, lithography, and etch processes.

cannot be ignored. Thick oxide films can cause bowing of the underlying substrate. Freestanding oxide membranes will buckle and warp, and thin oxides on silicon cantilevers will make them curl.

2.3.1.2 Chemical Vapor Deposition

Solid films, such as silicon dioxide, silicon nitride, and amorphous or polycrystalline silicon (polysilicon) can be deposited on the surface of a substrate by a CVD process, the film being formed by the reaction of gaseous species at the surface. The three most common types of CVD process are low-pressure CVD (LPCVD), plasma enhanced CVD (PECVD)—in which radio frequency (RF) power is used to generate a plasma to transfer energy to the reactant gases, and atmospheric pressure CVD (APCVD). For LPCVD, the step coverage (conformality), uniformity, and the composition and stress of the deposited layer are determined by the gases used and the operating temperature and pressure. For PECVD, the layer properties are affected additionally by the RF power density, frequency, and duty cycle at which the reactor is operated; and for APCVD, in which the deposition is mass transport limited, the design of the reactor is significant.

2.3.1.3 Polysilicon and Amorphous Silicon

Films deposited by LPCVD are used widely in the integrated circuit industry. Amorphous silicon and polysilicon, in particular, are usually deposited by LPCVD using silane. Although polysilicon can be deposited by PECVD, this is generally only done where large deposited areas are required or for thin-film transistor liquid crystal displays. The properties of LPCVD amorphous silicon and polysilicon layers depend on the partial pressure of silane in the reactor, the deposition pressure and temperature, and, if doped *in situ*, on the gas used for doping. If doped silicon is required, then diborane, phosphine, or arsine is included in the deposition process. The deposition temperatures range from 570°C for amorphous silicon to 650°C for polysilicon with the silicon grain size increasing with temperature. The final grain size for amorphous silicon is usually determined, however, by the temperature at which the film is annealed after deposition. For MEMS devices annealing can also be used to control the stress in amorphous and polysilicon films. The residual stress in as-deposited amorphous silicon and polysilicon films can be as much as 400 MPa and be either tensile or compressive depending on the deposition temperature. The transition from tensile to compressive stress is quite sharp and depends also on other deposition parameters, making it difficult to control the stress in the as-deposited film. The residual stress in polysilicon deposited at 615°C can be reduced to −10 MPa (compressive) by annealing for 30 minutes at 1,100°C in N_2 and that in amorphous silicon films deposited at 580°C is reduced to 10 MPa (tensile) by annealing for 30 minutes at 1,000°C in N_2. Perhaps more importantly, the residual stress gradient in these films is also reduced to near zero. An alternative method is to deposit alternating layers of amorphous silicon grown at 570°C and polysilicon grown at 615°C [4]. The amorphous silicon is tensile and the polysilicon is compressive. By adjusting the thickness and distribution in a multilayer film, it is possible to control both the stress and the stress gradient in an as-deposited polysilicon layer.

2.3.1.4 Epitaxy

Epitaxial silicon can be grown by APCVD or LPCVD. The ranges of temperatures at which this is done are 900°C to 1,250°C for APCVD and 700°C to 900°C for LPCVD. Epitaxy can be used to deposit silicon layers with clearly defined doping profiles that can be used as an etch stop, such as, for example, an electrochemical etch stop. It can also be used to thicken the SOI layers on Simox or UNIBOND wafers, for which the thickness of the original SOI layer is restricted by the manufacturing process. The most useful property of epitaxial silicon for MEMS applications, though, may be the fact that it can be grown selectively. Silicon dioxide or silicon nitride on wafers prevents the growth of epitaxial silicon, and a layer of amorphous silicon or polysilicon is normally deposited instead. However, this deposition process can be suppressed by the addition of HCl to the reaction gases. The HCl prevents spurious nucleation and growth of silicon on the silicon dioxide or nitride. An example of selective epitaxial growth is shown in Figure 2.5. This selective growth can be used to form useful microengineered structures. Epitaxial silicon reactors can also be used for depositing thick layers of polysilicon. Due to the growth time, polysilicon deposited by LPCVD is often no more than a couple of microns thick, whereas with the use of an epitaxial reactor, much thicker layers of more than 10 μm can be deposited. This type of polysilicon is referred to as epipoly.

2.3.1.5 Silicon Nitride

Silicon nitride is commonly deposited by CVD by reacting silane or dichlorosilane with ammonia. The film is in an amorphous phase and often contains a large amount of hydrogen. LPCVD silicon nitride is an exceptionally good material for masking against wet chemical etchants such as HF and hydroxide-based bulk silicon anisotropic etchants. The deposition temperature, however, which is in the range from 700°C to 850°C, prohibits its use on wafers with aluminum. Another limiting factor is the large intrinsic tensile stress, which is of the order of 1 GPa. Layers thicker than about 200 nm are likely to delaminate or crack, and freestanding structures are susceptible to fracture. For MEMS applications, low-stress LPCVD films can be deposited by increasing the ratio of silicon to nitrogen to produce silicon

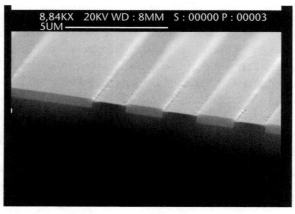

Figure 2.5 Epitaxial silicon grown selectively between bars of oxide.

rich nitride or by adding N_2O to the reaction gases, thereby depositing silicon oxynitride. Silicon nitride deposited by PECVD contains substantially more hydrogen than LPCVD nitride and is nonstoichiometric. Deposition temperatures are between 250°C and 350°C, thus making it possible to deposit it on wafers with aluminum interconnects. Stress in the films is a function of pressure, temperature, frequency, power, and gas composition and is in the range from −600 MPa (compressive) to +600 MPa (tensile). Films deposited at 50 kHz and 300°C are compressive, but at about 600°C the stress switches from compressive to tensile, making the deposition of low stress films possible. Unfortunately, this eliminates one of the advantages of PECVD, that is, low temperature deposition. Films deposited at 13.56 MHz are tensile and whereas most PECVD equipment operates at a fixed frequency, some equipment manufacturers have enabled their systems to be switched rapidly between high and low frequencies to obtain very low stress films. The step coverage of PECVD silicon nitride is conformal; however, the pinhole density and stress can be a problem if it is used as a masking material against wet chemical etchants. The exact film properties vary depending on the system, the gas purity, and the deposition conditions, yet, with the right conditions, low pinhole densities, conformal step coverage, and low stress layers can be obtained. Some properties of LPCVD and PECVD silicon nitride are shown in Table 2.3.

2.3.1.6   Silicon Dioxide

Silicon dioxide deposited by APCVD, LPCVD, and PECVD are all used in conventional semiconductor processing. In each case there are a number of different process conditions and gases used. A selection of the many different processes used with the properties of the deposited layers is shown in Table 2.4. APCVD films are generally deposited at temperatures below 500°C by reacting silane with oxygen or TEOS with ozone and are used as interlevel dielectrics between polysilicon and metal. Furthermore, with the addition of large quantities of dopants, these films can be flowed and reflowed at temperatures in excess of 800°C. Phosphorous doped oxide (phosphosilicate glass or PSG) reflows at decreasingly lower temperatures as the phosphorus content increases up to 8%. Although lower reflow temperatures are possible for higher dopant concentrations, it is inadvisable to go beyond this because of the possibility of corrosion of subsequently deposited aluminum. The addition of boron up to 4% to form borophosphosilicate glass (BPSG) reduces the

Table 2.3   Properties of Silicon Nitride

Deposition	PECVD	LPCVD
Process gases used	$SiH_4 + NH_4$ or $SiH_4 + N_2$	$SiH_4 + NH_4$ or $SiCl_2H_2 + NH_4$
Deposition temperature (°C)	250–350	700–850
Stress (GPa)	0.6 compressive to 0.6 tensile	1 tensile
Density (gcm^{-1})	2.4–2.8	2.9–3.1
Refractive index	1.85–2.5	2.01
Dielectric constant	6–9	6–7
Dielectric strength (10^6 Vcm^{-1})	5	10
Resistivity (Ω-cm)	10^6–10^{15}	10^{16}
Energy gap (eV)	4–5	5
Si/N ratio	0.8–1.2	0.75

Table 2.4   Properties of CVD Silicon Dioxide

	PECVD	APCVD	LPCVD	LPCVD	LPCVD
Process gases used	SiH_4+O_2 (or N_2O)	SiH_4+O_2	SiH_4+O_2	$TEOS+O_2$	$SiCl_2H_2+N_2O$
Deposition temp. (°C)	250	400	450	700	900
Stress (GPa)	0.3 compressive to 0.3 tensile	0.1 to 0.3 tensile	0.3 tensile	0.1 compressive	0.3 compressive
Dielectric strength (10^6 Vcm^{-1})	3–6	3–6	8	10	10
Dielectric constant	4.9	—	4.3	4.0	—
Refractive index	1.45	1.44	1.44	1.46	1.46
Density (gcm^{-3})	2.3	1–2	2.1	2.2	2.2

viscosity and enables reflow at even lower temperatures. The reflow process is illustrated in Figure 2.6. Although the addition of boron to PSG reduces the etch rate in solutions containing HF, these films etch very quickly and are therefore often utilized as sacrificial layers in surface micromachining. Because of the temperature constraints imposed by metal already on the wafer, the dielectric between each layer of metal, the interlevel metal dielectric, is deposited by LPCVD at 400°C or PECVD in the range from 250°C to 400°C. Other LPCVD processes working at temperatures up to 900°C have been developed to give conformal oxides with good uniformity. Silicon dioxide films deposited at temperatures below 500°C are of lower density than those deposited at higher temperatures or by thermal oxidation. Heating these oxides at temperatures above 700°C causes densification, a process in which the amorphous structure of the oxide is maintained but, due to a rearrangement of the SiO_4 tetrahedra, the density increases to that of thermal oxide. This is accompanied by a decrease in film thickness. The properties of densified oxides are similar to those of thermal oxides. For example, the etch rate in HF solutions is the same, whereas the etch rate of undensified oxides can be as much as an order of magnitude greater than densified oxides. The stress in deposited oxides is either compressive or

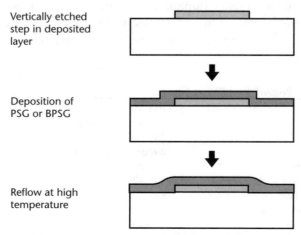

Vertically etched
step in deposited
layer

Deposition of
PSG or BPSG

Reflow at high
temperature

Figure 2.6   Illustration of the use of the reflow process to smooth the coverage over a vertical step.

tensile and is determined by the process. Typically this is up to 300 MPa. Control over this stress can only be exercised in PECVD deposition.

2.3.1.7 Metals

Although metals can be deposited by CVD, evaporation, e-beam evaporation, or plasma spray deposition, sputtering is the technique commonly used in integrated circuit processing. It is also safe to say that the metal predominantly used is aluminum, usually with a few percent silicon and/or copper added. The thickness of the metal is of the order of 1 μm and is usually deposited on thin layers, such as Ti, to improve adhesion, and barrier layers, such as TiN, to prevent diffusion. The stress in sputtered films is, in general, tensile, with the actual value depending on the pressure in the sputtering chamber and the temperature of the substrate.

2.3.1.8 Doped Silicon

Dopants are introduced into silicon either by ion implantation, during epitaxial growth, or by diffusion from solid or gaseous sources. Ion implantation is done by firing energetic ions directly into the silicon. After implantation, the silicon wafers have to undergo a thermal treatment, first, to anneal damage to the crystal caused by the impact of the energetic ions, and second, to move the dopant atoms into substitutional sites in the silicon crystal where they become electrically active. Doping during epitaxial growth is achieved by adding the appropriate gases, such as arsine, phosphine, or diborane, to the epitaxy growth chamber. Diffusion is done in a furnace at elevated temperatures in the range 800°C to 1,200°C. In all of these cases silicon dioxide can be used to create a two-dimensional spatially distributed pattern of doped silicon. The depth and the doping profile of the atoms introduced into the silicon depend on the exact conditions used. For MEMS mechanical sensors, ion implantation is usually used when a shallow doping profile is required as, for example, for piezoresistors. When a deeper doping profile is required—such as that required for the etch stop process discussed later in this chapter—then diffusion in a furnace is the obvious choice. Doping silicon to depths of up to ~10 µm can be achieved by diffusion. Beyond this, epitaxial growth of a doped layer of silicon is the only option.

2.3.2 Lithography

Lithography is the process by which patterns are formed in a chemically resistant polymer, applied by spinning it on to the silicon wafer. In optical lithography this polymer, called resist, is exposed to UV light through a quartz mask with an opaque patterned chrome layer on it to either break or link the polymer chains. The former is called positive resist and the latter negative resist. After exposure the soluble resist (the broken polymer chains in positive resist or the unlinked polymer chains in negative resist) is removed in developer and the remaining resist is baked in order to harden it against chemical attack. In integrated circuit processing the typical thickness of an optical resist is 1 μm and exposure is done with a wafer stepper. With state-of-the-art equipment, feature sizes of the order of 100 nm can be obtained.

The optical lithography process is illustrated in Figure 2.4. In electron beam lithography the resist is exposed to an energetic beam of electrons swept across the wafer. The beam is switched off and on to create a pattern in the resist, which again can be either positive or negative. E-beam resist is in general not as thick as optical resist, being of the order of 0.2 to $0.9\,\mu$m. Feature sizes are of the order of 10 nm. The minimum feature size that can be obtained with conventional lithography is not usually a concern for mechanical MEMS devices. However, other challenges have arisen as the lithography techniques used have expanded beyond the conventional limits. Double-sided and grayscale lithography, thick and laminated photoresists, liftoff processes, and the problems presented by large topographical features are all relevant examples.

2.3.2.1 Double-Sided Lithography

Many MEMS devices require double-sided processing; in the majority of cases this means that the patterns on either side of the wafer have to be aligned to each other. Although some workers have achieved this by etching completely through a wafer to form registration marks on the back side, the difficulties that this presents makes this a less than attractive option. Special alignment equipment is available for double-sided aligning. Some equipment uses an electronically captured image of crosshairs on a mask to which crosshairs on the back side of a wafer can be aligned. The front of the wafer is then exposed through the mask, which is clamped to the equipment. The alignment accuracy that can be achieved is of the order of 1 μm. Other equipment uses an infrared image converter to enable patterns on the backside of a wafer to be viewed on a monitor. The alignment accuracy in this case is limited to about 20 μm for a 4-inch wafer because the pattern on the wafer is separated from that on the mask by the thickness of the silicon wafer. This makes it impossible to focus sharply on both patterns simultaneously. In general, it is advisable to use double-sided polished wafers when using double-sided lithography.

2.3.2.2 Grayscale Lithography

This is a technique by which topographical features can be formed in photoresist. The amount of resist removed during the development cycle depends on the exposure in Joules per square meter, and a graph plotting the amount of resist removed against exposure is called a Gamma curve. The exposure at different pixel points on the resist can be controlled by having different gray levels on the mask. These gray levels are formed by arrays of submicron dots, and the gray level itself can be controlled by the number or size of the dots within the pixel. The important factor is that the dots themselves are not individually resolved by the mask aligner, but serve only to reduce the exposure. The number of gray levels that can be achieved with a times-five wafer stepper that can resolve 0.5-micron features is of the order of 300. In practice, 30 gray levels are sufficient for most applications. In principle, the features formed in the resist can be transferred to the underlying substrate by etching in, for example, an ion beam miller. One application of this technique is the fabrication of microlenses and microlens arrays as shown in the SEM photograph in Figure 2.7.

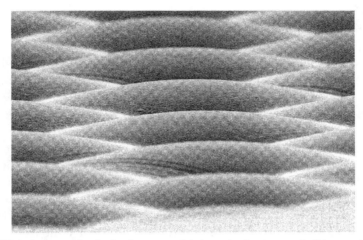

Figure 2.7 SEM photograph of microlens array fabricated using grayscale lithography.

2.3.2.3 Thick and Laminated Photoresists

There are a number of thick UV photoresists available and these have been used in a diverse range of applications. In conventional IC processing, the resist thickness spun on to the wafer is of the order of a micron thick, which means that 3 to 4 μm and above should be regarded as a thick resist. There are some thick resists, such as Shipley SPR 220-7, which will give a thickness of 7 μm if spun on to the wafer at the manufacturer's recommended speed. The thickness, however, can be increased by slowing the spin speed, and thicker layers of up to 60 μm can be obtained by repeating the process to give multiple layers. Other resists give thicker layers still, sometimes of the order of 500 μm in a single coating. Maintaining control over the thickness and uniformity becomes more difficult as the thickness increases. The thick resist most frequently reported on is the photoplastic polymer SU-8, which has been used as a micromold for injection molding or electroplating, as a mask for deep reactive ion etching (DRIE), as a structural MEMS component, and as a mechanical material. When cured, SU-8 forms a highly crosslinked matrix of covalent bonds giving it a wide range of elastic properties without plastic deformation. Thus, it has been used to make compliant structures such as springs and microgrippers [5]. There have been some reports on the difficulties associated with SU-8—for example, stress induced crack generation in mechanical structures—but by far the most frequently reported difficulty is the problem of removing it [6]. Both oxygen plasma [7] and hot NMP (1-methyl-2-pyrrolidinone) stripper [8] have been used, but in each case the removal has been either slow or incomplete. JSR manufactures a range of thick photoresists, which, it is claimed, can easily be stripped using the manufacturers own photoresist stripper and acetone [8]. Thicknesses of 1.4 mm have been reported for a double coating of JSR THB-430N. However, this resist has so far not been widely used in MEMS. A dry film photoresist, Ordyl P-50100, has been used successfully to form electroplating molds up to 100 μm thick, without any of the difficulties and limitations mentioned earlier [9]. An obstacle to using dry resists, however, is that application of the resist is done using a hot roll laminator, not normally found in silicon processing clean-rooms.

2.3.2.4 Liftoff Process

This is a simple method for patterning, usually metallic, layers. It is used for metals that are difficult to etch or where etching might damage other materials already on the substrate. A typical process is as follows. First, a resist is deposited and patterned with an image where the areas intended to have metal are cleared by the developer. Second, metal is deposited by evaporation or sputtering. Finally, the resist is removed in a solvent such as acetone that takes away the resist and lifts off the unwanted metal. For best results the developed pattern has undercut edges. This can be achieved by soaking the resist in chlorobenzene. Depending on the exposure time, this penetrates only a certain depth into the resist, causing the surface of the resist to develop at a slower rate than the resist in contact with the wafer. The process, however, is difficult to control and success is often only partial. A better approach is to use two different resists such as PMGI SF11 and a standard resist. In this process, illustrated in Figure 2.8, the PMGI SF11 is deposited and flood exposed before the application of a standard resist. After exposure with the pattern, the resist is developed. The PMGI SF11 develops at a faster rate than the standard resist, thereby leaving an overhang. Other materials can be used in a liftoff process. For example, the two layers of resist can be replaced by aluminum and polysilicon with orthophosphoric acid used both to create the overhang and to do the final liftoff. Providing that the layer to be patterned is not chemically attacked by orthophosphoric acid, the process will work.

2.3.2.5 Topography

Deep cavities etched into silicon are a common feature in MEMS devices, and ideally, the processing steps to produce these are done at the end of the process. However, the design of the device may not always allow this, for example, when contact

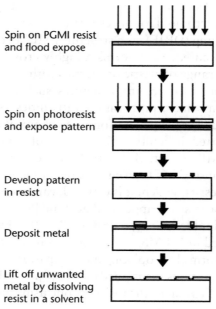

Spin on PGMI resist
and flood expose

Spin on photoresist
and expose pattern

Develop pattern
in resist

Deposit metal

Lift off unwanted
metal by dissolving
resist in a solvent

Figure 2.8 Process flow for liftoff.

to the silicon at the bottom of a cavity is required. In these cases difficulties arise, first, with step coverage and, second, with the minimum feature sizes that can be obtained. Resist coverage over a deep step is very nonuniform, with the resist thinning as it passes over the top edges and thickening at the bottom edges of a cavity leading to a disparity in the exposure and development conditions required for optimization. Typical resist profiles are illustrated in Figure 2.9. The thinner resist on the top edges requires short exposure and development times so that feature line widths are not reduced and the thicker resist at the bottom edges of the cavity requires long exposure and development times so as not to leave unwanted fillets of resist running around the bottom edges of the cavity. By using thicker resists and slower spin speeds the problem is reduced, although it can never be entirely eliminated, except by spray deposition. The bottom of the cavity will also be out of contact with the mask in a contact aligner and out of focus in a wafer stepper. However, most contact aligners have a sufficiently collimated beam for minimum line widths of 10 μm to be achieved at the bottom of a 400-μm deep cavity. Similar results can be obtained with a stepper.

2.3.3   Etching

Much of the early work on MEMS utilized micromachining using wet chemical etching; and although IC processing is dominated by dry etching, the majority of etch processing done in MEMS fabrication is still done using wet chemical etchants. In both wet and dry etching, consideration is given to the isotropy of the etch and the etch selectivity to the masking material and other exposed materials. The etch selectivity is defined as one film etching faster than another film under the same etching conditions.

Wet etchants used for etching silicon dioxide, silicon nitride, and aluminum are well known in the semiconductor industry. These are all isotropic etchants, which means they etch at the same rate in all directions. Wet etchants for silicon, on the other hand, may be either isotropic or anisotropic. The anisotropic silicon etchants etch crystalline silicon preferentially in certain directions in the crystal. For all the wet chemical etchants used in MEMS, the etchant and masking material can usually be chosen to give a highly selective etch.

Dry etching is done in a weakly ionized plasma at low pressure. Most dry etching is a combination of chemical and physical etching. Chemical etch processes give good selectivity and isotropic profiles are obtained, but physical etch processes have low selectivity and induce damage from ion bombardment. However, physical etch

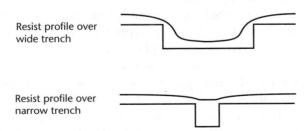

Resist profile over
wide trench

Resist profile over
narrow trench

Figure 2.9 Profiles of resist over wide and narrow trenches. Note the thinning of the resist near to the top edges and the thicker resist at the bottom edges of the wide trench.

processes give anisotropic etch profiles, which are extremely important for submicrometer semiconductor fabrication. By combining chemical and physical processes in a dry etch process, the optimum conditions for any particular process can be obtained.

The most common type of etching adapted for MEMS is deep etching into the silicon substrate; and this is often referred to as bulk micromachining. This bulk micromachining can be done either in a wet or dry process, and in each case it can be either isotropic or anisotropic. Other MEMS-specific etching is done on quartz or glass, using HF-based solutions or ammonium fluoride.

2.3.3.1   Silicon Wet Isotropic Etching

The most widely used isotropic etchant is a mixture of HNO_3, HF, and CH_3COOH, and this system proceeds by oxidation followed by dissolution of the oxide. Since the oxide is removed in the etch, masking materials such as silicon nitride, silicon carbide, or gold have to be used. The etch rate, surface roughness, and the geometrical aspects at the edges and corners of features depend on the precise composition of the etchant. All of these properties are difficult to control and even very small changes in temperature, agitation, and composition can cause large changes in the etch properties. Thus, the usefulness of this etchant is severely restricted. Etch rates as high as 1,000 μm/min have been reported, so the etch may be useful for removing large quantities of bulk silicon where precise definition is not required. Another potentially useful property is the dependence of the etch rate on the silicon dopant concentration. A solution of HF:HNO3:CH_3COOH mixed in a 1:3:8 ratio etches silicon doped at 10^{20} cm^{-3} 15 times faster than silicon doped at 10^{17} cm^{-3}, both for n- and p-type silicon. This provides an alternative etch stop to the usual etch stop method using anisotropic etchants described in the following section.

2.3.3.2   Silicon Wet Anisotropic Etching

There are many chemicals and mixtures that etch silicon anisotropically including the alkali metal hydroxides, simple and quarternary ammonium hydroxides, ethylenediamine mixed with pyrochatechol (EDP), hydrazine, and amine gallates. Many of these are still the subject of research and in practice only KOH, tetra methyl ammonium hydroxide (TMAH), and EDP are regularly used in MEMS manufacturing. The common properties of these etchants are that the etch rate is dependent on the crystal plane and that they selectively etch n-type or lightly p-doped silicon compared to heavily p-doped silicon. Without exception, the slowest etching planes are the {111} planes, but the fastest etching planes depend on the precise composition of the etchant. The other planes of interest are the {100} and {110} planes, which, although not the fastest etching planes, etch at a much faster rate than the {111} planes. Relative etch rates of 400 between the {100} and {111} planes are typical, for example, with KOH etching [10]. For all these etchants the etch rate drops significantly for heavily p-doped silicon. This property can be used to create etch stop layers, making it possible to fabricate a variety of structures, in which the structure is formed from the heavily doped material. The level of boron doping required for an etch stop layer is of the order of 5×10^{19} cm^{-3} and the etch rate selectivity is of the order of 1,000:1 [11]. An illustration of the use of etch stop using boron doping is shown in Figure 2.10.

The crucial difference between these etchants is in the etch rates of the masking and other materials that are deposited on the substrate. Suitable masks for KOH are silicon nitride or silicon carbide, which etch at negligible rates. Silicon dioxide, on the other hand, is not an ideal mask due to an etch rate that is typically 1/200 of the etch rate of {100} silicon. This may suffice in some circumstances, but for removing large amounts of silicon, the thickness of the oxide mask required is impractical. Another important consideration is that KOH is corrosive and therefore will damage metals such as aluminum. Refractory metals, such as gold and titanium, however, are not attacked. Silicon dioxide can be used as a mask when etching with TMAH, since the etch rate is negligible. This is a clear advantage. Another advantage is that it is possible to reduce the etch rate of aluminum to an acceptable level by the addition of silicon, polysilicic acid [12], $(NH_4)_2CO_2$, or $(NH_4)HPO_4$ to the etchant to lower the pH [13]. The drawback to this is that hillocks and rough surfaces are produced. These can be alleviated to some extent by the addition of an oxidizer such as ammonium peroxydisulfate [14]. Both oxide and nitride can be used as a mask for etching in EDP and, in addition, many metals are not attacked by EDP. One exception is aluminum, although the etch rate of aluminum for some formulations of the etchant can be reduced to useful proportions [15]. It is however extremely hazardous, very corrosive, carcinogenic, and has to be used in a reflux condenser. The surface roughness of the etched surface is also dependent on the etchant used. For a 30%wt KOH solution at 70°C, the mean surface roughness of the {100} plane is of the order of a few nanometers after etching ~200 μm. The smoothest surfaces obtained with TMAH are at concentrations above 20%wt where the mean surface roughness is of the order of 100 nm. Unfortunately, at these concentrations the pH is too high to make effective use of the methods used to reduce the aluminum etch rate mentioned above. A typical formulation for EDP is 750 ml ethylenediamine, 120g pyrochatechol, and 100 ml water used at 115°C. With this formulation surfaces comparable to KOH etched surfaces can be obtained. A comparison highlighting the main differences between these etchants can be found in Table 2.5.

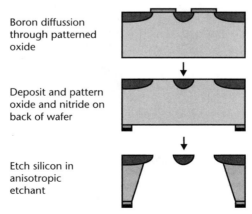

Boron diffussion through patterned oxide

Deposit and pattern oxide and nitride on back of wafer

Etch silicon in anisotropic etchant

Figure 2.10   Boron etch stop technique. In this illustration the technique is used to create freestanding structures such as cantilever beams.

Table 2.5   Comparison of Commonly Used Silicon Anisotropic Etchants

Etchant	Etches Aluminum	Etches Oxide	Silicon Surface	Advantages	Disadvantages
KOH	Yes	Yes	Very good	Easy to use and dispose of	Etches aluminum and oxide
EDP	Yes (but some formulations do not etch aluminum)	No	Good	Does not etch oxide	Hazardous, difficult to use, not clean-room compatible
TMAH	Yes	No	Good	Clean-room and IC process compatible	Etches aluminum
TMAH:Si	No	No	Poor	Does not etch aluminum	Poor surface finish

Etching silicon in these etchants results in three-dimensional structures bounded principally by {111} planes, but also by other planes. The simplest structures are made in {100} silicon. Illustrations showing the anisotropic etch property and the structures that are formed can be seen in Figure 2.11. The intersection between a {111} plane and the {100} surface of the silicon is in a {110} direction. Four such planes intersect the surface, such that the lines of intersection on the silicon surface are at right angles to each other. Each set of planes is inclined at an angle of 54.7 with respect to the surface. Etching is usually done through a window in a masking layer, and if the edge of the window is parallel to the intersection between a set of {111} planes and the surface, then the {111} facet that reaches the surface at this edge is

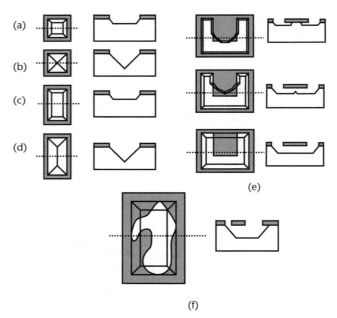

Figure 2.11   Illustration of wet anisotropic etching in {100} silicon showing plain views on the left-hand side and cross-sectional views on the right-hand side: (a) square opening in mask with the silicon etched for a relatively short time; (b) square opening in mask with the silicon etched until inverted pyramid forms; (c) rectangular opening in mask with the silicon etched for a relatively short time; (d) rectangular opening in mask with the silicon etched until V-groove is formed; (e) sequence showing undercutting to form a cantilever beam in the masking material; and (f) etching through arbitrarily shaped opening in mask.

gradually exposed as the etch proceeds. Using the anisotropic etch property, various shapes such as inverted pyramidal holes, V-shaped grooves, and flat bottomed trenches with sidewalls sloping at 54.7 can be formed. If the edge of the window is not in a {111} plane, then the mask is undercut and various crystal facets appear, although, by etching for a sufficient length of time, these crystal facets will eventually be eroded and a {111} plane will eventually be revealed. In addition to holes, silicon structures bounded by the {111} planes can also be formed. These are usually in the form of trapezoidal bosses bounded by the four {111} planes. In this case other crystal planes are exposed where the {111} planes meet at the corners of the structure, resulting in severe undercutting at these corners. By careful mask design, this undercutting can be avoided such that the corner of the boss is perfectly formed from two {111} planes. This technique is called corner compensation and a number of different patterns have been designed to achieve this [16, 17]. One of the simpler corner compensation techniques is shown in Figure 2.12. A particularly interesting feature (shown in Figure 2.13) that can be formed in KOH solutions is a vertical {100} face. This forms if the edge of the mask window lies in one of the {100} planes passing vertically through the wafer. However, as with other crystal facets this face is etched until two intersecting {111} planes are reached. Etching indefinitely through any arbitrarily shaped window will ultimately produce a rectangular feature bounded by four {111} planes that intersect in pairs. Conversely, etching indefinitely around any arbitrarily shaped island feature will ultimately remove the feature.

In addition to {100} silicon wafers, it is also possible to obtain wafers with other orientations, such as {110} and {111}. Although interesting features can be produced by anisotropic etching on these wafers, they are less versatile than {100} wafers. A pair of {111} planes pass vertically through {110} orientation wafers, which enables deep high aspect ratio grooves to be etched. The potential for

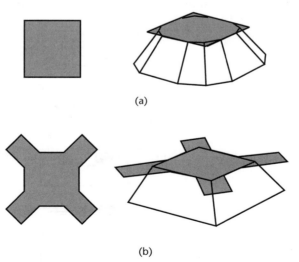

(a)

(b)

Figure 2.12   (a) Illustration showing the shape of a silicon boss formed beneath a square in the mask. Undercutting at the intersection of the {111} planes occurs at each corner of the square. (b) With simple compensation features added to the corners of the square it is possible to etch the structure such that the {111} planes meet perfectly at each corner. In this particular case the compensation feature at each corner is at an angle of 45° to the edge of the square and the width of the feature is twice the required etch depth.

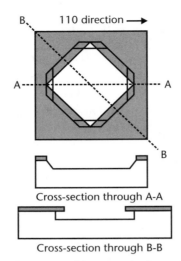

Figure 2.13 Illustration showing how vertical faces can be formed in {100} silicon. The edges of the opening in the mask are aligned to the <100> orientation. The vertical face is etched at the same rate as the horizontal surface.

producing useful anisotropically etched structures on {111} silicon is greater than on {110} silicon. By combining dry etching with anisotropic etching it is possible to form a variety of freestanding structures in the plane of the wafer; a trench is dry etched into {111} silicon in the shape of the structure to be formed; the sidewalls of the trench can be protected by, for example, oxidizing the silicon and if the bottom of the trench is then dry etched a little further, the silicon thus exposed can be etched in an anisotropic wet etch, which will remove the silicon laterally beneath the structure. The lower surface of the structure will be protected from the etchant by virtue of the fact that it is a slow etching {111} plane. An illustration of this process is shown in Figure 2.14.

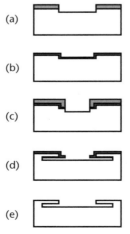

Figure 2.14 Process sequence for wet anisotropic etching of {111} silicon: (a) a trench is dry etched in the silicon; (b) silicon is oxidized; (c) a second trench is dry etched at the bottom of the first trench; (d) resist is removed and silicon is etched in wet anisotropic etch; and (e) oxide is removed.

Commercial software is available with which it is possible to simulate the results of anisotropic etching [18]. This is useful in predicting the outcome from employing various mask designs, and thereby it facilitates design of the layout.

2.3.3.3   Silicon Dry Isotropic Etching

Dry isotropic etches are not often used for bulk micromachining. However, there are a few examples. Etching in an SF_6 plasma has been used as an alternative to wet anisotropic etching. The advantage is that a resist mask can be used and the problem of protecting other materials on the wafers is easily overcome. The etch rates are comparable to wet etching, but it is considered to be slow because of the inability to process large numbers of wafers at a time. In another application the high selectivity of silicon dry isotropic etching in an SF_6/O_2 plasma against etching aluminum and silicon dioxide is utilized. By undercutting the aluminum after completion of a CMOS process, suspended structures can be made.

2.3.3.4   Silicon Dry Anisotropic Etching

Anisotropic etching of silicon has been used in the microelectronics industry for many years. The main applications have been in forming deep trench capacitors for memory devices and in constructing isolation trenches between active devices. However, for these applications the etch depth and aspect ratio used have been at most 10μm and 10:1, respectively. For MEMS applications there is a requirement for much deeper trenches, often through the thickness of the wafer, and in some cases higher aspect ratios are needed. The dry etch process that can achieve this is called DRIE and there are currently two different processes being used by equipment manufacturers. In each case the deep anisotropy is achieved by passivation of the sidewalls of the trench as it is etched. One process uses cryogenic cooling of the wafer to liquid nitrogen temperatures, which, it is believed, causes condensation of the reactant gases on to the silicon surface, thus passivating it. On horizontal surfaces, such as the bottom of trenches, this condensate is removed by ion bombardment and these surfaces are therefore etched. SF_6 is typically used because of the high etch rates that can be achieved. The passivation can be enhanced by the addition of oxygen to the plasma, which results in oxidation of the sidewalls. Possible problems with the cryogenic approach are in maintaining the temperature of structures during the etch process. Some structures may become thermally isolated resulting in adverse thermal stress. The other process is one patented by Bosch which uses alternate etch and passivation steps [19]. The passivation is achieved by deposition of a polymer using C_4F_8 as a source gas. Concurrent with this deposition step is some ion bombardment, and this prevents the formation of polymer on the bottom of the trench. The polymer on the bottom of the trench is, in any case, removed by energetic ions during the following etch step done in SF_6. The cycle time for this deposition/etch process is typically about 5 seconds with etch rates of between 1.5 and 4μm/min. Aspect ratios of more than 40:1 can be obtained. A limitation encountered with both DRIE processes is the etch rate dependence on trench width. The etch process is diffusion limited and for trench widths less than 60μm the etch rate becomes progressively slower as the trenches become narrower. This limitation can be overcome in design by avoiding large disparities in the feature sizes on the mask.

2.3.4 Surface Micromachining

Although the most popular sensor fabrication technology is bulk micromachining using deep wet or dry etching below the surface of the silicon, surface micromachining provides a complementary technique in which materials are added above the surface. These materials often act as spacers or sacrificial layers to be removed at a later stage to produce freestanding structures and moveable parts. A typical surface-micromachined structure, illustrated in Figure 2.15, uses silicon dioxide as the sacrificial layer and polysilicon for the structural layer [20]. In the most basic process the oxide is usually deposited by CVD because this etches more rapidly than thermally grown oxides. Holes are etched in the oxide to form anchor points for the structural layer. Polysilicon is then deposited and patterned and the oxide is etched laterally beneath the structure in a hydrofluoric acid etch. The structures thus formed can be designed to move either horizontally or vertically, in and out of the plane of the wafer. Complex structures can be made by stacking four or five alternating layers of polysilicon and silicon dioxide. Although other sacrificial and structural layer combinations, such as polysilicon and silicon nitride [21], nickel and copper [22], and copper and Ni/Fe [23], have been employed, the oxide and polysilicon combination has been by far the most prevalent. The challenges with surface micromachining are to control the mechanical properties of the structural layer to prevent the formation of internal residual stresses and to ensure that the released structures do not stick to the surface of the wafer after they are dried. Preventing stress in the polysilicon layer is done by carefully controlling the deposition and annealing conditions. Another method is to deposit alternate layers of amorphous silicon at 570°C, which is tensile, and polysilicon at 615°C, which is compressive [4]. In surface micromachining, structures are generally released by wet etching the sacrificial layer followed by rinsing in water. This gives rise to capillary forces as the wafers are dried causing the structures to stick to the underlying substrate. Many methods for preventing this stiction have been developed. One approach is to process the wafers through a series

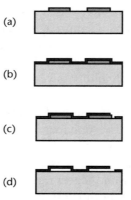

(a)

(b)

(c)

(d)

Figure 2.15 Typical surface-micromachined structure: (a) oxide deposited and etched; (b) polysilicon deposited; (c) polysilicon patterned and etched to create access holes through to the oxide; and (d) oxide etched selectively in HF to leave freestanding polysilicon structures.

of rinses such that the final rinse is in a hydrophobic liquid such as hexane or toluene [24]. Another approach relies on changing the phase of the liquid in which the wafers are finally rinsed, either by freezing or heating the liquid into a supercritical state. T-butyl can be frozen solid and sublimed at low vacuum pressures [25]. In the supercritical drying method the final rinse is done in a pressure vessel in liquid CO_2, which is then raised into a supercritical state. In this state, the interface between the liquid and gas phases is indistinguishable and there are no surface tension forces [26]. Thus, the CO_2 gas can be vented without affecting the structures. Other methods involve dry release of the structures. One such method is etching the oxide in an HF vapor [27], and another entails rinsing the wafers in acetone then adding photoresist, which fills the gaps beneath the structures after the acetone has evaporated. The resist can then be removed in an oxygen plasma. There are yet other methods, which have the added benefit of preventing the stiction of the structures when they are in use, that rely on modifying the surfaces of the structures using self-assembled monolayers formed, for example, from DDMS [$(CH_3)_2SiCl_2$] or ODTS [$(CH_3(CH_2)_{17}SiCl_3$] [28, 29]. In-use stiction can also be prevented by coating released structures in a fluorocarbon by PECVD [30].

2.3.5   Wafer Bonding

There are many wafer bonding processes currently available, and the choice of which is most suitable depends on the particular application and the materials involved. Bonding processes are as likely to be used at the beginning of a process sequence as at the end. For example, bonding is used in the fabrication of SOI wafers, but also in device fabrication processes, such as the bonding together of wafers to form the vacuum cavity of an absolute pressure sensor, as well as at the end of processes to package devices. In all the bonding processes described here, surface cleanliness is of paramount importance. Particulates trapped between wafers can lead to the formation of voids and ultimately failure of the bond. Also, as with most micromachining processes, attention has to be paid to the stress created by the process and this is particularly relevant to mechanical sensors. For this reason the materials bonded together and the material, if any, used to bond them should have a minimal thermal mismatch, otherwise temperature changes will result in strain being applied to devices. Bonds should also be stable over the life of the device. Any plastic flow, or creep, may alter the output of a device affecting its calibration and long-term stability. The bond should also be strong enough to withstand any strain the device is likely to be subjected to. If bonding is used for packaging devices it should provide, if possible, some degree of strain relief for the device.

2.3.5.1   Silicon Fusion Bonding

Silicon fusion bonding is a direct silicon-to-silicon bonding technique that does not require any melting alloys, glass layers, or polymer glues. As a result little or no stress due to thermal mismatch is introduced into the assembly, and the perfectly matched thermal expansion coefficients of the two wafers ensure that this low stress condition is preserved. The process requires the surfaces to be planar, clean, and hydrated. The hydration step can be carried out in a number of ways, either by boiling in nitric acid or ammonium hydroxide or simply by performing a standard RCA clean (so called because it was developed at the RCA Company). Two wafers can be joined together at room temperature, resulting in an immediate weak bond due to van der Waals forces. The bond is then strengthened by heat treating in a furnace or by RF or micro-wave heating at temperatures above 800°C [31]. At this temperature an hermetic seal is formed between the silicon wafers making it possible to fabricate sealed cavities. The exact nature of the bonding chemistry is not yet fully understood, but it is believed that Si-O-Si bonds and water molecules form at the interface as the tempera-ture is increased. The water molecules subsequently break and diffuse into the silicon. Silicon fusion bonding is also used to bond silicon to oxidized silicon wafers as in the manufacture of SOI wafers. The oxide thickness in this process can be as much as 4 μm, and typically temperatures of 1,100°C are used to obtain a permanent chemical bond. The formation of silicon dioxide to silicon dioxide fusion bonds at 1,100°C has also recently been reported [32]. Silicon-to-silicon bonding can also be done with a thin intermediate layer of sputtered, evaporated, or spun on glass. The wafers are clamped together and heated to the melting point of the glass, typically between 415°C and 1,150°C depending upon the glass. The assembly is then cooled and the glass solidifies and the process can be used to form a hermetic bond. The relatively thin layer of glass minimizes the residual stresses in the assembly.

2.3.5.2 Anodic Bonding

Otherwise known as electrostatic bonding, this process is used to bond silicon to glass. The method uses electrostatic attraction between the glass and the silicon to facilitate bonding, enabling this to be done at much lower temperatures than would otherwise be possible [33]. Circular glass substrates with a thermal expansion coeffi-cient closely matched to that of silicon are readily available. For example, Pyrex, which has a coefficient of thermal expansion of 3.25×10^{-6}/°C, is often used for this purpose. The process is carried out by placing the silicon on a grounded hotplate with the glass placed in contact with the silicon, as shown in Figure 2.16. The hot-plate is used to heat the silicon and glass to a temperature of between 350°C and 500°C and a negative potential of about 1 kV is applied to the glass. Extremely mobile positive ions, mainly sodium, in the glass drift towards the negative electrode leaving a negative charge on the glass side of the silicon-glass interface. A high electric field is generated between this fixed negative charge and positive charge in the silicon, thus pulling the glass and silicon together and facilitating the chemical bond. Hermetically sealed cavities can be obtained without difficulty. Residual stresses do occur and bonding temperatures as low as possible are recommended to keep these to a minimum. Narrow metal tracks, 100 to 200 nm thick, on the silicon do not compromise the hermiticity of the bond making it possible to run electrical feedthroughs into sealed cavities. Successful bonding even with a thin layer of silicon dioxide on the silicon (up to 100 nm thick) can be achieved using this technique.

Furthermore, the anodic bonding technique can be used to bond silicon to sili-con by sputtering or evaporating a thin layer of glass onto one of the surfaces. Alter-natively, glass layers can be spun on to wafers using a spin-on-glass. Wafers with sputtered glass layers of 0.5- to 4.0-μm thickness have been bonded to silicon [34]. The applied voltages required for this are much less than for bonding to glass sub-strates. In some cases bonding was achieved with as little as 30V applied. However,

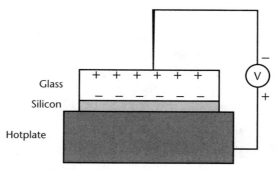

Figure 2.16 Setup for anodic bonding.

the deposition rate for glass sputtering is very low, and obtaining a uniform thickness as the layer grows is not a trivial task. Bonding to silicon with evaporated glass is also possible [35]. High compressive stress, much of which can be annealed out for layers up to $10\,\mu$m thick, can cause serious bowing of the wafers and control of the glass composition due to loss of sodium during evaporation is difficult. Spin-on-glass layers suitable for anodic bonding have been prepared [36]. One such preparation consists of a mixture of TEOS, MTEOS, and a potassium salt dissolved in ethanol with which layers up to $6\,\mu$m thick have been deposited. The layers are reported to have low intrinsic stress (30 MPa), are stable at temperatures above the typical bonding temperature (420°C), and have good uniformity across a 6-inch wafer ($\pm$20 nm) and low surface roughness (rms: 0.5 nm).

2.3.5.3 Eutectic Bonding

Eutectic bonding utilizes the eutectic properties of two materials combined, the combination having a lower melting point than each of the individual constituents. A common combination is silicon-gold for which the eutectic state occurs at a temperature of 363°C, the lowest bonding temperature for this system [37]. A typical composition is 97.1% Au and 2.85% Si by weight, which can be bonded at a temperature of 386°C. The process involves placing the gold in contact with the silicon and heating, causing the gold atoms to diffuse into the silicon. When the eutectic composition is reached, a liquid layer is formed at the interface and the eutectic alloy grows until the gold is exhausted. The alloy can then be cooled slowly, causing it to solidify and hence forming the bond. The gold can be deposited on one of the silicon surfaces by evaporation or sputtering, or a preform can be placed between the two surfaces to be bonded. The joints formed with this technique are hermetic. A drawback with this process is that the mismatch in thermal expansion coefficients results in high residual stresses within the alloy. In addition, these stresses change with time due to creep.

2.3.5.4 Adhesive Bonding

Micromachined components can be bonded together using a number of commercially available adhesives that possess a wide range of characteristics. There are, for example, numerous epoxies available with a wide range of thermal, electrical, and mechanical properties. Other adhesives include PMMA, polyamides, silicone rubbers, and negative photoresist [38]. Waxes can also be used as temporary adhesives during processing. Generally such bonds can be achieved at temperatures under 150°C and are relatively soft, providing some degree of stress relief for the wafers. They are, however, unsuitable for hermetic seals, can degrade over long periods of time, and can possess poor thermal stability [39].

2.3.5.5 Vacuum Bonding

A bonding stage may be carried out during a fabrication process to trap a vacuum in a cavity, which may, for example, contain a micromachined feature such as a resonator that requires a sufficiently high vacuum to operate. In this case the bonding process has to be carried out in a vacuum, with the component parts being brought together under vacuum. If anodic bonding is used, gas generation during the anodic bonding process and gas desorption from the inner surface of the sealed cavity necessitate the use of a getter within the cavity that is able to withstand the bonding temperature [40]. Vacuums of 4×10^{-5} torr have been achieved using this approach.

2.3.5.6 Bond Aligning

Some devices require accurate alignment between the two components being bonded together. Equipment is commercially available to enable wafers to be aligned and bonded to each other with an accuracy of a few micrometers. Glass-to-silicon bonding alignment is straightforward because of the transparency of the glass. For silicon-to-silicon bonding the aligners use infrared systems similar to those used in double-sided alignment in lithography. Equipment is available with various options so that anodic bonding, eutectic bonding, or silicon fusion bonding can be done in various environments including vacuum.

2.3.6 Thick-Film Screen Printing

Screen printing is one of the oldest forms of graphic art reproduction and involves the deposition of an ink (or paste) onto a base material (or substrate) through the use of a finely woven screen having an etched pattern of the desired geometry. The term "thick-film" can often be misinterpreted, so it is worth noting that it does not necessarily relate to the actual thickness of the film itself. The typical range of thicknesses for thick-film layers, however, is between 0.1 and 100 μm. The process is commonly used for the production of graphics and text onto items such as T-shirts, mugs, and pencils, and it is very similar to that used for microelectronic thick-films. The degree of sophistication for the latter is, however, significantly higher in order to provide high-quality, reproducible films for use in electronic systems.

The technology for manufacturing thick-film hybrid microelectronic circuits was introduced in the 1950s. Such circuits typically comprised semiconductor devices, monolithic integrated circuits, discrete passive components, and the thick-films themselves. In the early days of the technology, the thick-films were mainly resistors, conductors, or dielectric layers. Evidence of thick-film hybrids can still be found in many of today's commercial devices such as televisions, calculators, and telephones.

A typical mask, or screen, is made of a finely woven mesh of stainless steel, nylon, or polyester, which is mounted under tension on a metal frame and coated with a UV-sensitive emulsion. The desired pattern is exposed onto the screen photographically, leaving open areas through which a paste can be deposited. The pastes comprise a finely divided powder (typically 5-μm average particle size), a glass frit, and an organic carrier that gives the ink the appropriate viscosity for screen-printing. Typically, thick-film pastes are resistive, conductive, or dielectric in nature and are deposited onto substrates such as alumina or insulated stainless steels. Silicon has also been used as a base material to make devices such as micropumps, like the one described by Koch et al. [41].

The screen is held in position at a distance of around 0.5 mm away from and parallel to the substrate. The paste is poured onto the upper surface of the screen. A squeegee then traverses across the screen under pressure, forcing the ink through the open areas and onto the substrate leaving the required pattern. A wide variety of commercial screen printers, specifically developed for thick-film processing, are available for this task.

After screen-printing, the deposited films are dried in either a box oven or, more typically, in an infrared belt drier. This is usually achieved at a temperature of around 150°C. This stage of the processing removes the organic carriers that were present in the paste and produces a rigid film that can be handled or even over-printed with further layers. The final step is to fire (or sinter) the films to form a solid composite material. The glass frit melts during the annealing phase and bonds the film to the substrate and also binds the active particles together. This phase of processing is undertaken in a belt furnace at temperatures up to 900°C. The furnace operator has control over the peak temperature, throughput speed, and dwell time. After firing, the film is firmly attached to the substrate and additional screen printed layers can be added if needed, and the print, dry, fire cycle is then repeated. In addition to fabricating circuits, thick-film technology has been widely used as a means of making a variety of sensors [42].

2.3.7   Electroplating

Electroplating is used in many MEMS processes to obtain thick layers of a metal or alloy. Processes for depositing various different metals and alloys have been adapted for electroplating onto silicon. Those most commonly used are for depositing Ni, NiFe, Au, and Cu. A plating base, such as Ti, Ti/Pt, or Cr/Au has to be deposited onto the silicon. The uniformity, morphology, and composition of the deposited layer depend primarily on the design and operating parameters of the electroplating bath. To produce a patterned electroplated layer on silicon, a resist pattern (referred to as a mold) has to be applied. This resist pattern also influences the growth uniformity, morphology, and composition of the layer. Therefore, it is common in electroplating MEMS structures that some effort has to be put in on optimization to achieve a successful outcome. Even by MEMS standards, the deposited layers are unusually thick, sometimes of the order of 500 μm. There are photoresists (discussed earlier in this chapter) available with which it is possible to form molds of this thickness for electroplating. The process of electroplating using an optical photoresist as a mold is sometimes referred to as UV-LIGA in contrast with the X-ray LIGA process described in the next section. The highest aspect ratios that can be achieved with such UV-LIGA processes are much less than 10:1. For example, using the UV-sensitive negative resist, PMER N-CA3000, near vertical sidewalls with an aspect ratio of the order of 6:1 and resist several tens of microns thick have been obtained [43]. An example of electroplated nickel pillars using an optical resist as a mold is shown in Figure 2.17. Although not on the same scale as X-ray LIGA the UV-LIGA process is a simple practical process for MEMS. If higher aspect ratio structures are required, as is often the case, then X-ray LIGA must be used. This process, which combines X-ray lithography to form molds with electroplating, is more generally known as LIGA.

2.3.8   LIGA

In the LIGA process a resist layer several hundred microns thick is exposed through a mask to synchrotron X-ray radiation. By developing the exposed resist layer, a mold is formed that can be filled with metal by electroplating. After stripping the remaining resist, a metallic microstructure anchored to the substrate is obtained. To make devices with moving parts, the LIGA structure can be formed partly on a sacrificial layer, such as Ti, which can then be selectively removed to free part of the structure, with another part of the structure anchored to the substrate. The use of a highly collimated X-ray source enables structures with near vertical sidewalls and aspect ratios of more than 100:1 to be made. A multilevel LIGA process has also been developed for fabricating stacked electroplated structures. The LIGA process has been used in a wide variety of devices and applications including fluidic devices [44], optical components [45], gears [46], shock and acceleration sensors [23], and for making electrodischarge machining electrode arrays [47]. Although LIGA is not compatible with CMOS processes, a prototype flip-chip and selective bonding process has been developed to combine LIGA structures with IC substrates [48]. A drawback to the LIGA process is the cost, both of the masks and of access to the X-ray facilities.

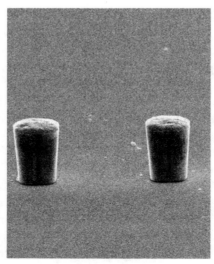

Figure 2.17   Electroplated nickel pillars formed through a photoresist mold. The pillars are approximately 20 μm high.

2.3.9 Porous Silicon

The etch rate of silicon in an electrochemical etch process can be controlled by applying an anodic potential to the silicon with reference to a platinum counter electrode. A typical setup for electrochemical etching is shown in Figure 2.18. The anodic potential causes holes to be drawn towards the surface of the silicon, which attract OH^- ions from the etching solution. These holes promote oxidation of the silicon, and if an HF/H_2O solution is used for the etchant, the oxide is rapidly etched, forming a smooth electropolished surface. By reducing the applied potential, the silicon is not completely oxidized and pores or voids are formed in place of the smoothly etched surface. These pores can penetrate to great depths in the silicon, forming a material known as porous silicon. The shape and size of the pores depend on many factors including the type and orientation of the silicon, the etchant used, and the current density. Porous silicon can be selectively grown, for example, depending on the doping concentration of the silicon, and selectively etched, either in a weak KOH solution or, since it is readily oxidized, by oxidation and etching in HF. This makes it useful as a sacrificial layer.

2.3.10 Electrochemical Etch Stop

The etch rate of silicon in electrochemical etching depends on the applied potential, and as this is made more positive a passivation potential is reached where SiO_2 is formed passivating the surface and thereby preventing etching. With KOH etching, this phenomenon can be used in an etch stop process where an n-type epitaxial layer has been grown on a p-type silicon substrate and the n-type layer is biased at its passivation potential. If the p-type substrate is not biased, the potential on it will float at its open circuit potential, which means that it will etch as normal. When the p-type substrate has etched through to the n-type layer, the passivation potential at which the n-type layer is held prevents further etching. This etch stop process is illustrated in Figure 2.19.

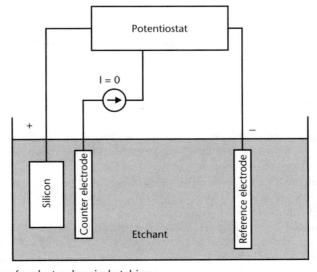

Figure 2.18 Setup for electrochemical etching.

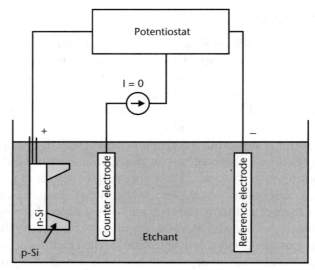

Figure 2.19   Setup for electrochemical etch stop.

2.3.11   Focused Ion Beam Etching and Deposition

Focused ion beam (FIB) technology enables localized milling and deposition of conductors and insulators with high precision. Milling can be accomplished by scanning a focused beam of energetic ions across a surface. Deposition occurs by a CVD reaction induced by the ion beam. A wide variety of materials can be etched or deposited by FIB technology and many different ions can be used. Feature sizes of the order of 1 μm with tolerances of 0.1 μm can be achieved. However, since it is a serial single substrate process, it can be slow and time consuming. Examples of the use of FIB technology are the release of micromechanical structures such as accelerometers and actuators [49] and the deposition of W and SiO_2 [50].

References

[1]   Hedlund, C., et al., "Anisotropic Etching of Z-Cut Quartz," *J. Micromech. Microeng.*, Vol. 3, 1993, pp. 65–73.

[2]   Kruelevitch, M., et al., "Thin Film Shape Memory Alloy Actuators," *J. Microelectromech. Syst.*, Vol. 5, 1996, pp. 270–282.

[3]   Kovacs, G., *Micromachined Transducers Sourcebook*, New York: WCB/McGraw-Hill, 1998.

[4]   Yang, J., et al., "A New Technique for Producing Large-Area As-Deposited Zero-Stress LPCVD Polysilicon Films: The Multipoly Process," *J. Microelectromech. Syst.*, Vol. 9, 2000, pp. 485–494.

[5]   Seideman, V., S. Bütefisch, and S. Büttgenbach, "Application and Investigation of In-Plane Compliant SU8-Structures for MEMS," *Transducers '01*, Munich, Germany, June 10–14, 2001, pp. 1616–1619.

[6]   Daniel, J., et al., "Large Area MEMS Fabrication with SU8 Photoresist Applied to an X-Ray Image Sensor," *Proc. SPIE*, Santa Clara, CA, September 2000, pp. 40–48.

[7]   Köser, H., et al., "A High Torque Density MEMS Magnetic Induction Machine," *Transducers '01*, Munich, Germany, June 10–14, 2001, pp. 284–287.

[8] Tseng, F.-G., and C.-S. Yu, "Fabrication of Ultrathick Micromolds Using JSR THB-430N Negative Photoresist," *Transducers '01*, Munich, Germany, June 10–14, 2001, pp. 1620–1623.

[9] Kukharenka, E., et al., "Electroplating Moulds Using Dry Film Thick Negative Photoresist," *Micromechanics Europe MME '02*, Sinaia, Romania, October 6–8, 2002, pp. 95–98.

[10] Seidel, H., et al., "Anisotropic Etching of Crystalline Silicon in Alkaline Solutions I: Orientation Dependence and Behavior of Passivation Layers," *Journal of the Electrochemical Society*, Vol. 137, No.11, 1990, pp. 3612–3626.

[11] Seidel, H., et al., "Anisotropic Etching of Crystalline Silicon in Alkaline Solutions II: Influence of Dopants," *Journal of the Electrochemical Society*, Vol. 137, No. 11, 1990, pp. 3626–3632.

[12] Hoffmann, E., et al., "3D Structures with Piezoresistive Sensors in Standard CMOS," *Proc. of IEEE Micro Electro Mechanical Systems*, Amsterdam, the Netherlands, January 29–February 2, 1995, pp. 288–293.

[13] Tabata, O., "pH-Controlled TMAH Etchants for Silicon Micromachining," *Sensors and Actuators*, Vol. A53, 1996, pp. 335–339.

[14] Klaassen, E., et al., "Micromachined Thermally Isolated Circuits," *Proc. of the 1996 Solid-State Sensor and Actuator Workshop*, Hilton Head Island, SC, June 3–6, 1996, pp. 127–131.

[15] Moser, D., *CMOS Flow Sensors*, Doctoral dissertation, Swiss Federal Institute of Technology, Zurich, Switzerland, 1993.

[16] Madou, M., *Fundamentals of Microfarication*, Boca Raton, FL: CRC, 1997.

[17] Bütefisch, S., A. Schoft, and S. Büttgenbach, "Three-Axes Monolithic Silicon Low-g Accelerometer," *J. Microelectromech. Syst.*, Vol. 9, 2000, pp. 551–556.

[18] Zielke, D., R. Lieske, and J. Will, "Automatic Transfer from Bulk-Silicon Technology Simulation into the FEM-Environment," *Transducers '01*, Munich, Germany, June 10–14, 2001, pp. 272–275.

[19] Lärmer, F., and P. Schlip, "Method of Anisotropically Etching Silicon," German Patent No. DE 4,241,045, 1994.

[20] Howe, R., "Surface Micromachining for Microsensors and Microactuators," *Journal of Vacuum Science and Technology B*, Vol. 6, 1988, pp. 1809–1813.

[21] Berenschot, N., et al., "Advanced Sacrificial Poly-Si Technology for Fluidic Systems," *Transducers '01*, Munich, Germany, June 10–14, 2001, pp. 624–627.

[22] Choi, Y.-S., et al., "Fabrication of a Solenoid-Type Microwave Transformer," *Transducers '01*, Munich, Germany, June 10–14, 2001, pp. 1564–1567.

[23] McNamara, S., and Y. Gianchandani, "A 19-Element Shock Sensor Array for Bi-Directional Substrate-Plane Sensing Fabricated by Sacrificial LIGA," *Transducers '01*, Munich, Germany, June 10–14, 2001, pp. 450–453.

[24] Scheepers, P., et al., "Investigation of Attractive Forces Between PECVD Silicon Nitride Microstructures and an Oxidized Silicon Substrate," *Sensors and Actuators*, Vol. A30, 1992, pp. 231–239.

[25] Takeshima, N., et al., "Electrostatic Parallelogram Actuators," *Transducers '91*, San Francisco, CA, June 24–27, 1991, pp. 63–66.

[26] Mulhern, G., D. Soane, and R. Howe, "Supercritical Carbon Dioxide Drying of Microstructures," *Transducers '93*, Yokahama, Japan, June 7–10, 1993, pp. 296–298.

[27] Lee, Y.-I., et al., "Dry Release for Surface Micromachining with HF Vapor-Phase Etching," *J. Microelectromech. Syst.*, Vol. 6, 1997, pp. 226–233.

[28] Kim, B., et al., "A New Organic Modifier for Anti-Stiction," *J. Microelectromech. Syst.*, Vol. 10, 2001, pp. 33–40.

[29] Ashurst, W., et al., "Dichlorodimethylsilane as an Anti-Stiction Monolayer for MEMS: A Comparison to the Octadecyltrichlosilane Self-Assembled Monolayer," *J. Microelectromech. Syst.*, Vol. 10, 2001, pp. 41–49.

[30] Man, P., B. Gogoi, and T. Harada, "Fabrication of an S-Shaped Microactuator," *J. Microelectromech. Syst.*, Vol. 6, 1997, pp. 25–34.

[31] Thompson, K., et al., "Si-Si Bonding Using RF and Microwave Radiation," *Transducers '01*, Munich, Germany, June 10–14, 2001, pp. 226–229.

[32] Wu, C.-H., et al., "Fabrication and Testing of Single Crystalline 3C-SiC Piezoresistive Pressure Sensor," *Transducers '01*, Munich, Germany, June 10–14, 2001, pp. 514–517.

[33] Wallis, G., and D. Pomerantz, "Field Assisted Glass-Metal Sealing," *J. Appl. Phys.*, Vol. 40, 1969, pp. 3946–3949.

[34] Esashi, M., et al., "Low Temperature Silicon-to-Silicon Anodic Bonding with Intermediate Low Melting Point Glass," *Sensors and Actuators*, Vol. A23, 1990, pp. 931–934.

[35] de Reus, R., and M. Lindahl, "Si-to Si Wafer Bonding Using Evaporated Glass," *Transducers '95*, Chicago, IL, June 16–19, 1997, pp. 661–664.

[36] Quenzer, H., et al., "Anodic Bonding on Glass Layers Prepared by a Spin-On Glass Process: Preparation Process and Experimental Results," *Transducers '01*, Munich, Germany, June 10–14, 2001, pp. 230–233.

[37] Valero, L., "The Fundamentals of Eutectic Die Attach," *Semiconductor International*, Vol. 7, 1984, pp. 236–241.

[38] den Besten, C., et al., "Polymer Bonding of Micro-Machined Silicon Structures," *Micro Electro Mechanical Systems '92*, Travemunde, Germany, February 4–7, 1992, pp. 104–108.

[39] Trigwell, S., "Die Attach Materials and Methods," *Solid Sate Technology*, April 1995, pp. 63–69.

[40] Henmi, H., et al., "Vacuum Packaging for Microsensors by Glass-Silicon Anodic Bonding," *Transducers '93*, Yokahama, Japan, June 7–10, 1993, pp. 584–587.

[41] Koch, M., et al., "A Novel Micropump Design with Thick-Film Piezoelectric Actuation," *Measurement Science and Technology*, Vol. 8, No. 1, 1997, pp. 49–57.

[42] White, N., and J. Turner, "Thick-Film Sensors: Past, Present and Future," *Measurement Science and Technology*, Vol. 8, No. 1, 1997, pp. 1–20.

[43] Wycisk, M., et al., "Low-Cost Post-CMOS Integration of Electroplated Microstructures for Inertial Sensing," *Sensors and Actuators*, Vol. A83, 2000, pp. 93–100.

[44] Kämper, K.-P., et al., "Microfluidic Components for Biological and Chemical Microreactors," *Proc. of 10th Annual Workshop of Micro Electro Mechanical Systems (MEMS '97)*, Nagoya, Japan, January 26–30, 1997, pp. 338–343.

[45] Mohr, J., "MOEMS Fabricated by the LIGA Technique—An Overview," *Proc. MOEMS '97*, Nara, Japan, November 1997, pp. 221–226.

[46] Mohr, J., et al., "Moveable Microstructures Manufactured by the LIGA Process as Basic Elements for Microsystems," in *Microsystems Technologies 90*, H. Reichl, (ed.), Berlin, Germany: Springer-Verlag, 1990, p. 529.

[47] Takahata, K., and Y. Gianchandani, "Batch Mode Micro-Electro-Discharge Machining," *J. Microelectromech. Syst.*, Vol. 11, 2002, pp. 102–110.

[48] Pan, L.-W., and L. Lin, "Batch Transfer of LIGA Microstructures by Selective Electroplating and Bonding," *J. Microelectromech. Syst.*, Vol. 10, 2001, pp. 25–32.

[49] Reyntjens, S., and R. Puers, "RASTA: The Real-Acceleration-for-Self-Test Accelerometer," *Transducers '01*, Munich, Germany, June 10–14, 2001, pp. 434–437.

[50] Reyntjens, S., D. De Bruyker, and R. Puers, "The NanoPirani—Presumably the World's Smallest Pressure Sensor," *Transducers '01*, Munich, Germany, June 10–14, 2001, pp. 490–493.

CHAPTER 3

MEMS Simulation and Design Tools

3.1 Introduction

Simulation of micromachined systems and sensors is becoming increasingly important. The motivation here is similar to that of the simulation of purely electronic VLSI circuits: before fabricating a prototype, one wishes to virtually build the device and predict its behavior. This allows for the optimization of the various design parameters according to the specifications. As it is a virtual device, parameters can be changed much more quickly than actually fabricating a prototype, then redesigning and fabricating it again. This considerably reduces the time to market and also the cost to develop a commercial device. Simulation software tools for electronic circuits are very mature nowadays, and the level of realism is striking. Often the first fabricated prototype of a novel circuit works in a very similar way as predicted by the simulation. In MEMS, however, this degree of realism cannot be achieved in many cases for two reasons. First, the simulation tools have not reached a similar maturity as their electronic equivalents; and second, and more importantly, simulation of MEMS devices is much more complex. A MEMS device typically comprises many physical domains such as mechanical, electrical, thermal, and optical. All these domains interact and influence each other, making the problem orders of magnitude more difficult.

Any MEMS simulation software uses either of two approaches:

- *System level (or behavioral or reduced order or lumped parameter) modeling:* This approach captures the main characteristics of a MEMS device. It provides a quick and easy method to predict the main behavior of a MEMS device. The requirement is that the device can be described by sets of ordinary differential equations and nonlinear functions at a block diagram level. This approach originated from control system engineering. The multidomain problem is avoided since, typically, the simulation tools are physically dimensionless—only the user interprets the input and output of the various blocks in a physically meaningful way.
- *Finite element modeling (FEM):* This approach originated from mechanical engineering where it was used to predict mechanical responses to a load, such as forces and moments, applied to a part. The part to be simulated is broken down into small, discrete elements—a process called meshing. Each element has a number of nodes and its corners at which it interacts with neighboring elements. The analysis can be extended to nonmechanical loads, for example, temperature. Additionally, finite element simulation techniques have been

successfully applied to simulate electromagnetic fields, thermodynamic problems such as squeeze film damping, and fluidics. FEM results in more realistic simulation results than behavioral modeling, but it is much more computationally demanding and hence it is difficult to simulate entire systems.

3.2 Simulation and Design Tools

3.2.1 Behavioral Modeling Simulation Tools

3.2.1.1 Matlab and Simulink

One of the most popular behavioral modeling tools is Simulink, which is a toolbox within Matlab [1]. It allows the user to perform system level simulation in the time domain. The user chooses blocks from a library that includes linear and nonlinear functions, which are either time continuous or discrete. Examples include gain, integrators and differentiators, z- and s-domain transfer functions, limiters, samplers, mathematical functions, switches, and many others.

Each block has a range of input and outputs. An input can be the output of another block or a source that can be an arbitrary waveform. Any output of a block can be visualized by different types of plots in the time or frequency domain; alternatively it can be stored as a variable to be analyzed or filtered further in Matlab. The software allows user-defined library and hierarchal modeling by defining parameterized subsystems. The software has a purely graphical interface; blocks are chosen by drag and drop and connected by wires drawn on the screen.

As an example, an accelerometer embedded in a force-feedback loop will now be described (for a description of the operating principle of such an accelerometer, see Chapter 8). The mechanical sensing element can be described to the first order by the differential equation of a mass-dashpot-spring system. Furthermore, it will be assumed that the proof mass is limited in its travel range by mechanical stoppers and that it has an optional deflection from its rest position at the start of the simulation. The input is an external inertial force and the outputs are the displacement, velocity, and acceleration of the proof mass as a response to the input force. The model of the sensing element is shown in Figure 3.1.

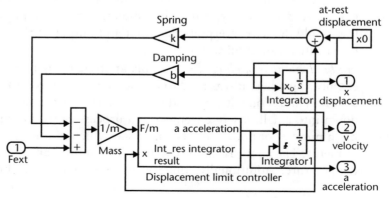

Figure 3.1 Simulink model of the sensing element of a micromachined accelerometer, which is a mass-dashpot-spring system including mechanical stoppers and initial deflection.

The model contains a user-defined submodel (displacement limit controller) that has two inputs: the input acceleration acting on the sensing element and the displacement of the proof mass. It models the nonlinear behavior of the sensing element in case the proof mass touches the mechanical stoppers (i.e., the displacement x exceeds a certain x_{max}). In this case the velocity of the proof mass is reduced to zero, hence Integrator1 in the figure is reset to zero until an acceleration in the direction away from the limit stopper is detected.

Another feature of the model is that a nonzero initial displacement of the proof mass can be set by x_0, which puts an initial condition on the second integrator. The summing block at the input sums up all external and internal forces acting on the proof mass.

The model of the sensing element is a subsystem in the overall sensor system model including the force-feedback control loop, and it is shown in Figure 3.2. Assuming further that the proof mass is embedded between two electrodes forming capacitors on either side, the displacement can be converted into a differential change of capacitance; this is modeled by a mathematical function block implementing the equations for parallel plate capacitances. The differential capacitance can be detected by an electronic position measuring circuit, which, to first order, can be represented by a gain block in the model. Followed by this are a comparator and sample and hold, which model the sigma-delta control system. In the feedback path the electrostatic forces on the proof mass are calculated if either of the two

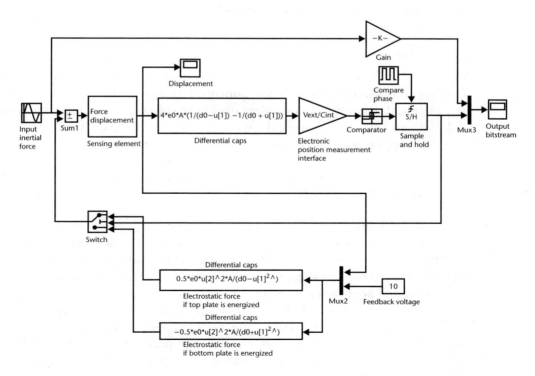

Figure 3.2   Simulink model of the entire sensor system. The model includes the sensing element dynamics, conversion from displacement to differential capacitance and, in turn, to a voltage, sigma-delta modulator control blocks, and the force-feedback arrangements.

electrodes is energized. These feedback forces are summed with any external inertial force acting on the proof mass.

The model allows the optimization of many design parameters such as the electrode area, spring constant, proof mass, the required electronic pick-off gain, and the sampling frequency. Predictions on the control loop stability can be made and the signal-to-quantization noise ratio can be derived. Additional effects such as inherent noise sources (Brownian or thermal noise) can be simulated by adding random number generators, or unwanted electrostatic forces due to the electrical excitation voltage required for the electronic interface circuitry can be added to the model and their influence on the performance of the sensor can be studied [2, 3]. Modeling of these second order effects obviously increases the simulation time considerably. On a modern computer a simulation run with the basic model presented in Figure 3.2 may only take seconds to a few minutes; if the other effects are added the simulation time may increase to a few hours. A typical methodology is to start with a basic model, capturing only first order effects, then adding various second order effects and evaluating their influence on the performance of the device. Those that have a negligible effect on the sensor can subsequently be discarded again to speed up the simulation.

The accuracy and merits of such an approach obviously rely on the analytical understanding of the underlying physics of the sensor to be simulated. The modeling process as such is done analytically by the designer, often by hand calculations. Certain FEM software tools automate this process by performing, for example, a full mechanical modal analysis, and then extracting a lump parameter model that is suitable for implementation in a system simulation tool.

3.2.1.2   Spice

Spice is typically an electronic circuit simulator. It can also, however, be used to simulate parts from another physical domain. Two approaches are possible: one can map electrical quantities to equivalent ones in the physical domain to be considered and build an equivalent electrical circuit. If a mechanical part is to be considered, then the mapping is as follows [4]:

> Mass == Inductance; Damping == Resistance; Stiffness == 1/Capacitance; Force == Voltage; Position == Charge.

A similar mapping process can be derived for other physical domains, for example, thermal processes. This allows the simulation of the dynamics of mechanical structures such as resonators, accelerometers, and pressure sensors. Even more complex phenomena such as squeeze film damping can be modeled in such a way [5].

The second approach is to make use of the analog behavioral library most commercial Spice packages include [6]. This library contains models for system level blocks such as integrator, transfer functions, look-up tables, summers, and gain blocks. It allows dynamic models of many physical sensors to be developed. In Figure 3.3, a system level model implemented in OrCad PSpice of a closed loop accelerometer is shown.

The main advantage of both approaches is obviously that in Spice the interface and control electronics of the sensor can be easily simulated.

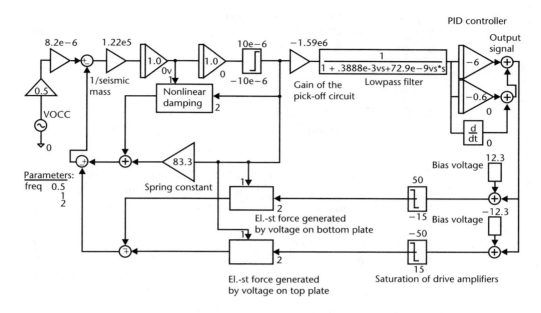

Figure 3.3 System level model of a closed loop micromachined accelerometer in Orcad PSpice.

3.2.1.3 Other System Level Simulators

A range of other system level simulators exists which are suitable for MEMS. Vis-Sim is a Windows-based program for the modeling and simulation of complex non-linear dynamic systems [7]. It is very similar in its capabilities to Simulink and hence will not be reviewed further here. Saber from Synopsis software simulates physical effects in different engineering domains (hydraulic, electric, electronic, mechanical) as well as signal-flow algorithms [8]. Saber is designed to perform simulations based on very few preconceptions about the target system. Consequently, the simulator can analyze designs containing multiple technologies, using the analysis units native to these technologies. The MEMS-relevant technologies include: electronic, electro-mechanical, mechanical, electro-optical, and controls systems.

3.2.2 Finite Element Simulation Tools

Finite element analysis (FEA) is a commonly used approach for simulating a broad range of engineering applications. The finite element method is well suited to the solution of differential equations with known boundary conditions, and it enables the analysis of complex geometries by subdividing them into a finite number of more simply shaped elements. Each element is defined by nodal points and can be specified with particular characteristics relevant to the engineering problem being solved. The solution involves approximating the required function over each finite element and, by considering element boundary conditions, obtaining nodal values of the function for each particular element. After considering interelement equilibrium and known global boundary conditions, a set of simultaneous equations is obtained. The solution of the simultaneous equations typically involves complex matrix algebra that requires the use of computers. Such computing power is readily accessible today, with even basic PCs being capable of solving complex FEA problems.

3.2.2.1 CoventorWare

CoventorWare [9] is a fully integrated design environment for MEMS design. The latest version is the CoventorWare 2003.1 running on Sun Solaris and Windows. It is process independent and consists of four bundles:

1. *Designer:* design, specify, and model MEMS structures, including two-dimensional layout creation and editing, process emulation, three-dimensional generation of solid models from two-dimensional masks, finite element meshing.
2. *Analyzer:* the specialty solvers creates electrical, mechanical, thermal, and fluidic solutions for MEMS-specific in-depth numerical analysis using mechanical simulation (FEM), electrostatic simulation (BEM), and coupled electromechanical simulation for complex multidomain.
3. *Integrator:* extracts reduced order models of physical effects (stiffness, damping, and inertia) found in most dynamic systems.
4. *Architect:* creates schematic models of MEMS designs and runs rapid simulations in other system-level simulators (Saber/Simulink).

CoventorWare generally follows a simulation and design procedure starting with the drawing of device layout, then the definition of the fabrication flow, generating a two- and three-dimensional solid model, meshing the structure, analysis, and detailed simulation, and optionally a reduced order system level model can be derived.

The first step is to create the two-dimensional layout of a MEMS design using Designer's layout editor, which is a full-featured two-dimensional mask-drawing tool capable of all-angle construction and curve creation for MEMS geometries, and parameterized layout generators. The layout editor supports true-curve structures and handles irregular MEMS solid components, which can be auto-meshed without partitioning, and it can also edit design subsections in any level of the hierarchy. Layout creates a .cat format file and supports the format used by other layout software such as GDS II, CIF, IGES, and DXF. Figure 3.4 is an example of the layout of a single-axial micromachined accelerometer with dimensions of $4.8 \times 4.0 \times 0.06$ mm.

In the second step the fabrication steps for a MEMS device are defined and emulated. The process editor supplies the information needed to create a three-dimensional MEMS model from the two-dimensional mask information provided by the layout editor. The depth information is defined by the various material layers in a sequence of deposit and etching steps with control of bulk and thin-film geometries. Materials for each process layer for the MEMS device are chosen from a material property database; the material properties include elasticity, stress, density, viscosity, conductivity, dielectric, piezoelectric, and thermal characteristics. The fabrication process parameters are defined by material thickness, deposition type (stacked, conformal, or planar), sidewall profiles of angular slope, mask perimeter offset, and mask polarity.

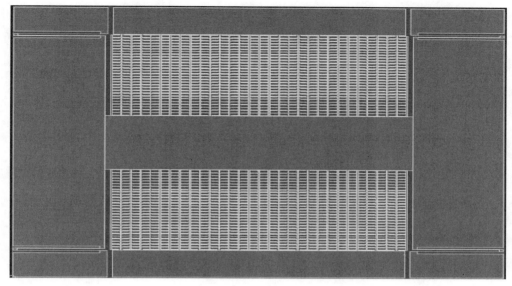

Figure 3.4   Layout of a micromachined accelerometer.

Once the layout and fabrication process flow have been generated, the *Solid Model* tool is used to build a three-dimensional model using the two-dimensional layout geometry from the mask files and the deposit/etch and thickness information from the fabrication process file.

The next step performs the finite element model creation and meshing of the device. The meshing tool creates a three-dimensional mesh based on the model created by the *Solid Model* tool and the process file created by the process editor. The user can choose between various mesh elements such as tetrahedral, bricks, and hexahedral according to the device geometry. The result is stored in a file containing all model input and output parameters such as the geometry and material properties, conductor and dielectric types, and mesh information used by the various simulation solvers.

Analyzer is the core of CoventorWare, consisting of various solvers such as MemCap, MemMech, CoSolveEM, SimMan (Simulation Manager), and some specialized solvers, which are briefly described next. All use finite element and boundary element techniques for solving the differential equations for each physical domain.

MemCap is the electrostatic solver that computes a charge matrix based on voltage conditions or a voltage matrix based on charge conditions for the MEMS design under investigation. Secondary effects such as fringe capacitances and the influence of a lossy media on the electric behavior can be simulated as well.

MemMech is the mechanical solver that analyses structural, displacement, modal, harmonic, stress, and contact steady-state thermomechanical properties. Figure 3.5 shows the modal analysis of the accelerometer shown in Figure 3.3, which computes and visualizes the natural frequencies of the mechanical structure.

CoSolveEM is a coupled electromechanical solver that combines the electrostatic and mechanical solution. It also can perform pull-in voltage and hysteresis analysis efficiently in one sweep.

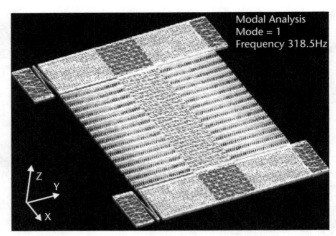

Figure 3.5 Modal analysis of a micromachined accelerometer using MemMech.

SimMan is a simulation manager that allows the users to set up a batch of solver runs to view trends or the sensitivity to various design or manufacturing parameters. Rather than manually adjust parameters and start individual solver runs, the Sim-Man provides a variety of ways to set parameter variations automatically and iterate solver results.

MemHenry is the inductance solver that computes the frequency-dependent resistance and inductance matrices for a set of conductors. This tool is aimed at magnetic sensor design, on-chip passive inductor analysis, and parasitic extraction for packaging analysis.

MemPackage is the package effects analyzer that computes the effects of package induced stresses and strains on a micromachined device mounted in a standard or user-defined package.

MemPZR analyses piezoelectric effects created by electric polarization due to mechanical stress and also addresses the converse, where strain in a piezoelectric material develops due to the application of an electric field. It can handle large displacements and other nonlinear effects introduced by electrical, mechanical, and thermal loading. Also, it includes transient analysis and voltage-driven harmonic analysis. Analyses necessary for obtaining S-parameters can be carried out, which are especially relevant for RF devices such as bulk acoustic resonators.

MemETherm is an electrothermal solver that computes the potential drop through a resistor resulting from a voltage and/or current flow and the resulting temperature distribution from joule heating.

AutoSpring is a spring constant extractor that allows the extraction of multidimensional, nonlinear spring behavior from complex tether designs. These values can then be used for system level models.

MemDamping is the damping solver that computes the squeeze-film damping and spring coefficients of a MEMS device using a hybrid Navier-Stokes-Reynolds approach. Many physical MEMS sensors are sensitive to squeeze-film damping effects—for example, this is one of most important features of the accelerometer illustrated above. Figure 3.6 shows a typical output graph plotting the damping coefficient of the accelerometer as a function of frequency.

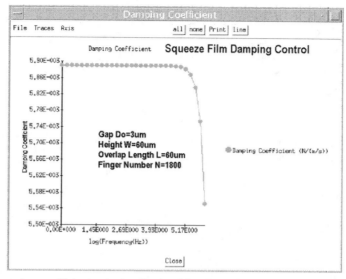

Figure 3.6   Damping coefficient calculated using MemDamping.

MemTrans is the transient analysis solver that computes transient thermal and mechanical deformation and stresses. A typical application is computing the failure conditions of a shock protection limiter of inertial sensors.

Furthermore, there are fluidic solvers for designing and simulating microfluidics devices and for integrating microfluidics and microarray technologies into lab automation and microarray products. MemCFD a general fluidic solver; NetFlow is·an electrokinetic fluidic solver; SwitchSim computes the electrokinetic response of a species in a fluid subjected to a switched electric field; DropSim provides full three-dimensional numerical simulation of droplet formation, transport, and impact; BubbleSim simulates the movement of bubbles and surrounding fluids in microchannels; and ReactSim is a chemical reaction solver. BioChip Developer is a development environment intended for users developing microscale biochemical analysis and synthesis systems, such as DNA sequencing, protein separation, and lab on-chip technologies. This software focuses on chip-scale implementation of biochemical systems.

CoventorWare also facilitates the simulation of a variety of MEMS and other microsystems-based RF components including switches, varactors, inductors, resonators, transmission lines, antenna elements, and waveguides. Additionally, CoventorWare contains tools for designing optical MEMS devices such as mirrors, shutters, fiber aligners used in optical switches, tunable optical filters, and lasers.

Finally, Integrator enables powerful macromodel extractions based on reduced order models of the most common physical effects (stiffness, damping, and inertia) found in most dynamic systems. This is done with three tools: (1) SpringMM extracts linear and nonlinear mechanical and electromechanical spring stiffness, (2) DampingMM provides a Stokes-equations flow solver that is applicable to arbitrary geometries, and (3) InertiaMM computes the mass inertia of the movable parts of the device. The extracted reduced order macromodels can be used in a system-level simulator such as Saber or SimuLink.

3.2.2.2 IntelliSuite

Intellisense Corning commercialized the MEMS CAD package IntelliSuite, current version 7.1 [10]. It also is a FEM-based simulation and design tool specifically developed for MEMS and runs on a standard PC under Windows. The user starts by drawing the masks in IntelliMask, which is a standard drawing package with typical features for mask designs such as multiple translations copy, layer control, and hierarchical cells. Each mask is drawn on a separate layer and saved in a different file. It is also possible to import and export the masks in GDS II of DXF file format. The next step is to define the fabrication process in a tool called IntelliFab. It contains a large database of silicon base materials, deposition steps of various materials, and etching steps for all commonly used materials in MEMS. The previously defined masks are used to define areas in which material is removed or added. Once the user has created the full process flow (referred to as Process Table) IntelliFab visualizes the fabricated device in an easy-to-use viewer that allows zooming, panning, and three-dimensional rotation of the virtual prototype. For standard MEMS processes, templates are available. The properties of a material used in any process step can be defined and altered in a powerful tool called MEMaterial. If, for example, one process step is to deposit silicon nitride (Si_3N_4) in a PECVD furnace, material properties include stress, density, thermal expansion coefficient, Young's modulus, and Poisson's ratio. If the material property is not a constant but depends on one or several fabrication conditions, their relationship may be graphically displayed. Stress of silicon nitride, for instance, depends on the deposition temperature, and their relationship is shown in the graph in the top-right window of Figure 3.7. In the lower

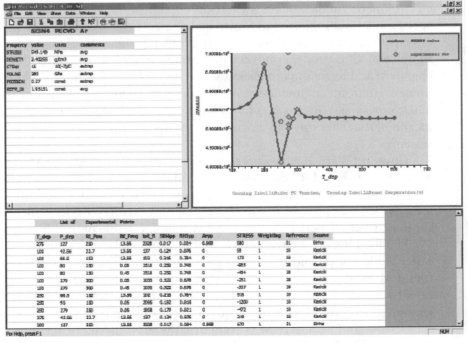

Figure 3.7 The MEMaterial tool within IntelliSuite. The user can graphically view and change material properties as a function of fabrication process parameters.

window the data points are given together with the literature source from which the information was taken.

The various simulation solvers, which are mechanical, electromagnetic, electromechanical, and electrostatic, can be run either from IntelliFab or directly. The mechanical solver meshes the device to be analyzed. The meshing process can be controlled by defining global or localized limits for the mesh of the certain areas of interest. Then it computes the natural mechanical resonant modes, which can be visualized in an animation. Furthermore, it allows the application of mechanical loads such as forces and moments to the different surfaces of the structure, but also thermal loads in form of heat convection. Thermal distribution generated by flow or current through materials with varying resistivity and their mechanical deformation caused by thermal strain can be simulated. Any analyses can be performed as a response to a static load or dynamically as a result of a time varying load.

The electrostatic solver uses a very similar meshing process and computes a capacitance matrix for the various layers and surfaces. Furthermore, it allows an analysis of the resulting charge density, electrostatic forces and pressures.

The electromechanical solver allows the user to apply various loads to the device under consideration such as electrostatic loads through applying voltages, temperature, pressure, acceleration, and displacements, and subsequently calculate the resulting mechanical reactions (such as stress distributions, deformations, and displacements) and electrical properties (such as capacitance, charge density, and electric field). As an example of an electromechanical simulation, Figure 3.8 shows

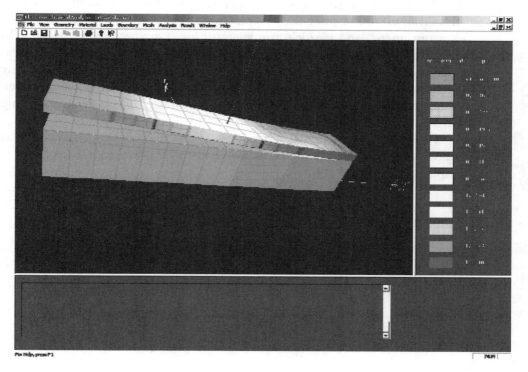

Figure 3.8   The result of a displacement simulation in the electromechanical solver. This particular example shows a micromechanical switch actuated by electrostatic forces.

the displacement of a beam as a result of an applied voltage giving rise to an attractive electrostatic force.

Another solver is the microfluidic analysis module. This tool allows the user to analyze thermal effects, concentrations, and flow within a fluid. It also simulates velocity and electric field distributions as a result of electrokinetic phenomena.

Another very useful tool is AnisE, an anisotropic etch process simulator. With AnisE, the user can use the layout of the microstructure to be prototyped to view a three-dimensional representation of it, access information about the etch rates of different etchants, and then simulate the etching under different time, temperature, and concentration parameters.

Finally, Intellisense contains a module called 3-D Builder, which can be called from any of the solvers or separately as a standalone application. This tool allows for building and meshing the three-dimensional geometry of MEMS structures with a graphical interface. The screen is divided into two areas: on the left is the two-dimensional layer window where the outline of different layers can be drawn; and on the right is the three-dimensional viewing window, which allows the user to visualize the device in three dimensions and includes zooming, rotating, and panning functions. Furthermore, the thickness of any layer can be changed. In this way, a MEMS device can be created without having to define the full fabrication process flow. The module produces a file that can be used for analysis in any of the solvers or, alternatively, a mask file that can be processed further by IntelliMask.

3.2.2.3 ANSYS (ANSYS Inc.)

The ANSYS FEA software is a commercially available simulation tool capable of structural, vibration (modal, harmonic, and transient), thermal, acoustic, fluidic, electromagnetic, and piezoelectric analyses (or combinations of these). While not specifically written for the simulation of MEMS, many of these analyses apply equally well in the microdomain, and as such, ANSYS has been widely used throughout the MEMS community. The software interface has evolved over many years, and the latest ANSYS Workbench environment is now relatively straightforward to use even for the novice.

The ANSYS Multiphysics software is of particular relevance to the simulation of MEMS and has the capability to simulate the following characteristics (shown graphically in Figure 3.9):

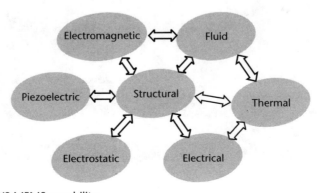

Figure 3.9 ANSYS MEMS capability.

- Structural (static, modal, harmonic, transient);
- Electrostatic effects;
- Piezoelectric films;
- Residual stresses;
- Fluidic damping;
- Microfluidics;
- Composite structures;
- Electrothermostructural coupling;
- Electromagnetic systems.

ANSYS can been used to simulate the vast majority of the MEMS physical sensors covered in this book, including those shown in Table 3.1. Given the nature of sensors, the ANSYS coupled field analyses are of particular interest.

The software also allows CIF files to be imported, thus enabling MEMS designs to be input from other software packages. By selecting the correct element (element 64), the anisotropic material properties of silicon can input in matrix form enabling accurate materials specification in the simulation. Other useful features include the optimization routine, which aims to minimize a specified objective variable by automatically varying the design variables. Taking finite element tools to the nanometer scale, the bulk material models used break down as quantum mechanical effects become dominant. The recent introduction of highly customizable, user programmable material models may, however, help to address the finite element analysis of some nanosystems.

ANSYS simulations are generally performed in three stages. The first is carried out in the preprocessor and defines the model parameters (i.e., its geometry, material properties, degrees of freedom, boundary conditions, and applied loads). Next is the solution phase, which defines the analysis type, the method of solving, and actually performs the necessary calculations. The final phase involves reviewing the results in the postprocessor. Different postprocessors are used depending upon the type of analysis (e.g., static or time based). The three stages are shown in Figure 3.10 along with the typical inputs required.

Several example MEMS simulations can be found on the Internet [11]. Example analyses performed by the authors are shown in Figures 3.11, 3.12, and 3.13. The

Table 3.1 Example MEMS Applications and Corresponding ANSYS Capabilities

MEMS Application	ANSYS Capability
Inertial devices: accelerometers and gyroscopes	Structural (static, modal, transient), coupled electrostatic-structural, coupled piezoelectric
Pressure transducers	Capacitance based: electrostatic structural coupling Piezoresistive based: electrostructural indirect coupling
Resonant microsensors (including comb and thermal drive)	Modal and prestressed modal analysis, electrostatic-structural coupling, thermal
Piezoelectric transducers	Piezoelectric-structural coupling
MEMS packaging	Structural and thermal analysis

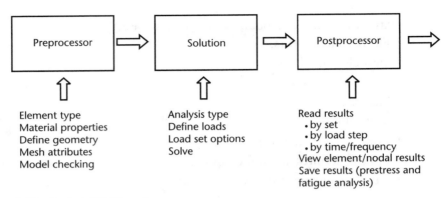

Figure 3.10 Typical ANSYS routine.

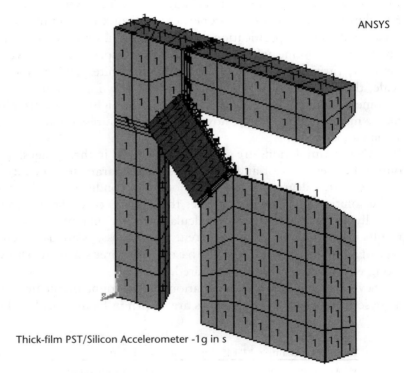

Figure 3.11 Finite element model of one-quarter of a PZT accelerometer.

first shows a model of one-quarter of a silicon accelerometer with a piezoelectric material deposited on the top surface of a beam supporting the inertial mass [12]. The device is a symmetrical structure, and therefore, only one-quarter needs to be modeled thus reducing solution time. The ANSYS coupled field piezoelectric analysis has been used to predict the sensor output from the piezoelectric material for a given acceleration. Modal and transient analyses were also performed to simulate the frequency response of the accelerometer. Figure 3.12 shows one-quarter of the

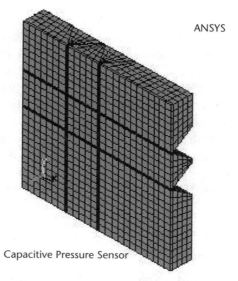

Capacitive Pressure Sensor

Figure 3.12   Element plot of one-quarter of a capacitive pressure sensor diaphragm.

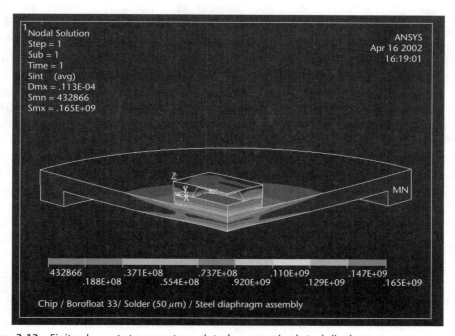

Figure 3.13   Finite element stress contour plot of a pressurized steel diaphragm.

diaphragm of a capacitive silicon pressure sensor [13]. The diaphragm was defined by anisotropically etched double corrugations designed in such a way that as the diaphragm deflects with applied pressure, it remains flat and parallel to the fixed electrode. This simplifies the linearization of the sensor output by removing the nonlinear component arising from the bending of the diaphragm. A straightforward ANSYS structural analysis was used to achieve a suitable corrugated geometry and to simulate the diaphragm's response to applied pressure. The third example, in Figure 3.13, shows a one-quarter model of a silicon resonant pressure sensor chip mounted on a glass support and bonded to a stainless steel diaphragm. A thermal analysis was performed to optimize the height of the glass support in order to minimize the effect of the thermal expansion coefficients of silicon and steel. In addition, sensitivity of the sensor to applied pressures was also simulated. The strains on the sensor chip arising from pressure applied to the underside of the steel diaphragm were applied to a separate model of the resonator. By performing a prestressed modal analysis, the frequency behavior of the resonator with applied pressure was determined.

3.2.2.4   MEMS Pro/MEMS Xplorer (MEMScAP)

MEMS Pro and MEMS Xplorer are PC and Unix-based CAD tools, respectively, and are supplied through MEMSCAP. The MEMS Pro package was developed originally by Tanner Research, Inc.

The basic MEMS Pro Suite is essentially an L-Edit based layout editor aimed at the MEMS designer. It contains libraries of standard MEMS components and some design functions specifically targeted at MEMS. It includes the MEMS Solid Modeler, which can produce three-dimensional models from the layout using user-designed fabrication processes. This feature supports both surface and bulk micromachining processes and enables visualization of the processed MEMS component. The model can also be exported into ANSYS, thereby enabling simulation of the function of the device. This link between the two software packages provides the complete MEMS CAD package, but it obviously requires the user to have access to both packages.

The MEMS Pro Verification Suite is the same as the basic suite but with the addition of a design rule checker, block place, and route function and user programmable interface with automated design tools. The next suite up is the MEMS Pro Design suite, which includes the T Spice Pro module, which enables simulation of both MEMS and electronic components. This provides an integrated system simulation utilizing an equivalent circuit approach and includes a library of MEMS components to facilitate modeling. It also includes a layout versus schematic (LVS) verification tool, which compares SPICE models extracted from both the layout and schematic editors. The top of the range MEMS Pro Complete suite also includes reduced order modeling (ROM) tools, which provide a behavioral model of the MEMS component from the FE results. This provides a link between the system and component designers. The Complete suite also accepts CIF files enabling layout files to be generated from an ANSYS three-dimensional model. ANSYS can also generate ROM components for use in the MEMS Pro environment. A schematic of the MEMS Pro Complete suite is shown in Figure 3.14.

Behavioral modeling of MEMS components is available in the MEMS Master software series developed by MEMScAP. MEMS Master is a prototyping and predimensioning environment that can be used in conjunction with MEMS Pro. Designs are carried out in the M2Architect tool and simulation is performed by the SMASH

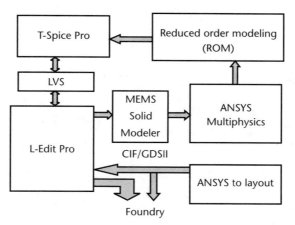

Figure 3.14   MEMS Pro Complete Suite.

VHDL-AMS simulator. The MEMS Master MemsModeler can generate VHDL-AMS models from ANSYS finite element models. A schematic of the MEMS Master software components and the links with MEMS Pro and ANSYS are shown in Figure 3.15.

The MEMS Xplorer suite offers a Unix-based design environment incorporating an IC design environment (Mentor/Cadence) and ANSYS FE tools. The architecture is shown in Figure 3.16. It uses some of the same modules described above but uses Cadence Virtuoso as the layout editor. This contains a MEMS library, MEMS design tools, and a three-dimensional model generator for integrating with ANSYS. The fabrication process simulation can be customized in the Foundry Process Manager, and this has the very useful capability of being linked to specific Foundry processes that enable precise simulation of the fabrication MEMS components.

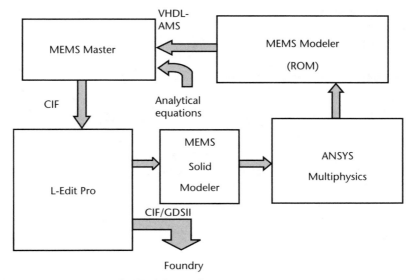

Figure 3.15   MEMS Master and MEMS Pro tools.

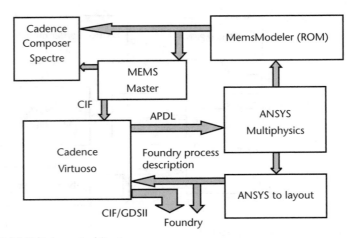

Figure 3.16 MEMS Xplorer Architecture.

References

[1]   http://www.matlab.com.

[2]   Mokhtari, M., et al., "Analysis of Parasitic Effects in the Performance of Closed Loop Micromachined Inertial Sensors with Higher Order SD-Modulators," *Proc. Micromechanics Europe (MME)*, Sinaia, Romania, October 2002, pp. 173–176.

[3]   Gaura, E., and M., Kraft, "Noise Considerations for Closed Loop Digital Accelerometers," *Proc. 5th Conf. on Modeling and Simulation of Microsystems*, San Juan, Puerto Rico, April 2002, pp. 154–157.

[4]   Marco, S., et al., "Analysis of Electrostatic Damped Piezoresistive Silicon Accelerometer," *Sensors and Actuators*, Vol. A37–38, 1993, pp. 317–322.

[5]   Veijola, T., and T. Ryhaenen, "Equivalent Circuit Model of the Squeezed Gas Film in a Silicon Accelerometer," *Sensors and Actuators*, Vol. A48, 1995, pp. 239–248.

[6]   Lewis, C. P., and M. Kraft, "Simulation of a Micromachined Digital Accelerometer in SIMULINK and PSPICE," *UKACC Int. Conf. on Control*, Vol. 1, 1996, pp. 205–209.

[7]   http://www.vissim.com.

[8]   http://www.analogy.com/products/mixedsignal/saber/saber.html.

[9]   http://www.coventor.com.

[10] http://www.corningintellisense.com.

[11] http://www.ansys.com/ansys/mems/index.htm.

[12] Beeby, S. P., J. N. Ross, and N. M. White, "Design and Fabrication of a Micromachined Silicon Accelerometer with Thick-Film Printed PZT Sensors," *J. Micromech. Microeng.*, Vol. 10, No. 3, 2000, pp. 322–329.

[13] Beeby, S. P., M. Stuttle, and N. M. White, "Design and Fabrication of a Low-Cost Microengineered Silicon Pressure Sensor with Linearized Output," *IEE Proc. Sci. Meas. Technol.*, Vol. 147, No. 3, 2000, pp. 127–130.

CHAPTER 4

Mechanical Sensor Packaging

4.1 Introduction

As with micromachining processes, many MEMS sensor-packaging techniques are the same as, or derived from, those used in the semiconductor industry. However, the mechanical requirements for a sensor package are typically much more stringent than for purely microelectronic devices. Microelectronic packages are often generic with plastic, ceramic, or metal packages being suitable for the vast majority of IC applications. For example, small stresses and strains transmitted to a microelectronics die will be tolerable as long as they stay within acceptable limits and do not affect reliability. In the case of a MEMS physical sensor, however, such stresses and strains and other undesirable influences must be carefully controlled in order for the device to function correctly. Failure to do so, even when employing electronic compensation techniques, will reduce both the sensor performance and long-term stability.

The need to control such external stresses is complicated by the simple fact that all MEMS sensors designed for physical sensing applications have to interact with their environment in order to function. The physical measurand must therefore be coupled to the sensor in a controlled manor that excludes, where possible, other undesirable influences and cross-sensitivities. In order to achieve this, the design of the sensor packaging is as important as the design of the sensor itself. The sensor packaging has a major influence on the performance of the device, especially with respect to factors such as long-term drift and stability. It is very important that the packaging of the sensor is considered at the outset and that the package design is developed in parallel with that of the sensor die itself. This is especially true when you consider that the cost of the package and its development can often be many times that of the sensor die.

The packaging of MEMS devices will often be specific to the application being addressed. Such a packaging solution will therefore involve a design, as well as the selection of materials and processes suitable for that particular application. Generic solutions suitable for a range of applications, such as is the case of microelectronic devices, are limited to simple, low-cost, high-volume MEMS applications. This chapter briefly describes the technologies developed for the packaging of integrated circuits before discussing the design considerations relating to the packaging of mechanical sensors. Typical problems encountered, and their potential solutions, are discussed in more detail. Example MEMS packaging solutions are given throughout the chapter in order to highlight some of the principles involved.

4.2   Standard IC Packages

From a cost point of view, it would certainly be advantageous if the mechanical sensor die could simply be mounted in one of the many standard IC packages available. These can be grouped into three types: ceramic, plastic, and metal. The functional requirements of microelectronics packages are to enclose the IC in a protective shell, to provide electrical connection from the IC to circuit board, and to enable adequate heat transfer. Key considerations in the design of an IC package are reliability (affected by packaging stresses and moisture ingress), heat flow, ease and cost of manufacture, and electrical characteristics such as lead resistance, capacitance, and inductance. For further information refer to Tammala et al. [1].

4.2.1   Ceramic Packages

Ceramic materials have been used to make a wide range of package types and, although more expensive than their plastic counterpart, possess an unrivaled range of electrical, thermal, and mechanical properties. Ceramics packages can be hermetically sealed and can be made very small with large numbers of reliable electrical interconnects. A wide variety of ceramic packages have been developed, including basic dual in-line packages (DIPs), chip carriers, flat packs, and multilayer packages. Such packages are used in high-performance applications where the increased cost can be justified. The most common ceramic materials used are alumina (Al_2O_3), alumina/glass mixtures, aluminum nitride (AlN), beryllium oxide (BeO), and silicon carbide (SiC).

Two approaches are used in the fabrication of ceramic packages. The first approach uses a mixture of ceramic and binders, which are molded into shape using a dry pressing process, and then sintered to form the finished component. A ceramic package is formed by sandwiching a metal leadframe between two such dry pressed ceramic components (the base and the lid). The three-layer package is held together hermetically by glass frit reflowed at temperatures between 400°C and 460°C. These pressed ceramic packages are lower in cost that the laminated multilayer package, but their simple construction limits the number of possible electrical features and interconnects. DIP packages fabricated in this manner are commonly known as CerDIPs.

The second approach is based upon a multilayer ceramic (MLC) structure. These are made from layers of unfired (green state) ceramics metallized with screen-printed tungsten patterns, which are then fired under pressure at high temperature (~1,600°C). Exposed metal features are electroplated with nickel and gold. Metal components, such as the contact pins, are attached using a copper-silver alloy braze. The laminated structure allows the package designer to incorporate electrical features into the package itself. Such MLC packages can be used for individual die or for mounting multiple die, known as multichip modules (MCMs). This approach can improve systems performance and can reduce the number of interconnects required at the circuit board level to a workable amount. Multilayer packages can now be produced with as many as 70 layers. MCMs can be used to package MEMS devices, and this is discussed further in Section 4.4.

Metallization can be realized on ceramic packages using either screen-printed thick-film or evaporated/sputtered thin-film technology. The thick-film approach deposits the metal, or indeed dielectric if required, in the pattern required, but it has traditionally been limited by poor resolution that yields typical line widths and spacing of 150 μm. Recent developments in photoimageable inks, however, allow line widths and spacing of 40 μm and 50 μm, respectively [2]. The thin-film approach, which involves subsequent lithographic and etching processes, is capable of even finer line widths and spacing (< 20 μm). The processing involved is not so straightforward and this approach is better suited to high-density, high-performance applications.

4.2.2   Plastic Packages

Molded plastic packages were developed in order to reduce the cost of IC packaging. At the center of a plastic package is a leadframe to which the die is attached and electrical connections are made. The leadframe material is typically a copper alloy, nickel-iron (the most widely used being alloy 42) or a composite strip (e.g., a copper clad stainless steel) and the leadframe geometry is obtained by stamping or chemical milling. The assembly is then encased in a thermoset plastic package using a transfer molding process. The molding resins used are a mixture of various chemicals. These have been developed in order to obtain the characteristics required by both the process and application. These characteristics include viscosity, ease of mold release, adhesion to leadframe, and low levels of ionic contamination. To prevent difficulties in packaging and future reliability problems, the component materials making up a plastic package must be chosen with care to avoid thermal expansion coefficient (TEC) mismatches, to allow adequate thermal conduction away from the IC, and to prevent moisture ingress.

4.2.3   Metal Packages

These are often used in military applications, since they offer the highest reliability characteristics, as well as in RF applications. Electrical connections are made using a metal feed-through and glass-to-metal seals. They are typically hermetically sealed by welding, soldering, or brazing a lid over the package, which prevents moisture ingress and resulting reliability difficulties (see Section 4.3.3). Common metals used in the construction are Kovar, cold rolled steel, copper, molybdenum, and silicon carbide reinforced aluminum. Hermetic seals can be formed. Common metal packages types are shown in Figure 4.1. Figure 4.2 shows a photograph of typical metal, ceramic, and plastic packages.

4.3   Packaging Processes

Irrespective of the type of package used, the assembly of the packaged device involves mounting the die, making electrical connections to the terminals provided, and sealing the assembled package. Several standard processes have been developed by the IC industry to meet these requirements, and these same processes are common to many MEMS packaging applications.

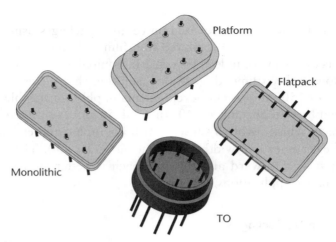

Figure 4.1   Common metal packages.

Figure 4.2   Photo of typical metal, ceramic, and plastic packages.

4.3.1   Electrical Interconnects

4.3.1.1   Wire Bonding [3]

Wire bonding uses thin wire to connect the bond pads on the die to the packaging interconnects. The attachment of the wire is achieved by using a combination of heat, pressure, and/or ultrasonic energy. The wire is brought into intimate contact with the surface of the pad, and the bonding process results in a solid phase weld via electron sharing or diffusion of atoms. The bonding pressure ensures intimate contact and aids the breakup of any surface contamination or oxidation, and this is further enhanced by the application of ultrasonic energy. Heat can be applied to accelerate atomic diffusion and therefore the bond formation. There are two wire bonding processes employed: ball and wedge bonding. These processes, and common wire and pad materials, are summarized in Table 4.1. Ball bonding most commonly uses relatively thin gold wire (< 75 μm) because it deforms readily under pressure and temperature, it resists oxide formation, and is well suited to the ball formation and cutting process. Gold wire is also attractive because it remains inert after bonding and does not require hermetic sealing. Ball bonding requires a pad pitch of more than 100 μm. Wedge bonding, on the other hand, can be used for both aluminum wire and gold wire bonding applications. Aluminum wire is bonded in an ultrasonic bonding process at room temperature. Gold wire wedge bonding uses a thermo-sonic bonding process. An advantage of wedge bonding is that it can be used on pads with a pitch of just 50 μm. It is however slower than thermo-sonic ball bonding. Aluminum ultrasonic bonding is the most common wedge bonding process because of the low cost and the low working temperature.

4.3.1.2   Tape Automated Bonding

In the case of tape automated bonding (TAB), the interconnections are first patterned on a multilayer polymer tape. The tape is positioned above the bare die so that the metal tracks on the polymer tape correspond to the contact pads on the die. Traditionally, the contact pads are located around the edge of the die, but a more recent innovation known as *area TAB* has contact pads in the form of metal bumps that are distributed over the entire surface of the die. This approach is able to support a greater number of connections to and from the die.

The TAB technology has several advantages over the wire bonding approach. These advantages include a smaller bonding pad and therefore increased I/O counts, smaller on-chip bonding pitch than for ball wire bonding (100 μm), an increased productivity rate, reduced electrical noise, suitability for higher frequency applications, lower labor costs, and lighter weight. The disadvantages of TAB technology include the time and cost of designing and fabricating the tape and the capital expense of the TAB bonding equipment. In addition, each die must have its own tape patterned for its pad and package configuration. For these reasons, TAB has typically been limited to high-volume production applications.

4.3.1.3   Flip Chip

Flip chip assembly, also called direct chip attach (DCA), involves placing the die face-down (hence, "flipped") onto the package or circuit board. The electrical connection is made by conductive bumps formed on the die bond pads. Flip chip assembly is predominantly being used for ICs, but MEMS devices are beginning to be developed in flip chip form. The advantages of flip chip include:

Table 4.1   Summary of Wire Bonding Processes

Wire Bonding Process	Technique	Pressure	Temp. (°C)	Ultrasonic	Wire	Pad	Speed (Wires/Sec)
Ball	Thermo-compression	High	300–500	No	Au	Al, Au	
Ball	Thermo-sonic	Low	100–150	Yes	Au	Al, Au	10
Wedge	Thermo-sonic	Low	100–150	Yes	Au, Al	Al, Au	4
Wedge	Ultrasonic	Low	25	Yes	Au	Al, Au	4

- Reduced package size;
- High-speed electrical performance due to the shortened path length;
- Greater flexibility of contact pad location;
- Mechanically rugged;
- Lowest cost interconnection method for high-volume production.

The disadvantages are similar to those associated with the TAB interconnects in that the package or substrate must be custom made for different die designs. Also, testing the quality of interconnects, repairing defects, and the relative complexity of the assembly process are drawbacks as well.

There are three stages in making flip chip assemblies: (1) bumping the die or wafer, (2) attaching the bumped die to the board or substrate, and (3) underfilling the remaining space under the die with an electrically insulating material. The conductive bumps can be formed from solder, gold, or conductive polymer. These bumps provide the electrical and thermal conductive path from chip to substrate and form part of the mechanical mounting of the die. They also act as a spacer preventing electrical contact between the die and substrate conductors. In the final stage of assembly, this space under the die is usually filled with a nonconductive adhesive that joins the surface of the die to the substrate. The underfill strengthens the assembly and prevents differences in thermal expansion between the package and the die from breaking or damaging the electrical connections. The underfill also protects the bumps from moisture and other potential hazardous materials. Figure 4.3 shows a cross-section of flip chip bonding.

A more recent innovation in flip chip assembly is the development of anisotropically conductive adhesives. These materials consist of conductive particles in an insulating matrix and are able to conduct in one axis (the z-axis) yet remain insulators in the x-y plane. This is achieved by trapping one or more conductive particles between conductive bumps on the flip chip and the pads on the substrate while preventing bridging between pads (see Figure 4.4). This requires precise control of the conductive filler loading, particle size distribution, and dispersion. The adhesive can be applied in the form of a paste or a film. This technique provides a simple method for forming conductive paths on flip chip assemblies and removes the need for subsequent underfilling. Studies have shown it to be highly reliable under optimized process conditions [4].

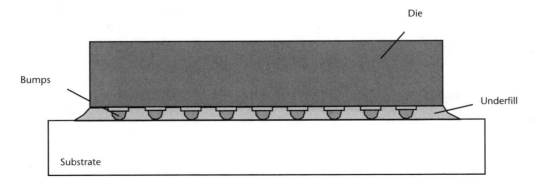

Figure 4.3   Cross-section of flip chip bonding.

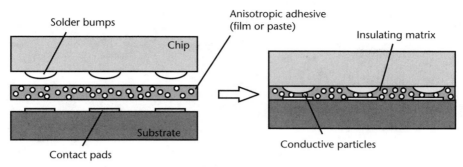

Figure 4.4   Anisotropic adhesive attachment.

4.3.2   Methods of Die Attachment

The process of mounting an IC to a substrate or package is known as die attach. The choice of attachment material is dictated by the size of the die, substrate material (e.g., ceramic, polymer, glass or metal), device requirements, and operating environment. Initial applications usually employed eutectic bonding or soldering on ceramics or metal substrates, but nowadays adhesives have become the predominant attachment medium. Glass frit techniques are rarely used. Other more recent techniques include the "Silicon-on-Anything" approach developed by Phillips. These methods and materials are described next, followed by a comparison of their relative merits shown in Table 4.2. These processes are discussed in relation to MEMS in Section 4.4.1.4.

4.3.2.1   Adhesive Die Attach

Adhesive bonding is achieved by depositing a film of epoxy thermoset, acrylic thermoplastic, or silicone resins between the chip and the substrate. The adhesives can be made electrically/thermally conducting (e.g., by loading with silver particles) or electrically isolating [5]. Adhesives are used in the Silicon-on-Anything technology, developed by Philips Research Laboratories, which enables circuits to be transferred to a range of insulating substrates resulting in greatly reduced parasitic capacitances. This particularly benefits high-frequency RF components. The process essentially involves the fabrication of active and passive bipolar silicon devices on the front surface of a SOI wafer. This wafer is then adhesively bonded face-down to a suitable dielectric substrate such as glass, and the back silicon wafer is then etched away to reveal the buried oxide layer and the inverted bond pads.

4.3.2.2   Soldering Die Attach

This approach uses solder alloys, typically in a thin-film preform placed between the die and the substrate. The assembly is heated up to the melting point of the solder (from 183°C for 63Pb-37Sn to 314°C for Pb-In-Ag solders), which then sets upon cooling. This approach mainly is used on high-power devices because of its good thermal/electrical conductivity and ability to absorb stresses due to expansion mismatch.

Table 4.2 Relative Merits of Die Attachment Methods

Process	Advantages	Disadvantages
Adhesive	Low cost	Outgases
	Easily automated	Contamination/bleed
	Low curing temperatures	Susceptible to voids
	Reduced die stress	Inferior thermal/electrical conductivity
	Special plated surfaces not required	Can require careful storage (e.g., −40°C) and mixing before use
	Rework possible	Not suited to harsh environments
Solder	Good electrical/thermal conductivity	Requires wettable metallized surfaces on the die and substrate
	Good absorption of stresses arising from of thermal expansion coefficients mismatches	Usually requires processing temperatures greater than 200°C
	"Clean"	Needs flux or an inert gas atmosphere
	Rework possible	Porosous
		Poor thermal fatigue resistance of some alloys
Eutectic	Good thermal conductivity	Poor absorption of stresses arising from of thermal expansion coefficients mismatches
	Electrically conducting	
	Good fatigue/creep resistance	High processing temperatures
	Low contamination	Die back metallization may be required
	"High" process/operating temperature capability	If bare die are used, a scrubbing action is required to break down surface oxide
		Rework difficult
Glass	Low void content	High processing temperature
	Good thermal/electrical conductivity	Glass requires an oxygen atmosphere, which can lead to oxidation of other plated systems
	Limited stress relaxation	
	Low contamination	Not commonly used
	High process/operating temperature resistance	

4.3.2.3   Eutectic Bonding

A eutectic bond typically uses gold and silicon, which, when heated, diffuse together at the interface. This diffusion continues until a suitable eutectic alloy is formed, which melts at a more workable temperature than would be the case for the base materials (for example, a 97Au-3Si eutectic melts at 363°C). The eutectic bond can be produced by heating the die then scrubbing it against a gold foil/metallization or by placing a eutectic foil preform at the interface.

4.3.2.4   Glass Die Attach

This process uses a glass layer between the die and the substrate. The glass can be either a solid frit placed beneath the die or be made into a screen printable paste and deposited onto the substrate. The assembly is then heated to typically between 350°C and 450°C until the glass softens to form a low viscosity liquid that will wet the die and substrate. The glass film solidifies upon cooling, thereby attaching the die. As with adhesive attachment, silver particles can be added to the glass to improve the thermal and electrical conductivity of the material. This is a more specialized process not commonly employed.

4.3.3 Sealing Techniques

Most types of IC plastic packages are sealed as part of the transfer molding process. Alternatively, premolded packages, in which the chip is placed in the package after the transfer molding process, require a lid to be placed over the package opening. Lids can be made from metal or preformed plastic and these are attached using a polymer adhesive. Premolded packages are the most common type of plastic package for microsensors. In either case, these packages are not hermetic and moisture will diffuse through the molding material and along the interface between the leadframe and the plastic. This moisture ingress is the main cause of failure in plastic packaged ICs, usually through corrosion of metallized features. Moisture resistance can be improved by encapsulating the die in silicone compounds prior to molding.

A variety of processes exist for sealing metal and ceramic packages once the die has been mounted and the electrical connections made [6]. The suitability of these processes will depend upon the nature of the package and the requirements of the application. The simplest method of sealing is to simply use a plastic seal to attach a lid to the package; this is generally known as epoxy sealing. This is a very inexpensive approach but does negate the hermetic nature of these packages. Hermetic packages require alternative sealing techniques that offer much greater levels of resistance to moisture. No material is truly impermeable, but metals, ceramics, and glasses possess permeability several orders of magnitude less than polymers.

Welding is the most reliable method for sealing hermetic metal packages and is widely used in military applications. The higher capital cost of the equipment is justified by the improved yields and reliability. The welding process involves the application of high current pulses resulting in localized heating of up to 1,500°C, thereby fusing the lid to the package. Other techniques include electron beam and laser welding, which is more attractive for larger packages and provides a noncontact sealing method. Welding is also more tolerant of uneven surfaces and the process does cause the outgassing of organic vapors, which can occur in soldering and glass frit sealing. Welding cannot be applied to ceramic lids and is not cost effective for high volume applications.

Alternative techniques, better suited to high volumes and suitable for use on both metal and ceramic packages, are soldering and brazing. In the case of ceramic packages a metal seal band should be incorporated on the substrate surface to facilitate the sealing process. Such a band can be formed by, for example, thick-film printing. When soldering and brazing, attention must be paid to the process temperature, which should be significantly lower than the temperatures necessary to melt the seal around contact pins and affect the die mount. Seals formed with a gold-tin eutectic braze are stronger and more reliable than their solder counterpart and also avoid the use of flux. The eutectic of choice is usually applied in a preform configuration that is placed between the lid and the package. Mechanical pressure is then applied via spring clips or weights and the assembly heated in a furnace. Flat surfaces are required on both the lid and package to ensure a reliable hermetic seal.

In addition to die mounting and the sealing of electrical interconnects, glass frits can also be used to seal packages. The attractions of glass frits include their inert nature, their electrical insulating properties, their impermeability to moisture and gases, and the wide range of available thermal characteristics. Their main disadvantages are their brittle nature and low strength. The seal design, choice of glass, and sealing process must be carefully considered to maximize the strength of the bond. Lead-zinc-borate glasses are often used and these require a process temperature below 420°C; and the TEC can be modified by the addition of suitable fillers to reduce stresses in the seals. The actual sealing process typically involves heating the package in a furnace to the required process temperature. The lid is normally pre-glazed with the appropriate sealing glass. Furnace profiles, and especially cooling rates, must be carefully controlled to reduce stresses and avoid reliability issues.

4.4 MEMS Mechanical Sensor Packaging

A MEMS sensor packaging must meet several requirements [7–9]:

- *Protect the sensor from external influences and environmental effects.* Since MEMS inherently include some microscale mechanical components, the integrity of the device must be protected against physical damage arising from mechanical shocks, vibrations, temperature cycling, and particle contamination. The electrical aspects of the device, such as the bond wires and the electrical properties of the interconnects, must also be protected against these external influences and environmental effects.
- *Protect the environment from the presence of the sensor.* In addition to protecting the sensor, the package must prevent the presence of the MEMS from reacting with or contaminating potentially sensitive environments [10]. The classic examples of this are medical devices that contain packaged sensors that can be implanted or used within the body; these must be biocompatible, non-toxic, and able to withstand sterilization.
- *Provide a controlled electrical, thermal, mechanical, and/or optical interface between the sensor, its associated components, and its environment.* Not only must the package protect both the sensor and its environment, it must also provide a reliable and repeatable interface for all the coupling requirements of a particular application. In the case of mechanical sensors, the interface is of fundamental importance since, by its nature, specific mechanical coupling is essential but unwanted effects must be prevented. A simple example would be a pressure sensor where the device must be coupled in some manner to the pressure but isolated from, for example, thermally induced strains. The package must also provide reliable heat transfer to enable any heat generated to be transmitted away from the MEMS device to its environment.

In the vast majority of cases, basic plastic, metal, or ceramic packages do not satisfy these requirements. While the requirements for electrical connections and heat transfers paths on sensor packages are typically much less than in the case of most ICs, it is the mechanical interface that complicates the package design. The mechanical interface must isolate the sensor from undesirable external stresses and provide relief from residual stresses in the assembly while enabling the desired mechanical effect arising from the measurand to be coupled to the sensor. In the vast majority of practical sensor applications, each packaging solution will be developed specifically for that particular application.

The sensor packaging can be broken down into two distinct components. *First order packaging* relates to the immediate mounting of the chip, and *second order packaging* refers to the mechanical housing surrounding the mounted sensor. The degree of engineering involved for each will depend upon the particular application. It is certainly common for the first order package, and often the case for the second order package, to perform an integral part of the device function.

The following sections present packaging solutions, both first and second order, that address the key requirements described above. Section 4.4.1 details methods of protecting the sensor die from its environment and includes a discussion of wafer level packaging techniques. Section 4.4.2 describes packaging techniques used to protect the environment from the presence of the sensor. Section 4.4.3 presents stress-relieving techniques used to isolate sensors from undesirable external stresses. It also includes an analysis of common packaging materials and bonding processes and discusses their influence on the behavior and performance of a packaged MEMS mechanical sensor. Finally, Section 4.4.4 discusses the latest developments and looks towards future packaging trends.

4.4.1 Protection of the Sensor from Environmental Effects

MEMS mechanical sensors require careful packaging in order to protect the inherently fragile mechanical components and to prevent undesirable external influences. Damage to the sensor chip can arise from chemical exposure, particulate contamination, mechanical shocks, and extremes of temperature [11]. Exposure to environmental media, either gases or liquids, can adversely affect MEMS in several ways. Corrosion of wire bonds, metal bond pads, or even the substrate material itself can lead to premature failure and reliability problems [12]. Water molecules can cause such effects. Another undesirable consequence is the occurrence of stiction, whereby surface machined components can become stuck to the substrate. Similarly, particle contamination will prevent mechanical components from functioning correctly, as well as potentially shorting electrical contacts. Excessive mechanical shocks can simply cause microstructures to fracture. Extremes of temperature will maximize packaging stresses arising from TEC mismatches, which can affect both performance and reliability, and possibly prohibit some forms materials and electronics. Finally, the electrical characteristics of interconnects and device electronics must also be protected. Such protection must be provided by the package as a whole, but packaging the device at wafer level provides the best level of protection. This approach ensures a robust sensor chip with some level of protection in place against the subsequent packaging processes.

4.4.1.1 Wafer Level Packaging

Wafer level packaging refers to any packaging step that can be performed using wafer-processing techniques and that act on all the devices simultaneously across the wafer. These packaging processes are carried out before dicing. Wafer level packaging is commonly used to provide some level of sensor isolation or stress relief (see Section 4.3.3) or to cap or seal part of or the whole die. The method of isolation and sealing will depend upon the application. The advantages of wafer level packaging compared to the normal packaging approach are:

- While wafer level packaging adds cost to the fabrication of the sensor, it simplifies subsequent packaging, leading to, in the majority of cases, a reduced overall cost. This is evidenced by the proliferation of low-cost, mass-produced accelerometers packaged in standard plastic encapsulations [13–15].
- The tight tolerances that can be achieved allow the cap over the device to perform a function such as over-range protection for inertial sensors.
- Wafer level capping can be used to trap a vacuum around a device. Such an approach has been used on numerous micromachined resonant sensors [16].
- Finally, the cap can protect the device during dicing, which is potentially both a damaging and contaminating process.

Wafer level sealing is typically achieved using glass or silicon capping wafers, and these can be joined together using anodic, organic adhesive, glass reflow, solder reflow, or silicon fusion bonding processes [17–19]. The suitability of each bonding process will depend upon the topology of the wafer, the materials involved, and the maximum permissible process temperature the devices can withstand. The suitability of the capping material will depend upon the application. Certain substrates materials, such a sapphire, offer improved resistance to corrosive media [20].

Micromachined accelerometers have been packaged at wafer level in this manner for many years, an example of which is shown in Figure 4.5 [17]. The piezoresistive accelerometer wafer is first bonded to a silicon supporting wafer. An etched silicon capping wafer is then bonded over the top, thereby sealing the accelerometer and forming a three-layer device. Due to the wafer topology, anodic or fusion bonding cannot be used in the final bonded step. As previously mentioned, these devices can then be placed in standard plastic packages and can even withstand the transfer molding process [13].

4.4.1.2 Electrical Interconnects for Wafer Level Packages

A negative aspect to wafer level capping is the complication of access to contact pads and on-chip electrical interconnects. Contact pads can be revealed by subsequent etching or sawing steps through the capping wafer [21]. On-chip electrical interconnects from the capped region of the die to the contact pads must not compromise the hermetic seal of the cap; they must possess low feedthrough resistance and remain

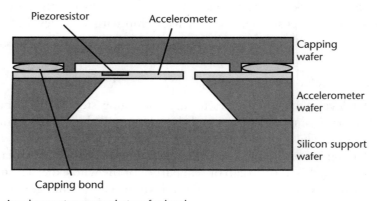

Figure 4.5 Accelerometer capped at wafer level.

electrically isolated from each other. Techniques for achieving such electrical inter-connects include [10]:

- P-n junction feedthrough;
- Buried electrode feedthrough;
- Sealed feedthrough channels;
- Thermomigration of aluminum.

Alternatively, through-wafer interconnects that allow contacts to be made on the underside of the sensor wafer are being developed [22]. Vertical vias have been etched through the thickness of the wafer using a DRIE process. Vias with diameters of up to 200 μm have been formed in this manner and successfully metallized along the length of the channel, thereby forming a low resistance conductive path between the front and back of the wafer. The underside contacts can be formed into solder bumps making this approach compatible with subsequent flip chip second order packaging (see Figure 4.6). The sealing of these underside contacts must be carefully carried out in order to preserve the hermetic nature of the sealed chamber. A similar technique that utilizes a 2-μm-thick polysilicon film heavily doped with phospho-rous deposited on the inside walls of the vias has also been presented [23]. The vias in this instance were just 20 μm in diameter and 400 μm long. A hermetic seal was insured by subsequently filling the vias with LPCVD oxide. Similar work has also been published by Chow et al. [24], and copper interconnects have been developed by Nguyen et al. [25].

In certain applications, wafer level capping alone may not be sufficient or wafer level processing may not be suitable. For example, the capping material may not be able to offer sufficient protection against corrosive media. In these instances, the capped sensor can be coated in a protective layer or the second order package must isolate and seal the device.

Protective coatings have been developed for a number of applications, and as with wafer level packaging, they can simplify second order packaging by removing the need to isolate the device. In wet applications polymer films such as Parylene and silicone gels have been successfully employed [26]. Despite the absorption of water molecules by these polymers, the adhesion of the film to the sensor prevents

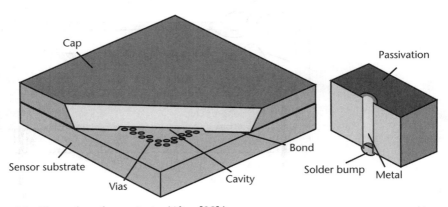

Figure 4.6   Through-wafer contacts. (*After:* [22].)

liquid water forming at the interface [27]. These polymers, however, offer poor levels of protection against alkaline solutions.

Protective silicon oxide and silicon nitride films possess a much greater resistance to the diffusion of water molecules. These films can be applied both at wafer level and on mounted chips using CVD processes. They must be free from cracks and pinholes, and in the case of mounted chips, the films must be deposited on all the exposed surfaces, including wirebonds and contact pads. The chemical resistance of these films is fundamentally important since they will only be deposited in thicknesses of a few microns. Even very low corrosion rates (27 angstroms/day) will remove a 1-micron-thick protective film after 1 year. Silicon carbide thin-films have been found to offer the most promising levels of chemical resistance [28]. A further consideration is the effects of thermal cycling, which can cause delamination of these films due to TEC mismatches.

If the second order package is required to protect the device, the sealing processes developed by the IC industry and described in Section 4.3.3 can be used. In the case of MEMS packaging, second order capping can be further complicated by the functionality of the device. The most common example of this is in pressure sensors where a stainless steel diaphragm in the second order package is used to provide media isolation [29]. Stainless steel offers excellent levels of chemical resistance and possesses good mechanical properties making it an ideal material for such a barrier diaphragm. This diaphragm must not only protect the sensor but transmit the media pressure to it. This is typically achieved by placing the sensing die in an oil-filled chamber behind the stainless steel diaphragm (see Figure 4.7). The pressure exerted on the stainless steel diaphragm is transmitted through the hydraulic oil to the sensor diaphragm. Both the stainless steel diaphragm and the oil used to fill the chamber will influence the behavior of the sensor. The corrugated steel diaphragm shown in Figure 4.7 is an example of a mechanical design used to minimize its influence on the behavior of the sensor. The thermal expansion of the oil will introduce another source of temperature cross sensitivity on the output of the sensor. This approach also places limitations on the minimum attainable size, increases the costs of the device, and restricts the number of applications.

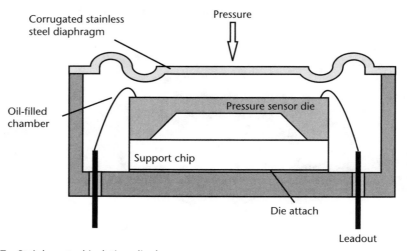

Figure 4.7   Stainless steel isolation diaphragm.

4.4.2   Protecting the Environment from the Sensor

The MEMS package must also protect the environment from the presence and function of the sensor. Application areas where this may be of particular concern include healthcare, food, beverage, and bioprocessing. These typically require the microsensor to be isolated from the chemical or biological media by a mechanical interface or sensor housing made from a suitable material. Types of interface materials include polymer membranes [30], ceramics, glass ceramics, and some metals. The duration of contact with the environment is a fundamental factor in choosing the material, and it must possess the following characteristics in the typical applications mentioned earlier:

- Biocompatible;
- Nontoxic;
- Able to withstand sterilization.

This is particularly important in biomedical applications, where the small size and performance characteristics of MEMS sensors make them highly attractive. Examples of such devices include catheter blood pressure sensors and chemical monitoring systems (e.g., glucose). These have been successfully used in both in vivo and in vitro applications.

Many of the applications discussed also impose space constraints upon the final packaged solution—catheter pressure sensors being the obvious example. Techniques such as flip chip assembly and wafer level packaging can be employed to reduce the packages' volumes. In extreme cases where more than one sensing die or separate ICs are to be incorporated, chip stacking can be employed to further reduce package size [31, 32]. Chip stacking introduces many potential difficulties including electrical interconnects, thermal issues, and packaging stresses [33]. Electrical interconnects have been realized using through-wafer techniques discussed above, purpose-made intermediate chips with a suitable track layout, and also by forming metal tracks on the outside of the stack [34]. A basic stacking approach can be used to reduce packaging stresses, and this is discussed in the following section.

Another interesting development that may suit some applications is the development of spherical semiconductors. A 1-mm-diameter spherical semiconductor has three times the surface area of a 1-mm-square chip [35]. Many sensing applications have been suggested for this form of device including medical sensors and accelerometers.

4.4.3   Mechanical Isolation of Sensor Chips

The mechanical isolation of the sensor chip is vital in avoiding unwanted cross sensitivities and the transmission of external stresses through the packaging to the sensor. Indeed, the package design itself must avoid introducing such undesirable effects and should provide relief from residual stresses trapped in the assembly during the packaging process. Factors such as the long-term stability of packaging materials and methods are of fundamental importance. A well-isolated sensor chip mounted on a carefully designed package will be less affected by changes in its environment, leading to improved long-term stability, resolution, and sensor accuracy.

While this applies equally to capacitive, piezoresistive, and resonant devices, the performance advantages offered by resonant sensing can only be achieved with capable package design. One of the major undesirable influences is the effect of temperature changes on the packaged sensor assembly. Uneven thermal expansion coefficients of the different materials making up the packaged assembly often induce stresses across the sensor chip. Similar packaging stresses can also be induced by the application of mechanical forces onto the second order packaging, changes in humidity, the presence of vibrations, or be in-built in the assembly during the packaging process.

The following techniques for providing mechanical isolation of a sensor chip have been applied to a simple pressure sensor. The pressure sensor in this case consists of an etched diaphragm with some form of strain-sensing mechanism fabricated on the top surface, as shown in Figure 4.8. This example assumes direct contact of the pressurized media with the sensor chip, and therefore, other packaging requirements, such as oil filling, are not considered in this case. Pressure sensors are discussed in more detail in Chapter 6.

The simplest and lowest cost form of sensor package is to bond the sensor chip directly to the second order packaging, in this a case a simple TO header as shown schematically in Figure 4.9. Coupling to the sensor diaphragm is facilitated by a pressure port formed in the header. Such an arrangement is based upon microelectronic device packaging and effectively has no first order packaging stage. As a result, mechanical stresses are transmitted directly to the sensor chip and the transducer housing is likely to be thermally incompatible with silicon due to TEC mismatches. The overall accuracy of the sensor will therefore be poor. Thermal stresses can be compensated for to some degree by the sensing electronics, but associated drift cannot be compensated for.

The above packaging solution is impractical in the vast majority of applications. Improved mechanical isolation can be achieved by the following range of techniques, the suitability of which will depend upon each application and its particular packaging requirements:

- Use of a first order packaging stage (i.e., placing an intermediate, or support chip, between the sensor chip and housing);
- Mechanical decoupling on the sensor or support chip;
- Displacing the sensor away from the second order packaging;
- Die attach using of soft ductile bond materials;

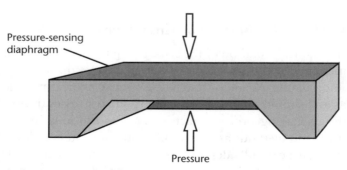

Figure 4.8 Typical pressure sensing die.

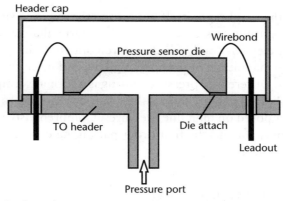

Figure 4.9   Basic packaging scheme.

4.4.3.1   Basic First Order Packaging Stage

A basic yet typical first order packaging arrangements is shown in Figure 4.10. The support chip used in Figure 4.10 can be fabricated from either thermally matched lead borosilicate glass, such as Pyrex 7740 or Schott Borofloat 33, or silicon itself. The glass constraint is typically anodically bonded to the silicon chip, providing and extremely strong molecular bond. This bond can be performed at wafer level, enabling all devices to be simultaneously mounted. If the glass constraints are not exactly matched to the silicon, some thermally induced stresses will occur because of the TEC mismatch. This drawback is exaggerated by the anodic bond, which is carried out at temperatures of around 400°C. As the bonded assembly cools, residual stresses will be inevitably introduced across the sensor chip. Thermal matching between the sensors and constraint will naturally be improved if the constraint is made from silicon [36, 37].

Another factor that should be considered in certain applications is that the presence of the support chip can alter the sensitivity of the sensor to the measurand. In the case of high-pressure sensors, for example, the pressure will not only be

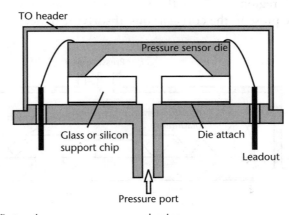

Figure 4.10   Basic first order pressure sensor packaging.

applied to the diaphragm itself but also to the exposed surfaces of the intermediate [38]. The resulting stresses induced in the intermediate will be transmitted in part to the sensing elements and will therefore contribute in some manner to the sensor output. The magnitude of the effect will depend upon the particular design and the application. Differential pressure sensing is another example application where this effect can be important, especially when attempting to detect small differential pressures imposed on high line pressures. Comprehensive modeling of the assemble sensor diaphragm and the first order packaging can be used in the design stage to predict this effect.

4.4.3.2 Mechanical Decoupling

Mechanical decoupling in the form of stress-relieving flexible regions may be incorporated on either the sensor or intermediate chip. The flexible regions take the form of micromachined corrugations that absorb stresses rather than transmit them to the sensing element within the assembly. This corrugated decoupling zone may be fabricated on the sensor chip itself, as shown in Figure 4.8 [39, 40]. The pressure-sensing diaphragm is located at the sensor of the chip and is supported by an inner rim. The sensor chip is fixed to its surroundings at an outer rim and the decoupling corrugations lie in between the two rims. The placement of the corrugations on the sensor chip could remove the need for any first order packaging (as depicted in Figure 4.11), but this does increase the overall size of the chip and reduces the number of devices that can be realized on each wafer. Also, the fabrication processes of the corrugations and the sensing mechanism employed on the sensor chip must be compatible. Another disadvantage is the difficulty in forming conductive paths over the corrugations to the outer rim, which would be the preferred location for the bond pads. This could be overcome by placing the bond pads on the inner rim or by providing planar paths, or bridges, over the corrugations [41].

Alternatively, the use of silicon intermediate support chips offers the opportunity of micromachining the stress-relieving regions on the constraint chip rather than the sensor chip itself. Finite element analysis employed to investigate various decoupling designs identified the structured washer style support chip, shown in Figure 4.12, as the most promising solution [41, 42]. The mechanical decoupling is provided by V-grooves etched into both sides of the constraint wafer, forming a thin corrugated region between the sensor chip and its mounting. When packaging stresses are present, the corrugations absorb the deflection rather than transmitting

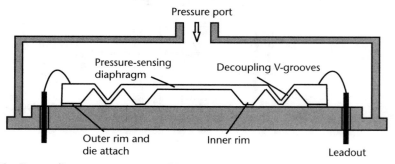

Figure 4.11   Decoupling zones on sensor chip.

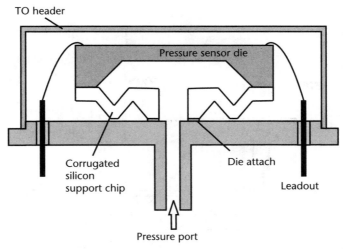

Figure 4.12   Corrugated silicon intermediate.

them to the sensor chip itself. In this manner, a 99% reduction in packaging stresses transmitted to the sensor chip is possible. The support chips can be fabricated and bonded to the sensor die at wafer level, therefore enabling the simultaneous processing of all the devices on a wafer.

This approach should improve the performance of the sensor and reduce the complexity and cost of the second order packaging. The disadvantages are the reduced strength of the assembly—because less area is available to bond the intermediate to the transducer housing—and the increased cost of the first order packaging due to the processing of the silicon intermediate. Also, as discussed previously, the presence of the support chips may influence the output of the sensor in certain applications and this may be further exaggerated by the corrugations.

The economic benefits of placing mechanical decoupling on the sensor chip or the silicon intermediate will depend upon the relative processing costs required by the sensor chip and the intermediate. If the sensor has a complex design requiring many processing steps, then it would be more economical to maximize the device density upon the wafer and incorporate the mechanical stress relief on the intermediate chip. If, on the other hand, the processing of the sensor is straightforward and not affected by incorporating the corrugations alongside the sensor structure, that approach could be favorable.

4.4.3.3   Displacing the Sensor from the Second Order Packaging

Other stress-relieving first order packaging designs involve removing the sensor as far away from the transducer housing as is practical. This can be achieved with both vertical and horizontal separation. Vertical separation can take the form of tall glass or silicon supports chips similar to the design shown in Figure 4.10. The packaging of the Druck resonant pressure sensor [43], described in Chapter 6, is an example of vertical separation. The package design is shown in Figure 4.13. The pressure-sensing diaphragm and resonator is mounted on a silicon support chip, which is in turn attached to a glass tube. The glass tube serves both to move the

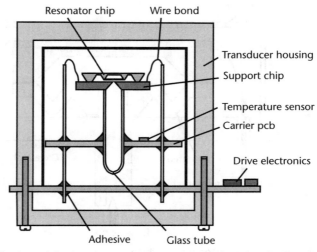

Figure 4.13 Packaging of the Druck resonant pressure sensor. (*After:* [43].)

sensor away from the transducer housing and, by sealing the end in a vacuum, trap a vacuum around the resonating element. This approach, however, is time consuming and expensive to assemble; wafer level vacuum encapsulation is greatly preferred.

Horizontal, or lateral, separation of the sensor chip away from transducer housing or supporting substrate is achieved by fixing the chip only at an insensitive part of the die (i.e., away from the location of the sensing elements) [44]. The sensing element is therefore separated from the substrate by a small gap, as shown in Figure 4.14, and packaging stresses will only be transmitted directly to insensitive regions of the sensor chip. This approach will not be suitable for many applications, but where it is applicable, experimental work has shown packaging stresses reduced by a factor of 10. This approach certainly offers a very simple isolating technique, but it may involve increasing the size of the sensor chip in order to include an insensitive region of sufficient area to enable robust mounting.

A similar approach has been employed in the packaging of a silicon high-pressure sensor designed for use in refrigeration and fluid power applications. The pressure-sensing membrane and associated piezoresistive elements are located at the end of a silicon *needle* [45]. This needle is housed within a metallic collar, and

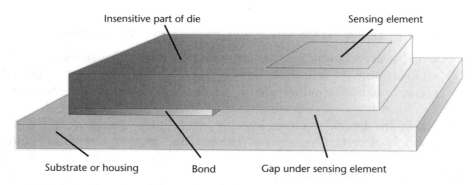

Figure 4.14 Lateral isolation of the sensing element.

the sensing elements protrude beyond the end of the collar and are in direct contact with the pressurized media. This low-cost packaging approach provides a good degree of mechanical isolation, but the drawbacks include the increased size of the sensor chip and the fact that it is in direct contact with potentially corrosive media.

4.4.3.4   Use of Soft Adhesives

The die mount material and the method of attachment will also have an influence on the mechanical isolation of the sensor and the level of in-built stresses trapped within the assembly. The various methods of die attach used in the IC industry and discussed in Section 4.3.2 are equally applicable to MEMS packaging. Typical parameters of these processes are shown in Table 4.3. The TECs of the bond materials, along with common packaging materials, are given in Table 4.4. The TEC of silicon varies with temperature and is listed against different temperatures in Table 4.5. The TEC of these materials is of fundamental importance to the MEMS designer since the stresses arising from TEC mismatches account for the majority of packaging-induced error.

The use of soft, ductile bond materials in the mounting of the die can provide a high degree of isolation from undesirable mechanical stresses. These soft adhesives absorb the stress in a manner similar to the mechanical decoupling structures described previously. In addition, the lower temperature die attach processes associated with typical soft adhesives are advantageous since the magnitude of thermally induced stresses trapped in the final assembly will be reduced. The drawbacks of soft adhesive typically relate to their bond strength, which is very weak compared to harder epoxies and especially solder and eutectic bonds. Soft adhesives are not suited to applications that place the sensor die under shear of tensile stress. Where harder, stronger bonds have to be used, trapped thermal stresses and the resulting temperature cross-sensitivity can be minimized by keeping the adhesive film as thin as possible.

Soft adhesives, such as RTV silicone, must be applied in a controlled thickness to achieve maximum benefit. Experimental analysis showed that the thermal behavior of the sensor shown in Figure 4.14 was improved by increasing the adhesive thickness up to 50 μm, but no further improvement was observed beyond this [46]. Glass spheres can be used in the assembly of the sensor to control this thickness, as shown in Figure 4.15.

Table 4.3   Typical Die Mounting Process and Material Parameters

Attachment Method	Adhesion Material	Process Temperature (°C)	Thermal Conductivity (W/m °C)	Young's Modulus (10^9 N/m²)
Eutectic	AuSi (97/3)	400	27.2	87
Solder	Pb/Sn	200	35	14
Glass	Pb glass	450–800	0.25–2	60
Anodic	Pyrex/ Borofloat 33	250–500	1.09	63
Epoxy	Epoxy (Ag loaded)	150	1.2	0.2–27
Thermoplastic	Thermoplastic	150	3	0.41
RTV silicone	RTV silicone	25	0.1	6.9 10^{-3}

Source: [47].

Table 4.4 Thermal Expansion Coefficients of Common Packaging Materials

Application	Material	TEC (10^{-6}/°C)
Die	Si	See Table 4.5
	GaAs	5.7
Lead frames	Copper	17
	Alloy 42	4.3–6
	Kovar	4.9
	Invar	1.5
Substrates/constraints	Alumina (99%)	6.7
	AIN	4.1
	Beryllia (99.5%)	6.7
	Pyrex 7740	3.3
Adhesives	Au-Si eutectic	14.2
	Pb-Sn	24.7
	Pb glass	10
	Ag loaded epoxy	23–40[1]
	Thermoplastic	30–54[1]
	RTV silicone	300–800[2]

[1]Below glass transition temperature.
[2]Above glass transition temperature.
Source: [48].

Table 4.5 Thermal Expansion Coefficient of Silicon Versus Temperature

Temperature (°C)	TEC (10^{-6}/°C)
–53	1.715
7	2.432
27	2.616
127	3.253
427	4.016

Source: [49].

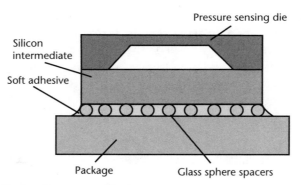

Figure 4.15 Soft adhesive die mount with glass spacers.

A commercial low absolute pressure sensor has also been successfully packaged using soft adhesives for use in space applications and in particular a mission to Mars [50]. The application requires the sensor to survive shocks of up to 100,000g, operate in temperatures as low as –80°C with fluctuations of 50°C and resolve 0.05 mbar over a 14-mbar range with an overall accuracy of 0.5 mbar. Given the size and weight restrictions, the sensor was packaged alongside the electronics using an MCM, as shown in Figure 4.16. The MCM incorporated epoxy-mounted ICs with thick-film tracks, surface mount capacitors, and thick-film resistors. The sensor and ICs were flush mounted to enable shorter wirebonds.

The mounting of the sensor die to the ceramic package is shown in Figure 4.17. The pressure sensor is bonded with a 25-μm-thick layer of soft adhesive (Silicone RTV 566) to a silicon support chip. Silicone RTV 566 was used because it has a glass transition temperature of –115°C, and therefore it maintains its ductile properties at the specified operating temperatures. The support chip is then bonded to the ceramic substrate using a much thicker layer of silicone (250 μm), which provides isolation from packaging and impact stresses. This layer of silicon could not be thicker than 250 μm because it would put the wirebonds under excessive strain. The support chip serves to isolate the sensor from the effects of the TEC mismatch between the silicon and the RTV silicone.

4.4.3.5 Summary of Techniques for Mechanically Isolating the Sensor Chip

Table 4.6 presents a summary of techniques for mechanically isolating the sensor chip.

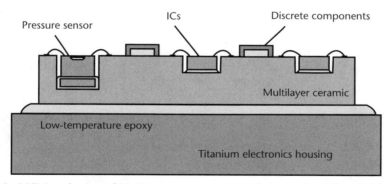

Figure 4.16 MCM packaging of Martian pressure sensor and electronics. (*After:* [50].)

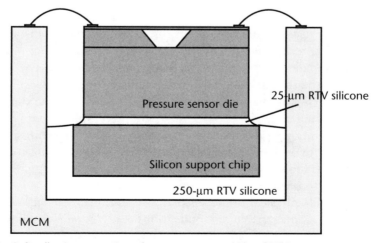

Figure 4.17 Soft adhesive mounting of pressure sensor. (*After:* [50].)

Table 4.6   Summary of Techniques for Mechanically Isolating the Sensor Chip

Technique	Advantages	Disadvantages
Pyrex intermediate (Figure 4.10)	Simplest first order package Low cost Bonded at wafer level Suitable for a wide range of applications	Pyrex not exactly matched with silicon Limited machining of Pyrex possible Relatively large first order assembly
Soft bond (Figures 4.15 and 4.17)	Simple Low cost No modifications to sensor chip required Can negate the need for, or be used, with first order packaging	Lower bond strength—reduced applications Bond material unsuitable for certain applications
Etched silicon intermediate (Figure 4.12)	Exact thermal match Bonded at wafer level Machining of intermediate possible High degree of isolation Smaller assembly size	Reduced bond area to second order packaging Critical alignment required over pressure port
On-chip decoupling (Figure 4.11)	Can negate the need for, or be used, with first order packaging	Increased chip area—fewer sensors per wafer Complicates sensor chip fabrication Not suitable for many applications
Vertical displacement–glass tube (Figure 4.13)	High level of isolation Allows simple evacuation of resonator surrounding	Labor-intensive assembly Sensors individually packaged–high cost Large assembly
Lateral displacement (Figure 4.14)	Simple Low cost Can negate the need for, or be used, with first order packaging	Limited applications Relatively poor degree of isolation Increased chip area—fewer sensors per wafer

4.5   Conclusions

It is clear that the packaging of the sensor is as important as the design of the sensor in determining the overall performance of the device. This is emphasized further by the fact that the packaging operation is likely to be more costly than the fabrication of the sensor itself. Many techniques for packaging microsensors can be taken from IC packaging techniques. However, microsensor packaging also requires that the sensor die remains well isolated from any undesirable stresses transmitted through, or arising from, the packaging while still transmitting the measurand to the sensor.

In order to minimize the total cost of the transducer, the simplest isolation techniques, utilizing wafer level processing, are preferable where possible. Of the basic isolation techniques, the use of soft adhesives is most promising, especially when combined with a glass or silicon constraint. It is interesting to note that the more complex, and more costly, isolating techniques involving machining stress-relieving structures into the silicon support or the sensor die have not been used in commercially available physical microsensors. The suitability of these isolation techniques will depend upon particular applications and the various design considerations involved. Important considerations include the temperature required by the various packaging process, the thermal compatibility of any the materials involved, and the long-term stability of the assembly.

References

[1] Tammala, R., E. J. Rymaszewski, and A. G. Klopfenstein, *Microelectronics Packaging Handbook*, New York: Wiley, 1997.

[2] Cazenave J. P., and T. R. Suess, "Fodel Photoimageable Materials—A Thick Film Solution for High Density Multichip Modules," *Proc. of the SPIE*, Vol. 2105, 1993, pp. 483–488.

[3] Harman, G. G., *Wire Bonding in Microelectronics: Materials, Processes, Reliability, and Yield*, New York: McGraw-Hill, 1997.

[4] Liu, J., and Z. Lai, "Reliability of Anisotropically Conductive Adhesive Joints on a Flip-Chip/FR-4 Substrate," *Transactions of the ASME, Journal of Electronic Packaging*, Vol. 124, No. 3, September 2002, pp. 240-245.

[5] Gilleo, K., and P. Ongley, "Pros and Cons of Thermoplastic and Thermoset Polymer Adhesives in Microelectronic Assembly Applications," *Microelectronics International*, Vol. 16, No. 2, 1999, pp. 34–38.

[6] Khanna, P. K., S. K. Bhatnagar, and W. Gust, "Analysis of Packaging and Sealing Techniques Microelectronic Modules and Recent Advances," *Microelectronics International*, Vol. 16, No. 2, 1999, pp. 8–12.

[7] Ko, W. H., "Packaging of Microsensors," *Proc. 4th Int. Conf. on Micro Opto Mechanical Systems and Components*, Berlin, Germany, October 19–21, 1994, pp. 477–480.

[8] Hsu, T.-R., "Packaging Design of Microsystems and Meso-Scale Devices," *IEEE Trans. on Advanced Packaging*, Vol. 23, No. 4, November 2000, pp. 596–601.

[9] Kelly, G., et al., "Microsystems Packaging: Lessons from Conventional Low Cost IC Packaging," *J. Micromech. Microeng.*, Vol. 7, 1997, pp. 99–103.

[10] Ko, W. H., "Packaging of Microfabricated Devices," *Materials Chemistry and Physics*, Vol. 42, No. 3, 1995, pp. 169–175.

[11] Sparks, D. R., "Packaging of Microsystems for Harsh Environments," *IEEE Instrumentation and Measurement Magazine*, September 2001, pp. 30–33.

[12] Maudie, T., and J. Wertz, "Pressure Sensor Performance and Reliability," *IEEE Industry Applications Magazine*, May/June 1997, pp. 37–43.

[13] Motta, V., et al., "Packaging for a Rotational Accelerometer: Is a Standard Plastic SOIC an Industrial Solution?" *Proc. of the SPIE*, Vol. 4174, 2000, pp. 377–387.

[14] Nysaether, J. B., et al., "Measurement of Package-Induced Stress and Thermal Zero Shift in Transfer Molded Silicon Piezoresistive Pressure Sensors," *J. Micromech. Microeng.*, Vol. 8, 1998, pp. 168–171.

[15] Li, G., and A. A. Tseng, "Low Stress Packaging of a Micromachined Accelerometer," *IEEE Trans. on Electronics Packaging Manufacturing*, Vol. 24, No. 1, January 2001, pp. 18–24.

[16] Chang-Chien, P. P. L., and K. D. Wise, "Wafer Level Packaging Using Localized Mass Deposition," *Transducers '01 / Eurosensors XV, 11th International Conference on Solid State Sensors and Actuators*, Munich, Germany, June 10–14, 2001, *Digest of Technical Papers*, Vol. 1, pp. 182–185.

[17] Sparks, D., et al., "Chip-Scale Packaging of Gyroscope Using Wafer Bonding," *Sensors and Materials*, Vol. 11, No. 4, 1999, pp. 197–207.

[18] Audet, S. A., and K. M. Edenfield, "Integrated Sensor Wafer Level Packaging," *Transducers '97, 1997 International Conference on Solid State Sensors and Actuators*, Chicago, IL, June 16–19, 1997, *Proceedings*, Vol. 1, pp. 287–289.

[19] Krassow, H., F. Campabadal, and E. Lora-Tamayo, "Wafer Level Packaging of Silicon Pressure Sensors," *Sensors and Actuators*, Vol. A82, No. 1–3, May 2000, pp. 229–233.

[20] Kimura, S., et al., "Stable and Corrosion Resistant Sapphire Capacitive Pressure Sensor for High Temperature and Harsh Environments," *Transducers '01/Eurosensors XV, 11th International Conference on Solid State Sensors and Actuators*, Munich, Germany, June 10–14, 2001, *Digest of Technical Papers*, Vol. 1, pp. 518–521.

[21] Renard, S., "Wafer Level Surface Mountable Chip Size Packaging for MEMS and ICs," *Proc. SPIE*, Vol. 4176, 2000, pp. 236–241.

[22] Neysmith, J., and D. F. Baldwin, "Modular Device Scale, Direct Chip Attach Packaging for Microsystems," *IEEE Trans. on Components and Packaging Technologies*, Vol. 24, No. 4, December 2001, pp. 631–634.

[23] Cheng, C. H., et al., "An Efficient Electrical Addressing Method Using Through-Wafer Vias for Two-Dimensional Ultrasonic Arrays," *Proc. IEEE Ultrasonics Symposium*, October 22–25, 2000, San Juan, PR, Vol. 2, pp. 1179–1182.

[24] Chow, E. M., et al., "Process Compatible Polysilicon-Based Electrical Through Wafer Interconnects in Silicon Substrates," *IEEE Journal of Microelectromechanical Systems*, Vol. 11, No. 6, December 2002, pp. 631–640.

[25] Nguyen, N. T., et al. "Through-Wafer Copper Electroplating for Three Dimensional Interconnects," *J. Micromech. Microeng.*, Vol. 12, 2002, pp. 395–399.

[26] Petrovic, S., et al., "Low-Cost Water Compatible Piezoresistive Bulk Micromachined Pressure Sensor," *Proc. of the Pacific Rim/ASME International Intersociety Electronic and Photonic Packaging Conference*, June 15–19, 1997, New York, Vol. 1, pp. 455–62.

[27] Dyrbye, K., T. R. Brown, and G. F. Eriksen, "Packaging of Physical Sensors for Aggressive Media Applications," *J. Micromech. Microeng.*, Vol. 6, 1996, pp. 187–192.

[28] Eriksen, G. F., and K. Dyrbye, "Protective Coatings in Harsh Environments," *J. Micromech. Microeng.*, Vol. 6, 1996, pp. 55–57.

[29] Terabe, H., et al., "A Silicon Pressure Sensor with Stainless Steel Diaphragm for High Temperature and Chemical Application," *Transducers '97, 1997 International Conference on Solid State Sensors and Actuators*, Chicago, IL, June 16–19, 1997, *Proceedings*, Vol. 2, pp. 1481–1484.

[30] Lee, N. K. S., et al., "A Flexible Encapsulated MEMS Pressure Sensor System for Biomedical Applications," *Microsystem Technologies*, Vol. 7, 2001, pp. 55–62.

[31] Gotz, A., et al., "Manufacturing and Packaging of Sensors for Their Integration in a Vertical MCM Microsystem for Biomedical Applications," *IEEE Journal of Microelectromechanical Systems*, Vol. 10, No. 4, December 2001, pp. 569–579.

[32] Heschel, M., et al., "Stacking Technology for a Space Constrained Microsystem," *IEEE Workshop on Microelectro Mechanical Systems, MEMS '98*, Heidelberg, Germany, pp. 312–317.

[33] Kelly, G., et al., "3-D Packaging Methodologies for Microsystems," *IEEE Trans. on Advanced Packaging*, Vol. 23, No. 4, November 2001, pp. 623–630.

[34] Barrett, J., et al., "Performance and Reliability of a Three-Dimensional Plastic Moulded Vertical Multichip Module," *Proc. 45th IEEE Electronics and Components Conf.*, May 21–24, 1995 Las Vegas, NV, pp. 656–663.

[35] Murzin, I., et al., "MO CVD Interconnects for Spherical Semiconductors," *Microelectronic Engineering*, Vol. 50, 2000, pp. 515–523.

[36] Germer, W., and G. Kowalski, "Mechanical Decoupling of Monolithic Pressure Sensors in Small Plastic Encapsulants," *Sensors and Actuators*, Vol. A23, 1990, pp. 1065–1069.

[37] Holm, R., et al., "Stability and Common Mode Sensitivity of Piezoresistive Silicon Pressure Sensors Made by Different Mounting Methods," *Proc. Int. Conf. Solid State Sensors and Actuators (Transducers '91)*, 1991, pp. 978–981.

[38] Matsuoka, Y., et al., "Characteristic Analysis of a Pressure Sensor Using the Silicon Piezoresistance Effect for High Pressure Measurement," *J. Micromech. Microeng.*, Vol. 5, 1995, pp. 25–31.

[39] Spiering, V. L., S. Bouwstra, and R. M. E. J. Spiering, "On Chip Decoupling Zone for Package Stress Reduction," *Sensors and Actuators*, Vol. A39, 1993, pp. 149–156.

[40] Spiering, V. L., et al., "Membranes Fabricated with a Deep Single Corrugation for Package Stress Reduction and Residual Stress Relief," *J. Micromech. Microeng.*, Vol. 3, 1993, pp. 243–246.

[41] Vaganov, V. L., "Construction Problems in Sensors," *Sensors and Actuators*, Vol. A28, 1991, pp. 161–172.

[42] Offereins, H. L., et al., "Stress Free Assembly Technique for a Silicon Based Pressure Sensor," *Tech. Digest Microsystems Technologies '90*, Berlin, Germany, September 10–13, 1990, pp. 515–520.

[43] Greenwood, J., and T. Wray, "High Accuracy Pressure Measurement with a Silicon Resonant Sensor," *Sensors and Actuators*, Vol. A37–38, 1993, pp. 82–85.

[44] Halg, B., and R. S. Popovic, "How to Liberate Integrated Sensors from Encapsulation Stress," *Sensors and Actuators*, Vol. A21–23, 1990, pp. 908–910.

[45] Birkelund, K., et al., "High Pressure Silicon Sensor with Low Cost Packaging," *Sensors and Actuators*, Vol. A92, 2001, pp. 16–22.

[46] Germer, W., and G. Kowalski, "Mechanical Decoupling of Monolithic Pressure Sensors in Small Plastic Encapsulations," *Sensors and Actuators*, Vol. A21–23, 1990, pp. 1065–1069.

[47] Reichl, H., "Packaging and Interconnection of Sensors," *Sensors and Actuators*, Vol. A25–26, 1991, pp. 63–71.

[48] Maudie, T., and J. Wertz, "Pressure Sensor Performance and Reliability," *IEEE Industry Applications Magazine*, May/June 1997, pp. 37–43.

[49] Lin, Y.-C., P. J. Hesketh, and J. P. Schuster, "Finite-Element Analysis of Thermal Stresses in a Silicon Pressure Sensor for Various Die-Mount Materials," *Sensors and Actuators*, Vol. A44, 1994, pp. 145–149.

[50] Reynolds, J. K., et al., "Packaging a Piezoresistive Pressure Sensor to Measure Low Absolute Pressures over a Wide Sub Zero Temperature Range," *Sensors and Actuators*, Vol. A83, 2000, pp. 142–149.

CHAPTER 5
Mechanical Transduction Techniques

There are many examples of micromachined mechanical transducers and these will be reviewed in detail in the following chapters. The purpose of this chapter is to present some of the fundamental concepts and techniques that are used in the design of mechanical microsensors and actuators. The most sensing-important mechanisms include the following effects: piezoresistivity, piezoelectricity, variable capacitance, optical, and resonant techniques. We will also review the main actuation methods, including: electrostatic, piezoelectric, thermal, and magnetic. The final section of this chapter includes a review of so-called intelligent (or smart) sensors.

5.1 Piezoresistivity

Piezoresistivity derives its name from the Greek word *piezin*, meaning "to press." It is an effect exhibited by various materials that exhibit a change in resistivity due to an applied pressure. The effect was first discovered by Lord Kelvin in 1856, who noted that the resistance of copper and iron wires increased when in tension. He also observed that iron wires showed a larger change in resistance than those made of copper. The first application of the piezoresistive effect did not appear until the 1930s, some 75 years after Lord Kelvin's discovery. Rather than using metal wires, these so-called strain gauges are generally made from a thin metal foil mounted on a backing film, which can be glued onto a surface. A typical metal foil strain gauge is depicted in Figure 5.1.

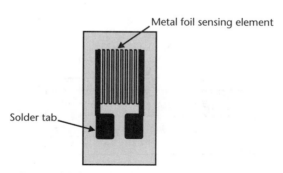

Figure 5.1 Illustration of a metal foil strain gauge.

The sensitivity of a strain gauge is generally termed the gauge factor. This is a dimensionless quantity and is given by

$$GF = \frac{\text{relative change in resistance}}{\text{applied strain}} = \frac{\Delta R / R}{\Delta L / L} = \frac{\Delta R / R}{\varepsilon} \qquad (5.1)$$

where R is the initial resistance of the strain gauge and ΔR is the change in resistance. The term $\Delta L / L$ is, by definition, the applied strain and is denoted as ε (dimensionless). For all elastic materials, there is a relationship between the stress σ(N/m^2) and the strain ε; that is, they obey Hooke's law and thus deform linearly with applied force. The constant of proportionality is the elastic modulus or Young's modulus of the material and is given by

$$\text{Young's modulus, } E = \frac{\text{Stress}}{\text{Strain}} = \frac{\sigma}{\varepsilon} (\text{N/m}^2) \qquad (5.2)$$

The Young's modulus of silicon is 190 GPa (1 Pa = 1 N/m^2), which is close to that of typical stainless steel (around 200 GPa). For a given material, the higher the value of Young's modulus, the less it deforms for a given applied stress (i.e., it is stiffer).

When an elastic material is subjected to a force along its axis, it will also deform along the orthogonal axes. For example, if a rectangular block of material is stretched along its length, its width and thickness will decrease. In other words, a tensile strain along the length will result in compressive strains in the orthogonal directions. Typically, the axial and transverse strains will differ and the ratio between the two is known as Poisson's ratio, ν. Most elastic materials have a Poisson's ratio of around 0.3 (silicon is 0.22). The effect on a rectangular block is depicted in Figure 5.2. The strains along the length, width, and thickness are denoted by ε_l, ε_w, and ε_t, respectively.

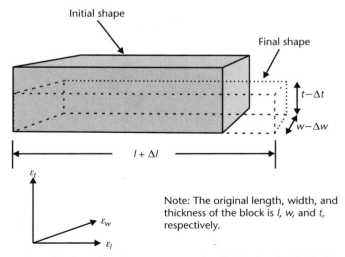

Figure 5.2 Illustration of Poisson's ratio on a rectangular, isotropic, elastic block. A longitudinal tensile strain results in deformation in the two orthogonal axes.

If it is assumed that the block is made of a resistive material, then its resistance, R, is given by

$$R = \frac{\rho l}{A} \qquad (5.3)$$

where ρ is the bulk resistivity of the material (Ωcm), l is the length, and A is the cross-sectional area (i.e., the product of width w and thickness t).

Hence,

$$R = \frac{\rho l}{wt} \qquad (5.4)$$

Differentiating the equation for resistance gives

$$dR = \frac{l}{wt} d\rho + \frac{\rho}{wt} dl - \frac{\rho l}{w^2 t} dw - \frac{\rho l}{wt^2} dt \qquad (5.5)$$

and hence

$$\frac{dR}{R} = \frac{d\rho}{\rho} + \frac{dl}{l} - \frac{dw}{w} - \frac{dt}{t} \qquad (5.6)$$

By definition, $\varepsilon_l = dl/l$, so the following equations apply on the assumption that we are dealing with small changes, and hence $dl = \Delta l$, $dw = \Delta w$, and $dt = \Delta t$:

$$\frac{dw}{w} = \varepsilon_w = -v\varepsilon_l \quad \text{and} \quad \frac{dt}{t} = \varepsilon_t = -v\varepsilon_l \qquad (5.7)$$

where v is Poisson's ratio. Note the minus signs, indicating that the width and thickness both experience compression and hence shrink. It is worth noting that the above example illustrates a positive Poisson's ratio.[1]

Therefore, from (5.6) and (5.7) we have

$$\frac{dR}{R} = \frac{d\rho}{\rho} + \varepsilon_l + v\varepsilon_l + v\varepsilon_l \qquad (5.8)$$

From (5.1) the gauge factor is therefore

$$GF = \frac{dR/R}{\varepsilon_l} = \frac{d\rho/\rho}{\varepsilon_l} + (1 + 2v) \qquad (5.9)$$

Equation (5.9) indicates clearly that there are two distinct effects that contribute to the gauge factor. The first term is the piezoresistive effect ($(d\rho/\rho)/\varepsilon_l$) and the second is the geometric effect $(1 + 2)$. As Poisson's ratio is usually between 0.2 and 0.3,

---

1.   Materials having a negative Poisson's ratio do exist. That is to say, as you stretch them, the width and thickness actually increase. Examples of such materials include special foams and polymers such as polytetrafluoroethylene (PTFE).

the contribution to the gauge factor from the geometric effect is therefore between 1.4 and 1.6. Sensors that exhibit a change in resistance as a result of an applied strain are generally termed strain gauges. Those in which the piezoresistive effect dominates are often referred to as piezoresistors. As Table 5.1 shows, different materials can have widely differing gauge factors.

So for a metal foil strain gauge or thin metal film, the geometric effect dominates the piezoresistive effect; whereas for a semiconductor the converse is true.

Semiconductor strain gauges possess a very high gauge factor. P-type silicon has a gauge factor up to +200, and n-type silicon has a negative gauge factor down to –125. A negative polarity of gauge factor indicates that the resistance decreases with increasing applied strain. In addition to exhibiting high strain sensitivity, semiconductor strain gauges are also very sensitive to temperature. Compensation methods must therefore be adopted when using semiconductor strain gauges.

A detailed account of the piezoresistive effect in silicon can be found in Middelhoek and Audet [1]; only a brief account will be given in this text. Essentially, the effective mobilities of majority charge carriers are affected by the applied stress. With p-type materials, the mobility of holes decreases and so the resistivity increases. For n-type materials, the effective mobility of the electrons increases and hence the resistivity decreases with applied stress. The effect is highly dependent on the orientation. If the geometric effect in semiconductor strain gauges is neglected, then the fractional change in resistivity is given by

$$\frac{d\rho}{\rho} = \pi_l \sigma_l + \pi_t \sigma_t \tag{5.10}$$

where π_l and π_t are the longitudinal and transverse piezoresistive coefficients and σ_l and σ_t are the corresponding stresses. The longitudinal direction is defined as that parallel to the current flow in the piezoresistor, while the transverse is orthogonal to it. The two coefficients are dependent on the crystal orientation and doping (p-type or n-type) and concentration. The temperature coefficient of piezoresistivity is around 0.25 %/°C in both directions.

Polysilicon and amorphous silicon are also piezoresistive, but because they comprise crystallites, the net result is the average over all orientations. The temperature coefficient of resistance (TCR), however, is significantly lower than that of single crystal silicon and is generally less than 0.05%/°C. By carefully choosing the doping levels, it is possible to reduce the TCR further.

Thin metal films behave in a similar manner to metal foil strain gauges and hence it not surprising that their gauge factors are very similar. Such films can be deposited directly onto the desired substrate (steel, ceramic, silicon) and therefore become an integral part of the system, thus removing the need for adhesives as with

Table 5.1 Gauge Factors of Different Materials

Material	Gauge Factor
Metal foil strain gauge	2–5
Thin-film metal	2
Single crystal silicon	–125 to +200
Polysilicon	±30
Thick-film resistors	10

metal foil strain gauges. The adhesives can contribute to a phenomenon called creep, whereby the gauge can effectively slip and therefore produce false readings as the adhesive softens with increasing temperature or over long periods of time.

Thick-film resistors, often used in hybrid circuits, have also been shown to be piezoresistive. Their gauge factor is around 10, and therefore, they offer a sensitivity between that of a semiconductor and foil strain gauge. The TCR is around 100 parts per million (ppm) per degree Celsius and matching between adjacent resistors is often less that 10 ppm/°C, making them well suited for use as active elements in Wheatstone bridge circuits, which reduce the overall temperature sensitivity.

An associated effect that has been observed in semiconductors is the so-called piezojunction effect, whereby a shift in the I-V characteristic of a *p-n* junction is observed as a result of an applied stress. Although this is an interesting physical effect, it has found little use in commercial micromachined devices.

5.2 Piezoelectricity

Certain classes of crystal exhibit the property of producing an electric charge when subjected to an applied mechanical force (direct effect). They also deform in response to an externally applied electric field (inverse effect). This is an unusual effect as the material can act as both a sensor and actuator. It was first discovered in quartz by Jacques and Pierre Curie in 1880. The physical origin of piezoelectricity arises because of charge asymmetry within the crystal structure. Such crystals are often termed noncentrosymmetric, and because of the lack of symmetry, they have anisotropic characteristics. Owing to its symmetric, cubic crystal structure, silicon is not, therefore, piezoelectric. Some crystals such as quartz and Rochelle salt are naturally occurring piezoelectrics, while others like the ceramic materials barium titanate, lead zirconate titanate (PZT), and the polymer material polyvinylidene fluoride (PVDF) are ferroelectric. Ferroelectric materials are those that exhibit spontaneous polarization upon the application of an applied electric field. In other words, ferroelectrics must be poled (polarized) in order to make them exhibit piezoelectric behavior. They are analogous to ferromagnetic materials in many respects. Figure 5.3 shows how an applied force gives rise to an electric charge (and hence voltage) across the faces of a slab of piezoelectric material.

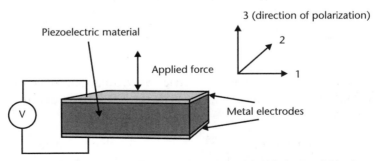

Figure 5.3 An illustration of the piezoelectric effect. The applied force results in the generation of a voltage across the electrodes.

If a ferroelectric material is exposed to a temperature exceeding the Curie point, it will lose its piezoelectric properties. Hence, there is a limit beyond which they cannot be used as sensors (or actuators). The Curie point of PZT type 5H is around 195°C, and its maximum operating temperature is generally lower than this value. In addition to this, the piezoelectric coefficients of the material also vary with temperature, and this is referred to as the pyroelectric effect. This can be exploited in its own right, and pyroelectric sensors based on modified PZT are often used as the basis of infrared sensor arrays.

Owing to the anisotropic nature of piezoelectric materials, a system of identifying each axis is required in order to specify its parameters. By convention, the direction of polarization is taken as the 3-axis, with the 1- and 2-axes being perpendicular. For example, the material shown in Figure 5.3 has the electrodes across the thickness of the material, and hence, this is the 3-axis. An important piezoelectric parameter is the charge coefficient d_{ij} (C/N). This relates the amount of charge generated on the surfaces of the material on the i-axis to the force applied on the j-axis. In the example given, the force applied and the charge generated are both across the thickness of the material, and hence, this charge coefficient is denoted as d_{33}. If a force, F_3, is applied to the piezoelectric sample, then the charge generated is given by

$$Q_3 = d_{33} F_3 \tag{5.11}$$

and so the voltage produced from a rectangular block of area A, thickness t, and relative permittivity ε_r is

$$V_3 = \frac{Q_3}{C} = \frac{d_{33} F_3 t}{\varepsilon_0 \varepsilon_r A} \tag{5.12}$$

where ε_0 is the permittivity of free space. For a 10×10-mm slab of PZT 5H (d_{33} = 600 pC/N, ε_r = 3,000) of thickness 1 mm, an applied force of 100N will produce an open circuit voltage of 22.6V. Strictly, the value of the relative permittivity is also dependent upon the direction in which it is used and the boundary conditions imposed upon the material. The nomenclature becomes a little cumbersome, however, and for the purpose of this text it should be assumed that the value quoted is for the direction in which the piezoelectric is being used.

Another important piezoelectric constant is the voltage coefficient denoted as g_{ij}. It is related to the d coefficient as shown here:

$$g_{ij} = \frac{d_{ij}}{\varepsilon_0 \varepsilon_r} \tag{5.13}$$

Owing to the inverse piezoelectric effect, an applied electric field will result in a deformation of the material. This gives rise to two definitions of the d and g coefficients:

$$d = \frac{\text{strain developed}}{\text{applied electric field}} \quad (\text{m/V}) \quad = \frac{\text{charge density}}{\text{applied mechanical stress}} (\text{C/N}) \tag{5.14}$$

and

$$g = \frac{\text{open circuit electric field}}{\text{applied mechanical stress}} \quad (\text{V}\cdot\text{m/N}) \quad = \frac{\text{strain developed}}{\text{applied charge density}}(\text{m/C}) \quad (5.15)$$

Table 5.2 shows some properties of various types of piezoelectric material. A search through the literature will reveal a wide variation in some of these values. In general, manufacturers of bulk piezoelectric materials quote a relatively wide tolerance (20%) on the values of the piezoelectric properties. Measurement of the properties of films deposited onto substrates is notoriously difficult, as the boundary conditions can grossly affect the measured value. Additionally, some materials, such as PZT, are available in a variety of compositions (4D, 5H, 5A, 7A) each exhibiting vastly different figures for their piezoelectric coefficients. The figures quoted in the table are only intended as a rough comparison.

Quartz is a widely used piezoelectric material that has found common use in watches and as a resonant element in crystal oscillators. There are no available methods to deposit it as a thin-film over a silicon substrate. PVDF is a carbon-based polymer material that is readily available in a light, flexible sheet form of typical thickness 9 to 800 μm. It is possible to spin-on films of PVDF onto substrates, but this must be polarized (poled) after processing in order to obtain piezoelectric behavior. Barium titanate and PZT are two examples of piezo ceramic materials and each of these can be deposited onto silicon using a variety of methods including sputtering, screen-printing, and sol-gel deposition. PZT is generally characterized by its relatively high value of d_{33} and is thus a desirable choice of piezoelectric material. Both zinc oxide and lithium niobate can be deposited as polycrystalline thin-films, but consistent data about their properties is not readily available.

In general, because of the relatively high voltages required for piezoelectric actuators to generate displacements in the micron range, they are not often used. For subnanometer movement, however, they provide an excellent method of actuation. Their high sensitivity to small displacements means that they offer many advantages as micromachined sensors. Devices such as surface acoustic wave sensors (SAWS) and resonant sensors utilize both modes of operation, meaning that only a single material is required for both the sensing and actuating mechanism.

An approximate electrical equivalent circuit of a piezoelectric material is depicted in Figure 5.4. Electrical engineers will recognize the circuit as a series-parallel resonant system. A plot of impedance against frequency is also shown.

The impedance exhibits both resonant and antiresonant peaks at distinct frequencies.

Table 5.2 Properties of Relevant Piezoelectric Materials

Material	Form	d_{33} (pC/N)	Relative Permittivity (ε_r)
Quartz	Single crystal	2	4
PVDF	Polymer	20	12
Barium titanate	Ceramic	190	2,000
PZT	Ceramic	300–600	400–3,000
Zinc oxide	Single crystal	12	12
Lithium niobate	Single crystal	6–16	30

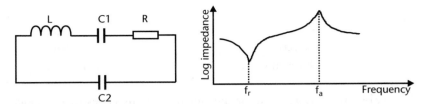

Figure 5.4 The equivalent electrical circuit of a piezoelectric material.

The mechanical resonance of the device is represented by the series inductor, capacitor and resistor (L, C_1, R) and these are the equivalent of mass, spring, and damper, respectively. Since the piezoelectric material is a dielectric with electrodes, it will have a shunt capacitance associated with it (C_2). The series resonant circuit is responsible for the resonant peak (f_r), and the parallel circuit gives rise to the anti-resonant behavior (f_a). The circuit behaves like a simple capacitor at frequencies below f_r and like an inductor between f_r and f_a. After f_a the impedance decreases with frequency, indicating typical capacitor behavior again. The two resonant frequencies are

$$f_r = \frac{1}{2\pi\sqrt{LC_1}} \quad \text{and} \quad f_a = \frac{1}{2\pi}\sqrt{\frac{C_1+C_2}{LC_1C_2}} \tag{5.16}$$

5.3 Capacitive Techniques

The physical structures of capacitive sensors are relatively simple. The technique nevertheless provides a precise way of sensing the movement of an object. Essentially the devices comprise a set of one (or more) fixed electrode and one (or more) moving electrode. They are generally characterized by the inherent nonlinearity and temperature cross-sensitivity, but the ability to integrate signal conditioning circuitry close to the sensor allows highly sensitive, compensated devices to be produced. Figure 5.5 illustrates three configurations for a simple parallel plate capacitor structure.

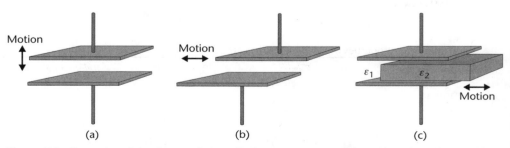

Figure 5.5 Examples of simple capacitance displacement sensors: (a) moving plate, (b) variable area, and (c) moving dielectric.

For a simple parallel plate capacitor structure, ignoring fringing fields, the capacitance is given by

$$C = \frac{\varepsilon_0 \varepsilon_r A}{d} \, (\text{F}) \qquad (5.17)$$

where ε_0 is the permittivity of free space, ε_r is the relative permittivity of the material between the plates, A is the area of overlap between the electrodes, and d is the separation between the electrodes. The equation shows that the capacitance can be varied by changing one or more of the other variables. Figure 5.5(a) shows the simple case where the lower electrode is fixed and the upper electrode moves. In this case the separation, d, is changing and hence the capacitance varies in a nonlinear manner. Figure 5.5(b) depicts a device where the separation is fixed and the area of overlap is varied. In this configuration, there is a linear relationship between the capacitance and area of overlap. Figure 5.5(c) shows a structure that has both a fixed electrode distance and area of overlap. The movement is applied to a dielectric material (of permittivity ε_2) sandwiched between two electrodes. A common problem to all of these devices is that temperature will affect all three sensing parameters (d, A, and ε_r), resulting in changes in the signal output. This effect must be compensated for in some manner, whether by additional signal conditioning circuitry or, preferably, by geometric design.

Figure 5.6 shows a differential capacitance sensor, which is similar in nature to a moving plate capacitor sensor except that there is an additional fixed electrode. Any temperature effects are common to both capacitors and will therefore be cancelled out, as the output signal is a function of the difference between the upper and lower capacitors. If we assume that the outer two electrodes (X and Z) are fixed and the inner electrode (Y) is free to move in a parallel direction towards X, then the gap between plates X and Y will decrease and that between Y and Z will increase. If the nominal gap distance is d and the center electrode is moved by a distance x, then the relationship between the differential output voltage and the deflection is given by

$$\left(V_2 - V_1\right) = V_s \left(\frac{x}{d}\right) \qquad (5.18)$$

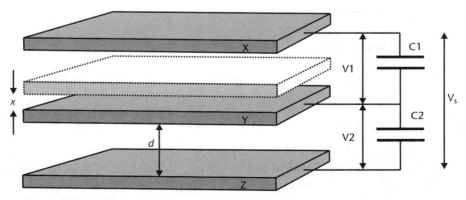

Figure 5.6   A differential capacitance sensor.

where V_s is the supply voltage. So this arrangement provides a linear relationship that is preserved over a range of $|x| < d$ and is capable of detecting displacement of a few picometers.

Capacitor structures are relatively straightforward to fabricate, and membrane-type devices are often used as the basis for pressure sensors and microphones. More elaborate structures, such as interdigitated capacitors, are also used, and the effects of the fringing fields cannot always be ignored. With such devices, the simple parallel plate capacitor equation only provides a crude estimate of the expected capacitance change.

Capacitive techniques are inherently less noisy than those based on piezoresistance owing to the lack of thermal (Johnson) noise. With micromachined devices, however, the values of capacitance are extremely small (in the range of femto- to attofarads), and the additional noise from the interface electronic circuits often exceeds that of a resistance-based system.

There are a variety of techniques for measuring capacitance changes including charge amplifiers (often used with piezoelectric devices), charge balance techniques, ac bridge impedance measurements, and various oscillator configurations. There are also a variety of commercially available ICs that can be used to measure capacitance changes of a few femtofarads in stray capacitances up to several hundred picofarads [2].

5.4   Optical Techniques

Optical sensing techniques primarily rely on modulating the properties of an optical frequency electromagnetic wave. In the case of optical sensors, the measurand directly modulates the properties of the electromagnetic wave. In the case of microsensors, which use optical interfacing, the miniaturized sensor interacts with the measurand. The microsensor then modulates a property of the optical signal in order to provide an indication of the measurand.

The following properties of the electromagnetic wave can be altered:

1. Intensity;
2. Phase;
3. Wavelength;
4. Spatial position;
5. Frequency;
6. Polarization.

The basic principles of each of these techniques will now be reviewed in turn.

5.4.1   Intensity

The primary advantage of intensity modulation is that intensity variations are simply detected because all optical detectors (e.g., photodiodes, phototransistors) directly respond to intensity variations. Therefore, if the microsensor can be arranged to vary the intensity of an optical signal, these variations can then be simply observed using a photodetector. A simple arrangement is for the microsensor to move in response to the measurand and for this movement to be arranged to block the path of the light beam incident on a photodetector. Figure 5.7 illustrates a simple transmissive arrangement, although reflected light is also used in some arrangements.

The optical source is shown as a light emitting diode (LED) since a coherent source is not required for intensity-based sensors. Alternative optical sources could be a laser, the output of an optical fiber, or simply an incandescent lamp.

The major difficulty with intensity-based systems is variations in intensity caused by factors not related to the measurand. For example, the output of an optical source can vary with time and temperature. For this reason intensity-based sensors often incorporate some form of reference measurement of the optical source intensity and a ratio taken between the optical intensity before and after modulation by the microsensor. This problem often negates the simplicity of intensity-based sensors. Variations in the sensitivity of the optical detector can also cause difficulties and complications.

A qualitative estimate of the resolution of intensity-based sensors can be obtained by estimation of the optical beam size. The minimum beam size is of the order of the wavelength of the optical source, so this gives an indication of the displacement required to give a 100% modulation of intensity.

5.4.2 Phase

As photodetectors do not respond directly to phase variation, it is necessary to convert a variation in phase to an intensity variation for measurement at the photodiode. This is usually achieved by using an interferometer to combine one or more optical beams that have interacted with the microsensor with one or more optical beams that are unaffected by the microsensor. A coherent source such as a laser diode is therefore typically used in phase-based optical sensing. The interaction with the microsensor has the effect of altering the optical path length of that optical beam and hence its phase. This can simply be achieved by reflecting the optical beam off the microsensor and the microsensor moving in response to the measurand so as to vary the optical path length.

A major advantage of phase-based systems is that subwavelength phase variations can be resolved, which equates to submicron displacement of the microsensor. Difficulties can be caused by the fact that the output of the interferometer is periodic; therefore, care has to be taken to establish the start point and the position relative to that. This can lead to complexity in the reference electrodes and errors in initializing the system.

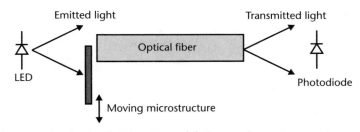

Figure 5.7 An example of a simple intensity modulation sensing system.

5.4.3 Wavelength

Wavelength-based sensing relies on the source spectrum being modulated by inter-action with the microsensor. Normally a source with a broad spectrum is used. The light returned from the microsensor is split into spectral segments and incident on a photodetector for measurement of its intensity. By a prior knowledge of the potential modulation mechanism present with the microsensor, one can identify the measurand and its magnitude. A good example of a wavelength-based sensor is one based on the gas absorption, which is highly wavelength specific according to the quantity of gas present.

The advantage of wavelength-based sensors is that they can be made insensitive to intensity variation since these affect the whole spectrum in the same way. There-fore, the measurement of a nonabsorbed wavelength can be used to reference the absorbed wavelength, therefore compensating for intensity variations. In addition, wavelength-based sensors often lend themselves to the measurement of multiple parameters since the light spectrum can be divided according to the particular wave-length corresponding to the measurand of interest.

5.4.4 Spatial Position

Figure 5.8 illustrates the principle of the modulation of special position by means of the movement of a microsensor. This technique is often known as triangulation.

This technique is simple to implement and has the advantage of immunity to source intensity variations. Its resolution is less then phase-based techniques.

5.4.5 Frequency

If optical radiation at a frequency f is incident upon a body moving a velocity v, then the radiation reflected from the moving body appears to have a frequency f_1, where

$$f_1 = \frac{f}{1 - \dfrac{v}{c}} \approx f\left[1 + \frac{v}{c}\right] \tag{5.19}$$

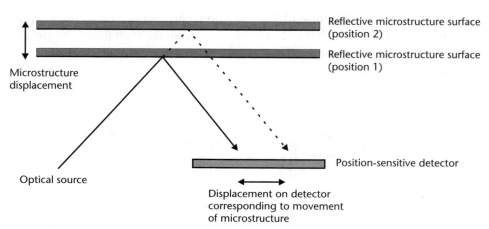

Figure 5.8 An example of a spatial position measurement system.

This Doppler frequency shift from a moving target can therefore be used as the basis of a detection technique of the velocity of the target. Laser Doppler velocimetry is a well-established field of research. Frequency variation is converted into intensity variation by interferometry by combining a nonfrequency-shifted reference beam with the shifted beam.

5.4.6   Polarization

Linear polarization is defined by the direction of the electric vector of the electromagnetic wave. Circular polarized light is defined by the direction of rotation of the electric field vector when viewed looking towards the source. Any polarization can be resolved into two orthogonal modes, and sensing can be achieved by altering the optical path length traversed by one mode with respect to the other. In practice this is normally achieved by a relative modification of the refractive index. A polarized light source such as a laser is required and the photodetector must be made polarization sensitive by including a polarizer.

Polarization-based interrogation of microsensors has not been widely investigated owing to the limited sensitivity available, as it is a differential technique. In addition, the method is susceptible to intensity changes in the source.

5.5   Resonant Techniques

A resonator is a mechanical structure designed to vibrate at a particular resonant frequency. Resonators can be fabricated from a range of single crystal materials with micron-sized dimensions using various micromachining processes. The resonant frequencies of such microresonators are extremely stable, enabling them to be used as a time base (the quartz tuning fork in watches, for example) or as the sensing element of a resonant sensor [3, 4]. The performance benefits of a well-designed resonant sensor compared with piezoresistive and capacitive techniques are shown in Table 5.3 [5]. The fabrication of such devices is, however, more complex and the requirement for packaging such devices more demanding.

A block diagram of a typical resonant sensor is shown in Figure 5.9 [6]. A resonant sensor is designed such that the resonator's natural frequency is a function of the measurand. The measurand typically alters the stiffness, mass, or shape of the resonator, hence causing a change in its resonant frequency. The other components of a resonant sensor are the vibration drive and detection mechanisms. The drive mechanism excites the vibrations in the structure while the detection mechanism senses these vibrations. The frequency of the detected vibration forms the output of

Table 5.3   Performance Features of Resonant, Piezoresistive, and Capacitive Sensing

Feature	Resonant	Piezoresistive	Capacitive
Output form	Frequency	Voltage	Voltage
Resolution	1 part in 10^8	1 part in 10^5	1 part in 10^4–10^5
Accuracy	100–1000 ppm	500–10,000 ppm	100–10,000 ppm
Power consumption	0.1–10 mW	$\approx$10 mW	<0.1 mW
Temperature cross-sensitivity	$-30 \times 10^{-6}/°C$	$-1,600 \times 10^{-6}/°C$	$4 \times 10^{-6}/°C$

Source: [5].

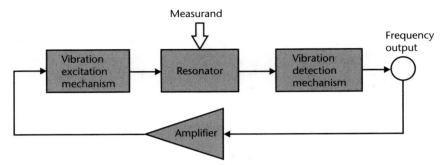

Figure 5.9   Block diagram of a resonant sensor.

the sensor and this signal is also fed back to the drive mechanism via an amplifier maintaining the structure at resonance over the entire measurand range.

In mechanical sensing applications, the most common mechanism for coupling the resonator to the measurand is to apply a strain across the structure. When used in such a manner the resonator effectively becomes resonant strain gauge. Coupling to the measurand is achieved by mounting the resonator in a suitable location on a specifically designed sensing structure that deflects due to the application of the measurand. The resonator output can be used to monitor the deflection of the sensing structure and thereby provide an indication of the magnitude of the measurand. When used as a resonant strain gauge, the applied strain effectively increases the stiffness of the resonator, which results in an increase in its natural frequency. This principle is commonly applied in force sensors, pressure transducers, and accelerometers (see Chapters 6 through 8 for detailed examples).

Coupling the measurand to the mass of the resonator can be achieved by surrounding the structure by a liquid or gas, by coating the resonator in a chemically sensitive material, or by depositing material onto the resonator. The presence of the surrounding liquids or gases increases the effective inertia of the resonator and lowers its resonant frequency. Density sensors and level sensors are examples of mass coupled resonant sensors. Coating the resonator in a chemically sensitive material is used in gas sensors. The sensitive material absorbs molecules of a particular gas, adding to the mass of the film and thereby reducing the frequency of the resonator.

The shape coupling effect is similar to the strain effect except changes in the measurand alter the geometry of the resonator, which leads to a shift in the resonant frequency. This is the least commonly used coupling mechanism.

5.5.1   Vibration Excitation and Detection Mechanisms

The piezoelectric nature of GaAs and quartz materials enables straightforward excitation and detection of resonant modes of vibrations [7]. Suitable electrode materials must be deposited and patterned on the surface of the resonator. The location and geometry of the electrodes should be carefully designed to maximize the electrical to mechanical coupling with the desired mode of operation (drive efficiency). Maximizing this coupling will promote the excitation of the desired mode and maximize the corresponding vibration detection signal.

The excitation and detection of resonance in silicon microresonators are not so straightforward because silicon is not intrinsically piezoelectric. Other mechanisms must therefore be fabricated on or adjacent to the resonator structure. There are many suitable mechanisms and these are all based on the sensing and actuating principles described in this chapter. For example, the resonators vibrations can be electrostatically excited and detected using implanted piezoresistors. Since the implanted piezoresistors could be used directly to measure the strain in the sensing structure, the added complexity of a resonant approach is only justifiable in high-performance sensing applications.

The various excitation and detection mechanisms used with silicon resonators are summarized in Table 5.4. Many of the mechanisms listed can be used to both excite and detect a resonator's vibrations, either simultaneously or in conjunction with another mechanism. Devices where a single element combines the excitation and detection of the vibrations in the structure are termed one-port resonators. Those that use separate elements are termed two-port resonators.

The suitability of these mechanisms for driving or detecting a resonator's vibrations depends upon a number of factors: the magnitude of the drive forces generated, the coupling factor (or drive efficiency), sensitivity of the detection mechanism, the effects of the chosen mechanism upon the performance and behavior of the resonator, and practical considerations pertaining to the fabrication of the resonator and the sensors final environment.

5.5.2   Resonator Design Characteristics

5.5.2.1   Q-Factor

As a structure approaches resonance, the amplitude of its vibration will increase, its resonant frequency being defined as the point of maximum amplitude. The magnitude of this amplitude will ultimately be limited by the damping effects acting on the system. The level of damping present in a system can be defined by its quality factor (Q-factor). The Q-factor is a ratio of the total energy stored in the system (E_M) to the energy lost per cycle (E_C) due to the damping effects present:

$$Q = 2\pi \left(E_M \Big/ E_C \right) \tag{5.20}$$

A high Q-factor indicates a pronounced resonance easily distinguishable from nonresonant vibrations, as illustrated in Figure 5.10. Increasing the sharpness of the resonance enables the resonant frequency to be more clearly defined and will improve the performance and resolution of the resonator. It will also simplify the operating electronics since the magnitude of the signal from the vibration detection

Table 5.4   *Summary of Excitation and Detection Mechanisms*

Piezoelectric	Piezoelectric
Magnetic	Magnetic
Electrothermal	Piezoresistive
Optothermal	Optical

Source: [8].

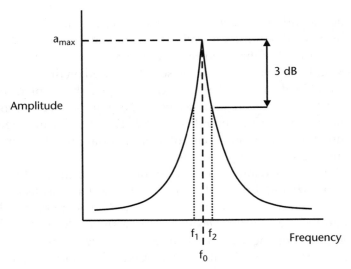

Figure 5.10 A typical characteristic of a resonant system.

mechanism will be greater than that of a low-Q system. A high Q means little energy is required to maintain the resonance at constant amplitude, thereby broadening the range of possible drive mechanisms to include weaker techniques. A high Q-factor also implies the resonant structure is well isolated from its surroundings, and therefore, the influence of external factors (e.g., vibrations) will be minimized.

The Q-factor can also be calculated from Figure 5.10 using

$$Q = \frac{f_0}{\Delta f} \tag{5.21}$$

where resonant frequency f_0 corresponds with a_{max}, the maximum amplitude, and Δf is the difference between frequencies f_1 and f_2. Frequencies f_1 and f_2 correspond to amplitudes of vibration 3 dB lower than a_{max}.

The Q-factor is limited by the various mechanisms by which energy is lost from the resonator. These damping mechanisms arise from three sources:

1. The energy lost to a surrounding fluid ($1/Q_a$);
2. The energy coupled through the resonator's supports to a surrounding solid ($1/Q_s$);
3. The energy dissipated internally within the resonator's material ($1/Q_i$).

Minimizing these effects will maximize the Q-factor as shown here:

$$\frac{1}{Q} = \frac{1}{Q_a} + \frac{1}{Q_s} + \frac{1}{Q_i} \tag{5.22}$$

Energy losses associated with $1/Q_a$ are potentially the largest, and therefore the most important, of the loss mechanisms. These losses occur due to the interactions of the oscillating resonator with the surrounding gas. There are several distinguishable loss mechanisms and associated effects. The magnitude of each depends primarily upon the nature of the gas, surrounding gas pressure, size and shape of the resonator, the direction of its vibrations, and its proximity to adjacent surfaces. Gas damping effects can be negated completely by operating the resonator in a suitable vacuum, and this is used in most micromechanical resonator applications.

Molecular damping occurs at low pressures of between 1 and 100 Pa when the surrounding gas molecules act independently of one another [9]. The damping effect arises from the collisions between the molecules and the resonator's surface as it vibrates. This causes the resonator and molecules to exchange momentum according to their relative velocities. The magnitude of the loss is directly proportional to the surrounding fluid pressure, and also close proximity of the oscillating structure to adjacent surfaces will exaggerate the damping effects. Viscous damping predominates at pressures above 100 Pa where the molecules can no longer be assumed to act independently and the surrounding gas must be considered as a viscous fluid. Viscous drag occurs as the fluid travels over the surface of the resonator. The formation of boundary layer around the resonator can also result in the vibrations forming a transverse wave, which travels into the fluid medium. Other damping mechanisms associated with surrounding fluids are acoustic radiation and squeezed film damping.

Structural damping, $1/Q_s$, is associated with the energy coupled from the resonator through its supports to the surrounding structure and must be minimized by careful design of the resonant structure. Minimizing the energy lost from the resonator to its surroundings can be achieved by a designing a balanced resonant structure, supporting the resonator at its nodes, or by employing a decoupling system between the resonator and its support.

The coupling mechanism between the resonator and its support can be illustrated by observing a fixed-fixed beam vibrating in its fundamental mode. Following Newton's second law that every action has an equal and opposite reaction, the reaction to the beam's vibrations is provided by its supports. The reaction causes the supports to deflect and as a result energy is lost from the resonator.

The degree of coupling of a fixed-fixed beam can be reduced by operating it in a higher-order mode. For example, the second mode in the plane of vibrations shown above will possess a node halfway along the length of the beam. The beam will vibrate in antiphase either side of the node, and the reactions from each half of the beam will cancel out at the node. There will inevitably still be a reaction at each support, but the magnitude of each reaction will be less than for mode 1. The use of such higher order modes is limited by their reduced sensitivity to applied stresses and the fact there will always be a certain degree of coupling.

Balanced resonator designs operate on the principle of providing the reaction to the structure's vibrations within the resonator. Multiple-beam style resonators, for example, incorporate this inherent dynamic moment cancellation when operated in a balanced mode of vibration. Examples of such structures are the double-ended tuning fork (DETF), which consists of two beams aligned alongside each other, and the triple-beam tuning fork (TBTF), which consists of three beams aligned alongside each other, the center tine being twice the width of the outer tines. Figure 5.11 shows these structures and their optimum modes of operation.

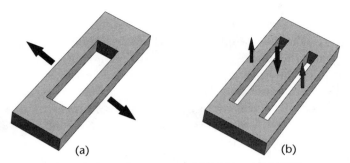

Figure 5.11   Examples of two balanced resonators: (a) DETF and (b) TBTF.

$1/Q_s$ is of fundamental importance since it not only affects the Q-factor of the resonator, but provides a key determinant of resonator performance. A dynamically balanced resonator design that minimizes $1/Q_s$ provides many benefits [10]:

- High resonator Q-factor and therefore good resolution of frequency;
- A high degree of immunity to environmental vibrations;
- Immunity to interference from surrounding structural resonances;
- Improved long-term performance since the influence of the surrounding structure on the resonator is minimized.

The Q-factor of a resonator is ultimately limited by the energy loss mechanisms within the resonator material. This is illustrated by the fact that even if the external damping mechanisms $1/Q_a$ and $1/Q_s$ are removed, the amplitude of its vibrations will still decay with time. There are several internal loss mechanisms by which vibrations can be attenuated. These include the movement of dislocations and scattering by impurities, phonon interaction, and the thermoelastic effect.

5.5.2.2   Nonlinear Behavior and Hysteresis

Nonlinear behavior becomes apparent at higher vibration amplitudes when the resonator's restoring force becomes a nonlinear function of its displacement. This effect is present in all resonant structures. In the case of a flexurally vibrating fixed-fixed beam, the transverse deflection results in a stretching of its neutral axis. A tensile force is effectively applied and the resonant frequency increases. This is known as the hard spring effect. The magnitude of this effect depends upon the boundary conditions of the beam. If the beam is not clamped firmly, the nonlinear relationship can exhibit the soft spring effect whereby the resonant frequency falls with increasing amplitude. The nature of the effect and its magnitude also depends upon the geometry of the resonator.

The equation of motion for an oscillating force applied to an undamped structure is given by (5.23) where m is the mass of the system, F is the applied driving force, ω is the frequency, y is the displacement, and $s(y)$ is the nonlinear function [11].

$$m\ddot{y} + s(y) = F_0 \cos \omega t \qquad (5.23)$$

In many practical cases $s(y)$ can be represented by (5.24), the nonlinear relationship being represented by the cubic term.

$$s(y) = s_1 y + s_3 y^3 \tag{5.24}$$

Placing (5.24) in (5.23), dividing through by m, and simplifying gives

$$\ddot{y} + s_1/m\left(y + s_3/s_1 y^3\right) = F_0 \cos \omega t \tag{5.25}$$

where s_1/m equals ω_{or}^2 (ω_{or} representing the resonant frequency for small amplitudes of vibration) and s_3/s_1 is denoted by β. The restoring force acting on the system is therefore represented by

$$R = -\omega_{or}^2\left(y + \beta y^3\right) \tag{5.26}$$

If β is equal to zero, the restoring force is a linear function of displacement; if β is positive, the system experiences the hard spring nonlinearity; a negative β corresponds to the soft spring effect. The hard and soft nonlinear effects are shown in Figure 5.12. As the amplitude of vibration increases and the nonlinear effect becomes apparent, the resonant frequency exhibits a quadratic dependence upon the amplitude, as shown in

$$\omega_r = \omega_{or}\left(1 + \tfrac{3}{8}\beta y_0^2\right) \tag{5.27}$$

The variable β can be found by applying (5.27) to an experimental analysis of the resonant frequency and maximum amplitude for a range of drive levels.

The amplitude of vibration is dependent upon the energy supplied by the resonator's drive mechanism and the Q-factor of the resonator. Driving the resonator too hard or a high Q-factor that results in excessive amplitudes at minimum practical drive levels can result in undesirable nonlinear behavior. Nonlinearities are undesirable since they can adversely affect the accuracy of a resonant sensor. If a resonator is driven in a nonlinear region, then changes in amplitude—due, for

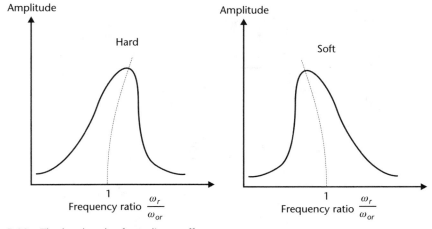

Figure 5.12   The hard and soft nonlinear effects.

example, to amplifier drift—will cause a shift in the resonant frequency indistinguishable from shifts due to the measurand. The analysis of a resonator's nonlinear characteristics is therefore important when determining a suitable drive mechanism and its associated operating variables.

A nonlinear system can exhibit hysteresis if the amplitude of vibration increases beyond a critical value. Hysteresis occurs when the amplitude has three possible values at a given frequency. This critical value can be determined by applying

$$y_0^2 > \frac{8h}{3\omega_{or}|\beta|} \qquad (5.28)$$

where h is the damping coefficient and can be found by measuring the Q-factor of the resonator at small amplitudes and applying

$$Q = \frac{\omega_{or}}{2h} \qquad (5.29)$$

5.6 Actuation Techniques

In Chapter 1 we defined an actuator as a device that responds to the electrical signals within the transduction system. Specifically, a mechanical actuator is one that translates a signal from the electrical domain into the mechanical domain. In the ideal case, we would like the conversion to be 100% efficient. Of course, any real system cannot achieve a figure anywhere near this, owing to internal and external losses. Typical micromechanical actuators offer an efficiency between 5% and 35%. Other factors such as ease of fabrication, robustness, resistance to external effects (i.e., temperature, humidity), and range of motion, result in a series of trade-offs for selecting the appropriate mechanism.

For the purpose of this text, four fundamental approaches for actuator design will be discussed. Other techniques such as chemical and biological actuation are not covered here.

5.6.1 Electrostatic

Electrostatic actuators are based on the fundamental principle that two plates of opposite charge will attract each other. They are quite extensive as they are relatively straightforward to fabricate. They do, however, have a nonlinear force-to-voltage relationship. Consider a simple, parallel plate capacitor arrangement again, having a gap separation, g, and area of overlap, A, as shown in Figure 5.13. Ignoring fringing effects, the energy stored at a given voltage, V, is

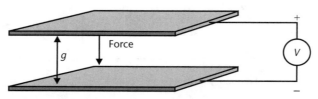

Figure 5.13 A simple planar capacitor electrostatic actuator.

$$W = \frac{1}{2}CV^2 = \frac{\varepsilon_0 \varepsilon_r A V^2}{2g} \tag{5.30}$$

and the force between the plates is given by

$$F = \frac{dW}{dg} = \frac{\varepsilon_0 \varepsilon_r A V^2}{2g^2} \tag{5.31}$$

It is therefore clear that the force is a nonlinear function of both the applied voltage and the gap separation. Use of closed loop control techniques can linearize the response.

An alternative type of electrostatic actuator is the so-called comb-drive, which is comprised of many interdigitated electrodes (fingers) that are actuated by applying a voltage between them. The geometry is such that the thickness of the fingers is small in comparison to their lengths and widths. The attractive forces are therefore mainly due to the fringing fields rather than the parallel plate fields, as seen in the simple structure above. The movement generated is in the lateral direction, as shown in Figure 5.14, and because the capacitance is varied by changing the area of overlap and the gap remains fixed, the displacement varies as the square of the voltage.

The fixed electrode is rigidly supported to the substrate, and the movable electrode must be held in place by anchoring at a suitable point away from the active fingers. Additional parasitic capacitances such as those between the fingers and the substrate and the asymmetry of the fringing fields can lead to out-of-plane forces, which can be minimized with more sophisticated designs.

Electrostatic actuation techniques have also been used to developed rotary motor structures. With these devices, a central rotor having surrounding capacitive plates is made to rotate by the application of voltages of the correct phase to induce rotation. Such devices have been shown to have a limited lifetime and require lubrication to prevent the rotor from seizing. The practical use has therefore been limited, but they are, nevertheless, the subject of intensive research.

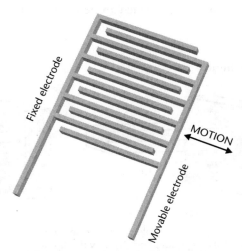

Figure 5.14   An illustration of the electrostatic comb-drive actuator.

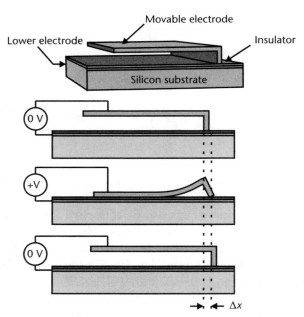

Figure 5.15 Illustration of the principle of operation of the electrostatic scratch drive actuator as described by Akiyama and Katsufusa. (*After:* [12].)

Another interesting type of electrostatic actuator is the so-called scratch drive actuator (SDA) as described by Akiyama and Katsufusa [12]. The device comprises a flexible, electrode plate and a small bushing at one end. It is depicted in Figure 5.15, which also illustrates the principle of operation. The free end of the electrode in the actual device is usually supported by a thin beam, but this is not shown in the figure. When a voltage is applied between the electrode plate and the buried electrode layer on the substrate, the plate buckles down and so causes the bushing to "scratch" along the insulator, thereby resulting in a small forward movement. When the voltage is removed, the plate returns to its original shape, thereby resulting in a net movement of the plate. The cycle can be repeated for stepwise linear motion.

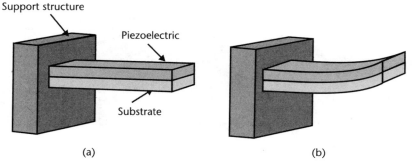

(a)   (b)

Figure 5.16 An example of a simple cantilever beam with a deposited piezoelectric layer: (a) the structure with no applied voltage; and (b) how the tip of the beam moves upon the application of an applied voltage.

5.6.2   Piezoelectric

As we have already seen, piezoelectric devices can be used for both sensor and actuator applications. An applied voltage across the electrodes of a piezoelectric material will result in a deformation that is proportional to the magnitude of the voltage (strictly electric field). The displacement across a bulk sample of PZT with an actuation voltage of several hundred volts, for example, is only a small fraction of a micron. When such a system is scaled down to that of a typical MEMS actuator, a displacement of several orders of magnitude less is obtained! For this reason, some form of mechanical amplification is needed in order to generate useful displacements. Such a device can be fabricated by depositing a piezoelectric film onto a substrate in the form of a cantilever beam as shown in Figure 5.16. This type of structure is referred to as a piezoelectric *unimorph*. The deflection at the free end of the beam is greater than that produced in the film itself.

Piezoelectric actuators are often used in micropumps (see Chapter 9) as a way of deflecting a thin membrane, which in turn alters the volume within a chamber below. Such a structure is depicted in Figure 5.17. The device comprises two silicon wafers bonded together. The lower wafer comprises an inlet and outlet port, which have been fabricated using bulk micromachining techniques. The upper wafer has been etched to form the pump chamber. The shape of the ports gives rise to a preferential direction for the fluid flow, although there is a degree of flow in the reverse direction during pumping. So the ports behave in a similar manner to valves. An alternative structure comprises cantilever-type flaps across the ports, but these often suffer from stiction during pumping. When a voltage is applied to the piezoelectric material, this results in a deformation of the thin membrane and hence changes the volume within the chamber. This is depicted in Figure 5.17(b). Typical flow rates are in the range of nanoliters to microliters per minute, depending on the dimension of the micropump.

5.6.3   Thermal

Thermal actuation techniques tend to consume more power than electrostatic or piezoelectric methods, but the forces generated are also greater. One of the basic

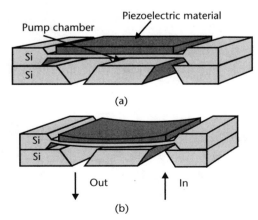

(a)

(b)

Figure 5.17   A simple micropump with a piezoelectric actuator: (a) the situation with no applied voltage; and (b) the effect of applying a voltage to the piezoelectric layer.

approaches is to exploit the difference in linear expansion coefficients of two materials bonded together. Such structures are often referred to as thermal bimorphs and are analogous to the familiar bimetallic strips often used in thermostats. One layer expands by a different amount to the other, resulting in thermal stresses at the interface leading to bending of the structure. The amount of bending depends on the difference in thermal coefficients of expansion and also on the temperature. An illustration of a thermal bimorph is shown in Figure 5.18. If an electric current is passed through the aluminum layer, it heats up (Joule heating), thereby causing the free end of the beam to move. These devices are relatively straightforward to fabricate and in addition to consuming relatively large amounts of power, they also have a low bandwidth because of the thermal time constant of the overall structure (i.e., beam and support).

An example of a commercial device based on thermal actuation is the so-called fluistor from Redwood Microsystems in California. This device is comprised of a cavity with a sealed fluid that can be heated and thus expanded. The heat is applied to the fluid via a thin-film resistive element. If one section of the cavity, such as a wall, is made more compliant than the other sections, then it will deform under pressure, thereby generating a mechanical force. The cavity is formed by bulk micromachining in silicon and is sealed using a Pyrex wafer, containing the heating element, anodically bonded to the silicon. Strictly, this is a thermopneumatic actuator and the commercial device is often used as a microvalve in applications such as medical instrumentation, gas mixers, and process control equipment. Such actuators may require up to 2W of power to operate.

Another thermal effect that can be exploited in thermal actuators is the shape-memory effect, which is a property of a special class of metal alloys know as shape-memory alloys. When these materials are heated beyond a critical transition temperature, they return to a predetermined shape. The SMA material has a temperature-dependent crystal structure such that, at temperatures below the transition point, it possesses a low yield strength crystallography referred to as a *Martensite*. In this state, the alloy is relatively soft and easy to deform into different shapes.

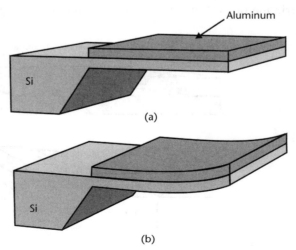

Figure 5.18 A simple thermal bimorph actuator (a) before and (b) after the application of electric current.

It will retain this shape until the temperature exceeds the phase transition temperature, at which point the material reverts to its parent structure known as *Austenite*. One of the most widely used SMA materials is an alloy of nickel and titanium called Nitinol. This has excellent electrical and mechanical properties and a long fatigue life. In its bulk form, it is capable of producing up to 5% strain. The transition temperature of Nitinol can be tailored between −100°C and +100°C by controlling the impurity concentration. The material has been used in MEMS by sputter depositing TiNi thin-film layers [13].

5.6.4   Magnetic

If a current-carrying element is placed within a magnetic field, an electromagnetic force (Lorentz force) will occur in a direction perpendicular to the current and magnetic field. The magnitude of the force is proportional to the current, length of the element, and the magnetic field. The availability of permanent magnetic materials, which are compatible with MEMS processing, is very limited, and thus it is common for the magnetic field to be generated externally. Discrete magnetic actuators often comprise coils, but such structures are not currently achievable with conventional MEMS processing and planar coils must be used.

Another approach that can be used as the basis of a magnetic actuator is the magnetostrictive effect. Magnetostriction is defined as the dimensional change of a magnetic material caused by a change in its magnetic state. Like the piezoelectric effect, it is reversible, and an applied stress results in a change of magnetic state. All magnetic materials exhibit varying degrees of magnetostriction. J. P. Joule discovered the effect in 1847 by observing the change in length of an iron bar when it was magnetized. A popular modern-day magnetostrictive material is Terfenol-D, an alloy of terbium, dysprosium, and iron. The magnetostriction of Terfenol-D is several orders of magnitude greater than that of iron, nickel, or cobalt and gives rise to strains in the region of 2×10^{-3}. Bulk Terfenol-D produces much larger strains than those achievable with piezoelectric materials. Research has been undertaken to investigate the feasibility of depositing thin and thick-films of magnetostrictive material onto substrates such as silicon, glass, and alumina; the magnetostriction achievable, however, is inferior to that of the bulk material.

Figure 5.19 shows an example of a magnetic actuator as described by Judy et al. [14]. The device comprises a 7-μm-thick layer of Permalloy, which was electroplated onto a polysilicon cantilever. The root of the beam is thin and narrow and acts as a spring, thereby allowing the tip to deflect over a wide angular range. The magnetic field is applied externally to the device, and this causes a deflection of the actuator in the direction of the plane of the substrate. The device is made using polysilicon surface micromachining techniques. Deflections exceeding 90° were achieved with this configuration.

5.7   Smart Sensors

Advances in the area of microelectronics in recent years have had a major effect on many aspects of measurement science. In particular, the distinction between the

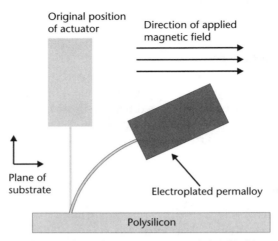

Figure 5.19 An example of an in-plane magnetic actuator. (*After:* [14].)

sensor and the instrument may not be apparent. Many of today's commercial devices have some form of electronic processing within the main sensor housing; perhaps simple electronic filtering or more sophisticated digital signal processing. The terms *intelligent* and *smart* sensor have been used, almost interchangeably, over the past 20 years or so to refer to sensors having additional functionality provided by the integration of microprocessors, microcontrollers, or application specific integrated circuits (ASICs) with the sensing element itself. The interested reader is encouraged to read the texts by Brignell and White [15], Gardner et al. [16], and Frank [17], for a deeper insight into the field of smart sensor technologies. For consistency in this text, we will adopt the term *smart sensor* to refer to a microsensor with integrated microelectronic circuitry.

Smart sensors offer a number of advantages for sensor system designers. The integration of sensor and electronics allows it to be treated as a module, or black-box, where the internal complexities of the sensor are kept remote from the host system. Smart sensors may also have additional integrated sensors to monitor, say, localized temperature changes. This is sometimes referred to as the sensor-within-a-sensor approach and is an important feature of smart sensor technology. An example of a smart sensor system is depicted in Figure 5.20.

Many physical realizations of smart sensors may contain some or all of these elements. Each of the main subsystems will now be described in more detail.

The *sensing element* is the primary source of information into the system. Examples of typical sensing techniques have already been outlined in this chapter. The smart sensor may also have the ability to stimulate the sensing element to provide a self-test facility, whereby a reference voltage, for example, can be applied to the sensor in order to monitor its response. Some primary sensors, such as those based on piezoelectrics, convert energy directly from one domain into another and therefore do not require a power supply. Others, such as resistive-based sensors, may need stable dc sources, which may benefit from additional functionality like pulsed excitation for power-saving reasons. So excitation control is another distinguishing feature found in smart sensors.

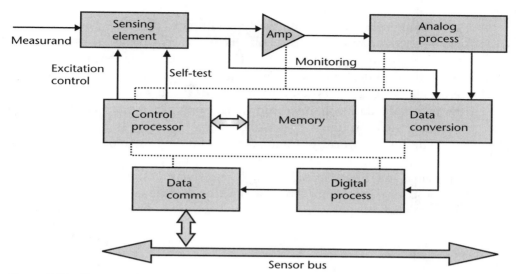

Figure 5.20   Elements of a smart sensor.

Amplification is usually a fundamental requirement, as most sensors tend to produce signal levels that are significantly lower than those used in the digital processor. Resistive sensors in a bridge configuration often require an instrumentation amplifier; piezoelectric devices may need a charge amplifier. If possible, it is advantageous to have the gain as close as possible to the sensing element. In situations where a high gain is required, there can often be implications for handling any adverse effects such as noise. In terms of chip layout, the sharp transients associated with digital signals need to be kept well away from the front-end analog circuitry.

Examples of *analog processing* include antialiasing filters for the conversion stage. In situations where real-time processing power is limited, there may also be benefits in implementing analog filters.

Data conversion is the transition region between the continuous (real-world) signals and the discrete signals associated with the digital processor. Typically, this stage comprises an analog-to-digital converter (ADC). Inputs from other sensors (monitoring) can be fed into the data conversion subsystem and may be used to implement compensation, say for temperature. Note that such signals may also require amplification before data conversion. Resonant sensors, whose signals are in the frequency domain, do not need a data conversion stage as their outputs can often be fed directly into the digital system.

The *digital processing* element mainly concerns the software processes within the smart sensor. These may be simple routines such as those required for implementing sensor compensation (linearization, cross-sensitivity, offset), or they may be more sophisticated techniques such as pattern recognition methods (such as neural networks) for sensor array devices.

The *data communications* element deals with the routines necessary for passing and receiving data and control signals to the sensor bus. It is often the case that the smart sensor is a single device within a multisensor system. Individual sensors can communicate with each other in addition to the host system. There are many examples of commercial protocols that are used in smart sensor systems, but we will not go into detail here. It is sufficient to be aware that the smart sensor will often have to deal with situations such as requests for data, calibration signals, error checking, and message identification. Of course, it is feasible in some applications that the data communications may simply be a unit that provides an analog voltage or current signal.

The *control processor* often takes the form of a microprocessor. It is generally the central component within the smart sensor and is connected to most of the other elements, as we have already seen. The software routines are implemented within the processor and these will be stored within the memory unit. The control processor may also issue requests for self-test routines or set the gain of the amplifier.

References

[1]   Middelhoek, S., and S. A. Audet, *Silicon Sensors*, New York: Academic Press, 1989.

[2]   http://www.qprox.com.

[3]   Tudor, M. J., and S. P. Beeby, "Resonant Sensors: Fundamentals and State of the Art," *Sensors and Materials*, Vol. 9, No. 3, 1997, pp. 1–15.

[4]   Langdon, R. M., "Resonator Sensors—A Review," *J. Phys. E: Sci. Instrum.*, Vol. 18, 1985, pp. 103–115.

[5]   Greenwood, J. C., "Silicon in Mechanical Sensors," *J. Phys. E: Sci. Instrum.*, Vol. 21, 1988, pp. 1114–1128.

[6]   Stemme, G., " Resonant Silicon Sensors," *J. Micromech. Microeng.*, Vol. 1, 1991, pp. 113–125.

[7]   Eernisse E. P., R. W. Ward, and R. B. Wiggins, "Survey of Quartz Bulk Resonator Sensor Technologies," *IEEE Trans. Ultrasonics Ferroelectrics and Frequency Control*, Vol. 35, No. 3, May 1988, pp. 323–330.

[8]   Prak, A., T. S. J. Lammerink, and J. H. J. Fluitman, "Review of Excitation and Detection Mechanisms for Micromechanical Resonators," *Sensors and Materials*, Vol. 5, No. 3, 1993, pp. 143–181.

[9]   Newell, W. E., "Miniaturization of Tuning Forks," *Science*, Vol. 161, September 1968, pp. 1320–1326.

[10] Beeby, S. P., and M. J. Tudor, "Modeling and Optimization of Micromachined Silicon Resonators," *J. Micromech. Microeng.*, Vol. 5, 1995, pp. 103–105.

[11] Andres, M. V., K. H. W. Foulds, and M. J. Tudor, "Nonlinear Vibrations and Hysteresis of Micromachined Silicon Resonators Designed as Frequency Out Sensors," *ElectronicsLetters*, Vol. 23, No. 18, August 27, 1987, pp. 952–954.

[12] Akiyama, T., and S. Katsufusa, "A New Step Motion of Polysilicon Microstructures," *Proc MEMS '93*, 1993, pp. 272–277.

[13] Walker, J. A., K. J. Gabriel, and M. Mehregany, "Thin-Film Processing of TiNi Shape Memory Alloy," *Sensors and Actuators*, Vol. A21–23, 1990, pp. 243–246.

[14] Judy, J. W., R. S. Muller, and H. H. Zappe, "Magnetic Microactuation of Polysilicon Flexure Structures," *Tech. Dig. Solid State Sensor and Actuator Workshop*, Hilton Head, SC, 1994, pp. 43–48.

[15] Brignell, J. E., and N. M. White, *Intelligent Sensor Systems*, Bristol, England: IOP Publishing, 1994.

[16] Gardner, J. W., V. K. Varadan, and O. O. Awadelkarim, *Microsensors, MEMS and Smart Devices*, Chichester: John Wiley and Sons, 2001.

[17] Frank, R., *Understanding Smart Sensors*, 2nd ed., Norwood, MA: Artech House, 2000.

CHAPTER 6

Pressure Sensors

6.1 Introduction

The application of MEMS to the measurement of pressure is a mature application of micromachined silicon mechanical sensors, and devices have been around for more than 30 years. It is without doubt one of the most successful application areas, accounting for a large portion of the MEMS market. Pressure sensors have been developed that use a wide range of sensing techniques, from the most common piezoresistive type to high-performance resonant pressure sensors.

The suitability of MEMS to mass-produced miniature high-performance sensors at low cost has opened up a wide range of applications. Examples include automotive manifold air and tire pressure, industrial process control, hydraulic systems, microphones, and intravenous blood pressure measurement. Normally the pressurized medium is a fluid, and pressure can also be used to indirectly determine a range of other measurands such as flow in a pipe, volume of liquid inside a tank, altitude, and air speed. Many of these applications will be highlighted in this chapter, demonstrating MEMS solutions to a diverse range of requirements.

This chapter will first introduce the basic physics of pressure sensing and discuss the influence of factors such as static and dynamic effects as well as media compressibility. Following that is a section on the specifications of pressure sensors, which serves to introduce the terms used and the characteristics desired in a pressure sensor. Before describing the many MEMS developments that have occurred in the field of pressure sensing, there is brief discussion on traditional pressure sensors and diaphragm design. The MEMS technology pressure sensor section then looks at silicon diaphragm fabrication and characterization, applied sensing technologies, and example applications.

Pressure is defined as a force per unit area, and the standard SI unit of pressure is N/m^2 or Pascal (Pa). Other familiar units of pressure are shown in Table 6.1 along

Table 6.1 Units of Pressure and Conversion Factor to Pa (to Two Decimal Places)

Unit	Symbol	No. of Pascals
Bar	bar	1×10^5
Atmosphere	atm	1.01325×10^5
Millibar/hectopascal	Mbar/hPa	100
Millimeter of mercury	mmHg/torr	133.32
Inch of mercury	inHg	3,386.39
Pound-force per square inch	lbf/in^2 (psi)	6,894.76
Inch of water	inH$_2$O	284.8

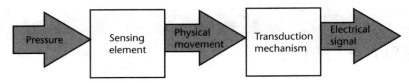

Figure 6.1 Block diagram of key pressure sensor components.

with the conversion factor to Pascals. The chosen mechanism for measuring pressure depends upon the application. Typically, pressure is measured by monitoring its effect on a specifically designed mechanical structure, referred to as the sensing element. The application of pressure to the sensing element causes a change in shape, and the resulting deflection (or strain) in the material can be used to determine the magnitude of the pressure. A block diagram of this process is shown in Figure 6.1. A range of sensing elements designed to deform under applied pressures can be fabricated using micromachining techniques, the most common by far being the diaphragm. The transduction mechanisms suitable for measuring strain or displacement described in Chapter 5 can be used to measure the resulting deflection of the sensor element. Other techniques such as using micromachined airflow sensors to measure pressure will also be discussed later in this chapter.

6.2 Physics of Pressure Sensing

The pressure at a given point within a static fluid occurs due to the weight of the fluid above it. The pressure at a given point depends upon the height of the fluid above that point to the surface, h, the density of the fluid, ρ, and the gravitational field g (see Figure 6.2). The pressure, p, is given by [1]

$$p = h\,\rho g \qquad (6.1)$$

This pressure acts in all directions, which leads us to Archimedes' principle, which states that when a body is immersed in a fluid it is buoyed up (i.e., appears to lose weight) by a force equal to the weight of the displaced fluid. Figure 6.3 shows a block of material area A and thickness t submerged in a fluid. The buoyancy pressure acting upwards is given by (6.2). The net pressure, shown in (6.3), is given

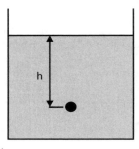

Figure 6.2 Pressure in a static fluid.

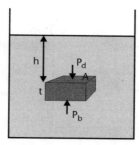

Figure 6.3   Pressures on a submerged block.

by the downwards pressure on the top face of the block, p_d [given by (6.1)], minus this buoyancy pressure, is given by

$$p_b = (h+t)\rho g \tag{6.2}$$

$$p_d - p_b = t\,\rho g \tag{6.3}$$

This is the basic principle by which objects float in liquids. If the weight of a displaced liquid exceeds the weight of the object, then it has positive buoyancy and will float on the surface. Conversely, if the weight of the object exceeds the weight of the liquid it will have negative buoyancy and sink. Neutral buoyancy is obtained by when the weight of the object equals the weight of displaced liquid, and therefore $P_b = P_d$. Objects with neutral buoyancy will remain suspended in the liquid at whatever depth they are located. Submarines, for example, typically operate at neutral buoyancy and change depth by angling fins and moving forward.

Atmospheric pressure is related to the above case. The fluid in question is the Earth's atmosphere, which extends to a height of 150 km. The calculation of atmospheric pressure is complicated by the fact that the density of the atmosphere varies with height due to the Earth's gravitational field and the compressible nature of gases. Liquids, on the other hand, are nearly incompressible and therefore this complication does not occur. The atmospheric pressure at the Earth's surface is referred to as 1 atmosphere (numerous equivalent units of pressure were given in Table 6.1).

The incompressible nature of liquids enables them to be used in hydraulic systems. Pascal's principle states that a liquid can transmit an external pressure applied in one location to other locations within an enclosed system. By applying the pressurizing force on a small piston and connecting this to a large piston, mechanical amplification of the applied force can be achieved, as shown in Figure 6.4. The distance moved by the larger piston will be less than that moved by the smaller piston, as shown in (6.4). This principle is used in hydraulic car jacks and presses.

$$d_2 = \frac{F_1}{F_2}\,d_1 = \frac{A_1}{A_2}\,d_1 \tag{6.4}$$

The rules applying to static pressures described above no longer apply when pressure measurement is carried out in moving fluids. Bernoulli's investigations of the forces present in a moving fluid identified two components of the total pressure of the flow: static and dynamic pressure. Bernoulli's equation, one form of which is

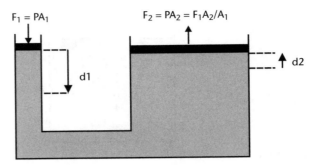

Figure 6.4 Hydraulic force multiplication [1].

shown in (6.5), states that for an inviscid (zero viscosity), incompressible, steady fluid flow of velocity v with negligible change in height, the static pressure (p) plus dynamic pressure equals the total pressure (p_t), which is a constant.

$$p + \frac{\rho v^2}{2} = p_t \qquad (6.5)$$

The dynamic pressure is given by the second term. This principle is used in measurement of airspeed using a Pitot tube as shown in Figure 6.5. The tube incorporates a center orifice that faces the fluid flow and a series of orifices around the circumference of the tube that are perpendicular to fluid flow. The perpendicular orifices measure static pressure, p_s, while the center orifice measures the total pressure at the stagnation point. Equation (6.5) can be rearranged to calculate velocity v, as shown by

$$v = \sqrt{\frac{2(p_s - p_t)}{\rho}} \qquad (6.6)$$

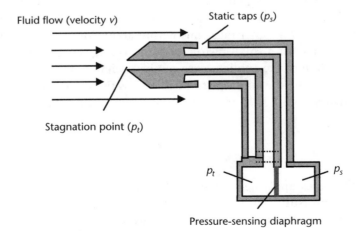

Figure 6.5 Pitot tube arrangement.

Many of the principles discussed so far rely on fluids being incompressible. Gases, however, such as the Earth's atmosphere mentioned above, are compressible. Boyle's law relates pressure to volume, V, as shown by

$$p \propto \frac{1}{V} \quad \text{or} \quad pV = \text{constant} \tag{6.7}$$

The value of the constant depends upon the mass of the gas and the temperature. This is shown by

$$pV = nRT \tag{6.8}$$

where n equals the mass of the gas divided by the molar mass, and R is the universal molar gas constant (8.31 J mol^{-1} K^{-1}). The relationship between pressure, volume, and temperature can be shown graphically in Figure 6.6.

6.2.1 Pressure Sensor Specifications

A wide variety of pressure sensors have been developed to measure pressure in a huge range of applications over many years. In order to select the correct type of sensor for a particular application, the specifications must be understood (i.e., what makes a good pressure sensor?). The fundamental specification is the operating pressure range of the sensor. Other specifications are also obvious: cost, physical size, and media compatibility. Specifications relating to performance, however, are not so obvious. and this is exacerbated by subtle differences in definitions used by manufacturers. The performance will depend upon the behavior of the sensor element, the influence of the material from which it is made, and the nature of the transduction mechanism. Common performance specifications are therefore explained next.

6.2.1.1 Zero/Offset and Pressure Hysteresis of Zero

Zero or offset is defined as the sensor output at a constant specified temperature with zero pressure applied. Pressure hysteresis of zero is a measure of the repeatability of the zero pressure reading after the sensor is subjected to a specified number of full pressure cycles. This is typically expressed as a percentage of full-scale output (% fs).

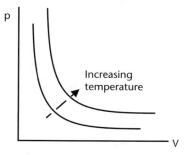

Figure 6.6   Pressure versus volume for a compressible gas.

6.2.1.2 Linearity

A linear sensor response to pressure over the entire operating range is highly desirable. This greatly simplifies subsequent signal processing. In practice, this is unlikely to be the case. Pressure sensors of the MEMS variety tend to be based on micromachined diaphragms and typically exhibit a declining rate of increased output with increases in applied pressure [2]. Linearity (also referred to as nonlinearity) can be defined as the closeness to which a curve fits a straight line. There are generally three definitions of linearity used in the specification of pressure sensors [3], and these are shown in Figure 6.7:

- *Independent linearity:* the maximum deviation of the actual measurement from a straight line positioned so as to minimize this deviation (a best fit straight line);
- *Terminal based linearity:* the maximum deviation of the actual measurement from a straight line positioned to coincide with the actual upper and lower range values;
- *Zero-based linearity:* the maximum deviation of the actual measurement from a straight line positioned to coincide with the actual lower range value and minimize the maximum deviation.

6.2.1.3 Hysteresis

Hysteresis is a measure of the repeatability of the sensor output over the operating pressure range after one or more cycles. Elastic behavior at low stresses suggests the sensor element will deflect by a constant amount for the same pressure after any number of cycles. In reality, the sensor output as pressure increases from zero to full scale will be different to the output as pressure falls from full scale to zero. This is shown in Figure 6.8. The measure of hysteresis is the difference between ascending and descending readings usually at mid-scale. It is normally expressed as a percentage of full scale. It is due to molecular effects such as molecular friction causing the

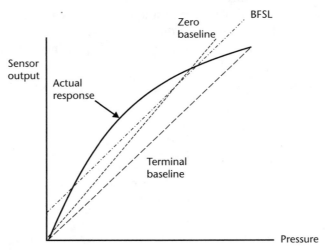

Figure 6.7 Linearity baselines.

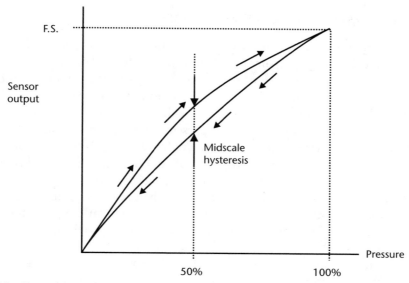

Figure 6.8   Hysteresis.

loss of energy to entropy. This is more commonly a problem associated with traditional metal sensor elements rather than single crystal materials such as silicon. Single crystal materials exhibit negligible hysteresis effects.

6.2.1.4   Sensitivity

This is the ratio of the sensor output to the applied pressure, and the units by which
it is expressed vary depending upon the manufacturers preferred units and the transduction mechanism employed in the sensor.

6.2.1.5   Long-Term Drift

This is a measure of the change in sensor output over a specified period of time. Sensor output at zero or full scale may be used. Drift over time is commonly associated
with the effects of temperature and pressure cycling on the sensor and its mounting.
The relaxation of adhesives, for example, is a common cause of drift.

6.2.1.6   Temperature Effects

The specified operating temperature range of the sensor can have many negative
effects on the sensor performance. Span temperature hysteresis is the difference in
span readings after application of minimum and maximum operating temperatures.
It is expressed as a percentage of full scale. Temperature coefficient of zero relates
sensor output at zero pressure over the specified operating temperature range. This
is commonly specified to fall within a percentage of full scale anywhere within the
temperature range. Temperature hysteresis of zero provides a measure of the repeatability of the zero pressure reading after temperature cycling. Again this is specified
as a percentage of full scale.

6.2.2 Dynamic Pressure Sensing

Dynamic pressure sensing covers applications where the user is interested in monitoring changes in pressure over small time intervals. This can provide additional information such as rate of change and the pattern of change. An example where such additional information is of use is blood pressure monitoring where it provides more detailed information about the health of the cardiovascular system. In addition to the requirements of a static pressure sensor, the frequency response of the measurement system must be considered. Frequency response is defined as the ability of a measurement system (the packaged transducer, its assembly, and electronics) to accurately reflect dynamic pressure changes. All the components of the measurement system must be considered. Within the packaged transducer and its assembly this includes the response of the mechanical element coupling the pressure to the sensing mechanism and the response of the pressurized media within the package and assembly.

The mechanical element will behave like a spring mass system and therefore its dynamic response will depend upon its stiffness, mass, and the degree of damping present. The natural frequency of such mechanical elements will be specified by the sensor manufacturer. Operation close to this frequency must be avoided. In addition, the correct level of damping must be applied for the transducer to be suitable for dynamic sensing. Underdamping will cause amplification of the pressure wave and dynamic error in the measured pressure. Overdamping will attenuate the pressure wave.

The dynamic response of miniature pressure sensors is discussed in more detail in Section 6.5.1, but broadly speaking, due to their small size and the elastic properties of single crystal silicon, resonant frequencies in the megahertz range are possible. This gives them excellent inherent dynamic response characteristics. Typically, however, a stainless steel barrier diaphragm is employed between the pressure sensor and the pressurized media to ensure media compatibility. The volume between the stainless steel diaphragm and the silicon sensor is filled by hydraulic oil that transmits the pressure to the sensor die. The presence of the barrier diaphragm and the hydraulic oil will both serve to lower the resonant frequency of the transducer as a whole. Hydraulic over range protection mechanisms also limit dynamic response since these tend to overdamp the system rendering the transducer unsuitable for dynamic pressure sensor applications.

The frequency response of the pressurized media within the fluid channels and sensor cavity is often the most limiting factor. The natural frequency of such a fluidic system depends upon the volume of the sensor cavity, the length and diameter of the channels, and the speed of sound in the fluid to be measured. As with the natural frequency of the mechanical element, dynamic pressure measurements at the natural frequency of the fluidic system are not recommended. This would cause severe distortion and amplification of the pressure waveform. The frequency at which tolerable distortion occurs will depend on the damping in the system. Assuming the worst case where damping levels are low, as a rule of thumb the maximum usable frequency for any given fluidic system is generally taken to be one-fifth or one-seventh of its natural frequency.

The electronics associated with a pressure sensor provide power to the sensing mechanism and perform signal conditioning on the output signal. Signal conditioning can include amplification, filtering, and compensation. The frequency response of the electronics is likely to be a limiting factor only when used with very high frequency sensors as described above. Sensors requiring ac excitation (e.g., capacitive) will be limited in particular by the frequency of this driving signal.

6.2.3   Pressure Sensor Types

Pressure can be measured relative to vacuum, atmosphere, or another pressure measurand.

- *Absolute pressure sensors* are devices that measure relative to a vacuum and therefore must have a reference vacuum encapsulated within the sensor. Atmospheric pressure is measured using absolute sensors.
- *Gauge pressure sensors* measure relative to atmospheric pressure, and therefore, part of the sensor must be vented to the ambient atmosphere. Blood pressure measurements are taken using a gauge pressure sensor. Vacuum sensors are a form of gauge pressure sensor designed to operate in the negative pressure region.
- *Differential pressure sensors* measure the difference between two pressure measurands. The design of differential sensors often represents the greatest challenge since two pressures must be applied to the mechanical structure. The specifications for such devices can also be exacting since it is often desirable to detect small differential pressures superimposed on large static pressures.

6.3   Traditional Pressure Sensors

Traditional macroscale pressure sensors have been developed that are based on a wide range of mechanical sensing elements and transduction principles. These are discussed briefly in this section to illustrate the development of pressure sensors.

6.3.1   Manometer

This is a simple yet accurate method for measuring pressure based upon the influence of pressure on the height of a column of liquid. The best-known form is the U-tube manometer shown in Figure 6.9. If pressure is exerted to one side of the U-tube as shown, the liquid is displaced, causing the height in one leg to drop and the other to rise. The difference in height h between the fluid-filled legs indicates the pressure. The measurement is usually taken visually by reading the height from the scale incorporated into the instrument. Resolution can be improved by inclining one leg, allowing more precise reading of the scale. Often a liquid reservoir is incorporated onto one side, making the drop in fluid height on that leg negligible. The unit of pressure will depend upon the liquid (e.g., inch of water, inch of mercury). Manometers can be used both as a gauge sensor with one side vented to atmosphere and as differential sensors with pressure applied to both legs. The disadvantages associated with manometers include their slow response (they are not suitable for dynamic applications) and the limited range of pressures for which they are suitable.

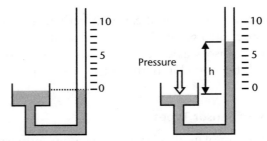

Figure 6.9   U-tube manometer.

6.3.2   Aneroid Barometers

Aneroid barometers essentially consist of an evacuated metal capsule with flexible top and bottom faces. The shape of the capsule changes with variations in atmospheric pressure and this deformation is usually mechanically amplified via a series of levers or gears. The pressure capsule can be fabricated in the form of bellows to provide further deflection. The measurement of deflection is done visually by a pointer connected to the levers aligned to an appropriate scale. Alternatively, they can be connected to a plotter for recording pressure against time (known as a Barograph). These devices were first practically realized in the mid-nineteenth centaury as an alternative to the manometer and, while not as accurate, offered the advantages of ruggedness, compactness, and no liquid (aneroid meaning without liquid). Dynamic response of these devices is poor due to the mechanical mass of the sensor element, and they are not suited for dynamic pressure sensing applications. They are still widely used today.

6.3.3   Bourdon Tube

Bourdon tubes operate on the same principle as the aneroid barometer, but instead of an evacuated capsule or bellows arrangement, a C-shaped or helical tube is used (see Figure 6.10). The tubes are closed at one end and connected to the pressure at the other end, which is fixed in position. The tube has an elliptical cross-section, and when pressure is applied, its cross-section becomes more circular, which causes the tube to straighten out until the force of the fluid pressure is balanced by the elastic resistance of the tube material. Different pressure ranges are therefore

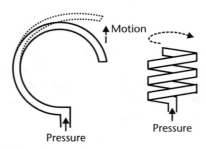

Figure 6.10   Bourdon tube sensor elements.

accommodated by using different materials such as phosphor bronze or stainless steel. Changes in pressure move the closed end of the tube to which a linkage arm and a gear and pinion assembly are attached. These rotate a pointer around a graduated scale, providing visual reading of the pressure. Bourdon tubes are usually used by gauge pressure sensing applications, but differential sensing is possible by connecting two tubes to one pointer. By correctly arranging the linkages, the pointer can be made to measure the pressure difference between the tubes. Helical tubes are more compact, reliable, and offer performance advantages over the more traditional C-shaped devices. Bourdon tubes are used throughout the industry and are available in a wide range of pressure specifications.

6.3.4 Vacuum Sensors

Vacuums are pressures that are below atmospheric. Vacuums are difficult to measure by mechanical means, and therefore, different sensing techniques have been developed. These techniques are suited for different vacuum levels. Within the low vacuum region (atmosphere to $\sim 10^{-3}$ mbar) the Pirani gauge is normally used. This is based upon a heated wire, the electrical resistance of which is proportional to its temperature. At atmospheric pressure convection occurs and heat energy is lost from the wire. As the vacuum increases, gas molecules are removed and less convection occurs, causing the wire to heat up. As it heats up, its electrical resistance increases, and this can be calibrated against pressure to provide a reasonably accurate measure of the vacuum.

Below the range of the Pirani gauge, Ion gauges (also known as Penning or Cold Cathode gauges) are used. These consist of a filament, a grid, and a collector. The filament produces thermonic emission of electrons, and a +ve charge on the grid draws the electrons away from the filament. The electrons circulate around the grid, which has a fine structure enabling the electrodes to pass through many times until they eventually collide. Any gas molecules present around the grid may collide with circulating electrons, which results in the gas molecule being ionized. The collector inside the grid is −ve charged and attracts these +ve charged ions. The number of ions collected is directly proportional to the number of molecules inside the vacuum system, and therefore, the collected ion current gives a direct reading of the pressure.

6.4 Diaphragm-Based Pressure Sensors

Diaphragms are the simplest mechanical structure suitable for use as a pressure-sensing element. They are used as a sensor element in both traditional and MEMS technology pressure sensors. In the case of MEMS, due to the planar nature of many established fabrication processes, the diaphragm is the main form of sensor element developed. This section will first review basic diaphragm theory before analyzing in more detail particular aspects relating to MEMS pressure sensors. This review of traditional diaphragm theory is particularly relevant in the packaging of MEMS technology pressure sensors. Stainless steel diaphragms are routinely incorporated into the package to isolate the sensor from the media. The behavior of the stainless steel diaphragm will affect the performance of the sensor and must be considered when designing the device as a whole at the outset. For a more detailed analysis of diaphragm behavior, the author recommends the work of Di Giovanni [3].

Pressure applied to one (or both) side(s) of the diaphragm will cause it to deflect until the elastic force balances the pressure. The pressure range of a given diaphragm will depend upon its dimensions (surface area and thickness), geometry, edge conditions, and the material from which it is made. Traditional metal diaphragm pressure sensors are made from a range of materials such as stainless steels 316L, 304, 17-4 PH, PH 15-7 Mo, titanium, nickel alloys, and beryllium copper. The metals are characterized by good elastic properties and media compatibility.

In the case of traditional sensors, diaphragms are the simplest sensor element to manufacture, they are the least sensitive to vibrations, they offer the best dynamic response, and they are compatible with simple forms of overload protection. However, the deflection associated with diaphragms is much less than, for example, Bourdon tubes. Therefore, electromechanical transduction mechanisms may be employed to measure the deflection rather than the mechanical linkages associated with Bourdon tubes.

Metal diaphragms are typically circular and may incorporate corrugations to modify diaphragm characteristics. The behavior of a diaphragm will depend upon many factors, such as the edge conditions and the deflection range compared to diaphragm thickness. The edge conditions of a diaphragm will depend upon the method of manufacture and the geometry of the surrounding structure. It will vary between a simply supported or rigidly clamped structure, as shown in Figures 6.11(a) and 6.12(a). Simply supported diaphragms will not occur in practice, but the analytical results for such a structure may more accurately reflect the behavior of a poorly clamped diaphragm than the rigidly clamped analysis. At small deflections (<~10% diaphragm thickness) the pressure-deflection relationship will be linear. As the pressure increases, the rate of deflection decreases and the pressure-deflection relationship will become nonlinear. As a rule of thumb, a deflection of 12% of diaphragm thickness will produce a terminal nonlinearity of 0.2%; a deflection of 30% produces a nonlinearity of 2% [3]. The suitability of the deflection range will depend upon the desired specification of the sensor and the acceptable degree of compensation.

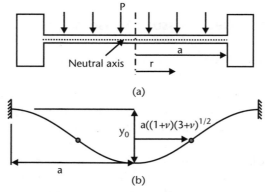

Figure 6.11 (a) Rigidly clamped diaphragm and (b) its associated displacement under uniform pressure.

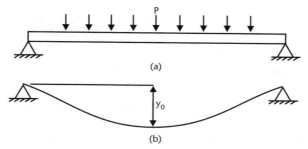

Figure 6.12   (a) Simply supported diaphragm and (b) its associated displacement under uniform pressure.

The following two sections present analytical equations for the deflection and resulting stress of rigidly clamped and simply supported metal diaphragms. These have been grouped according to the degree of deflection in relation to thickness, denoted small deflection diaphragms, medium deflection diaphragms, and membranes. The following equations assume the following assumptions:

- The diaphragm is flat and of uniform thickness.
- The material is homogenous and isotropic (silicon will be covered later).
- Pressure is applied normally to the plane of the diaphragm.
- The elastic limit of the material is not exceeded.
- The thickness of the diaphragm is not too thick (e.g., maximum 20% of diaphragm diameter).
- Deformation is due to bending, the neutral axis of the diaphragm experiences no stress.

6.4.1   Analysis of Small Deflection Diaphragm

For small deflection diaphragms the maximum deflection is 30% of diaphragm thickness. The deflection y at radial distance r of a round diaphragm under a uniform pressure P, rigidly clamped as shown in Figure 6.11(a), is given by

$$y = \frac{3(1-v^2)P}{16Eh^3}(a^2 - r^2)^2 \tag{6.9}$$

where h is the diaphragm thickness, E and v are the Young's modulus and Poisson's ratio of the diaphragm material, respectively, and a is the radius of the diaphragm. The maximum deflection y_0 will occur at the diaphragm center where $r = 0$. Assuming a common value for metals of $v = 0.3$, the maximum deflection is given by

$$y_0 = \frac{0.1709Pa^4}{Eh^3} \tag{6.10}$$

The deflection of a rigidly clamped diaphragm is shown in Figure 6.11(b). As mentioned previously, the measurement of the deflection associated with diaphragm pressure sensors typically requires the use of electromechanical transducers rather than mechanical linkages. Electromechanical effects can be used to measure displacement directly or to measure the stress/strain induced in the diaphragm material. Therefore, it is also useful to provide an analysis of the stress distribution across a pressurized diaphragm.

The stress distribution will vary both across the radius and through the thickness of the diaphragm. For example, the neutral axis [shown in Figure 6.11(a)] experiences zero stress while the maximum stress occurs at the outer faces. At any given distance r from the center of the diaphragm, one face will experience tensile stress while the other experiences compressive stress. There are two stress components associated with a circular diaphragm: radial and tangential. The radial stress, σ_r, at distance r from the center of the diaphragm is given by (6.11). The maximum radial stress that occurs at the diaphragm edge ($r = a$) is given by (6.12).

$$\sigma_r = \pm \frac{3}{8} \frac{Pa^2}{b^2} \left[(3+v)\frac{r^2}{a^2} - (1-v) \right] \tag{6.11}$$

$$\sigma_{r_{max}} = \pm \frac{3}{4} \frac{Pa^2}{b^2} (1+v) \tag{6.12}$$

Radial stress is equal to zero at a value of r given by $a((1 + v)(3 + v))^{1/2}$ (shown in Figure 6.12). This equals 0.628 if $v = 0.3$.

The tangential stress, σ_t, at distance r from the center of the diaphragm is given by (6.13). The maximum tangential stress that occurs at the diaphragm center ($r = 0$) is given by (6.14).

$$\sigma_t = \pm \frac{3}{8} \frac{Pa^2}{b^2} \left[(3v+1)\frac{r^2}{a^2} - (1+v) \right] \tag{6.13}$$

$$\sigma_{t_{max}} = \pm \frac{3}{8}(1+v)\frac{Pa^2}{b^2} \tag{6.14}$$

The inflection circle for tangential stress is removed from that of radial stress and is given by $a((1 + v)(3v + 1))^{1/2}$. This equals 0.827 if $v = 0.3$.

In the case of simply supported diaphragms [as shown in Figure 6.12(a)], for a round diaphragm under a uniform pressure P, the deflection y at radial distance r is given by (6.15). The maximum deflection occurs at the diaphragm center and, assuming $v = 0.3$, is given by (6.16).

$$y = \frac{3}{16} \frac{(1-v^2)P(a^2 - r^2)}{Eb^3} \left(\frac{5+v}{1+v} a^2 - r^2 \right) \tag{6.15}$$

$$y_0 = \frac{0.695 Pa^4}{Eb^3} \tag{6.16}$$

The deflection of a simply supported diaphragm is shown in Figure 6.12(b). The radial is given by (6.17). The maximum radial stress that occurs at the diaphragm center ($r = 0$) is given by (6.18).

$$\sigma_r = \pm \frac{3}{8} \frac{Pa^2}{h^2} \left[(3+v) \left(1 - \frac{r^2}{a^2} \right) \right] \tag{6.17}$$

$$\sigma_{r_{max}} = \pm \frac{3}{8} \frac{Pa^2 (3+v)}{h^2} \tag{6.18}$$

The tangential stress, σ_t, at distance r from the center of the diaphragm is given by (6.19). The maximum tangential stress occurs at the diaphragm center and is equal to the radial stress given by (6.18):

$$\sigma_t = \pm \frac{3}{8} \frac{P}{h^2} \left[a^2 (3+v) - r^2 (1+3v) \right] \tag{6.19}$$

6.4.2 Medium Deflection Diaphragm Analysis

The operation of diaphragms at deflections beyond 30% of thickness as covered in Section 5.4.2 may be required in certain designs. In such a case, both tensile and bending stresses must be considered. The characteristic equation, assuming the material remains within the elastic limit, in such a case is given by [3]

$$P = \frac{16}{3(1-v^2)} \frac{Eh^3}{a^4} (y) + \frac{7-v}{3(1-v^2)} \frac{Eh}{a^4} (y^3) \tag{6.20}$$

This may be written as a cubic equation form $P = cy + dy^3$, where

$$c = \frac{16}{3(1-v^2)} \frac{Eh^3}{a^4} \quad \text{and} \quad d = \frac{7-v}{3(1-v^2)} \frac{Eh^3}{a^4} \tag{6.21}$$

These represent the linear and nonlinear terms of the characteristic equation.

6.4.3 Membrane Analysis

Membranes can be considered as very thin diaphragms with large deflection $(y_0/h > 5)$ [3]. In theory, a membrane has no flexural rigidity and experiences tensile stress, but no bending stress. The characteristic equation for a membrane is given by [4]

$$\frac{Pa^4}{Eh^4} = 2.86 \frac{y_0^3}{h^3} \tag{6.22}$$

Radial stress in a membrane at radius r is given by (6.23); the maximum stress occurs at the diaphragm center and, assuming $v = 0.3$, is given by (6.24). Tangential stress is given by (6.25). Maximum tangential stress occurs at the center of the membrane and is equal to the maximum radial stress.

$$\sigma_r = \frac{Ey_0^2}{4a^2} \left(\frac{3-v}{1-v} - \frac{r^2}{a^2} \right) \tag{6.23}$$

$$\sigma_r = 0.96 \frac{E y_0^2}{a^2} \tag{6.24}$$

$$\sigma_t = \frac{E_0^2}{4a^2} \left(\frac{3-v}{1-v} - 3\frac{r^2}{a^2} \right) \tag{6.25}$$

Other factors such as tensioned membranes or the inclusion of rigid centers are beyond the scope of this book.

6.4.4 Bossed Diaphragm Analysis

A bossed diaphragm is a flat diaphragm with a thicker center portion, which increases the rigidity in that location [see Figure 6.13(a)]. The inclusion of the center section, or boss, affects the behavior of the diaphragm under pressure. A bossed diaphragm, for example, will exhibit higher stresses for a given deflection, which is attractive in the case of a traditional bonded strain gauge pressure sensor. They are particularly well suited to sensing low pressures and exhibit improved linearity characteristics compared with flat diaphragms. The boss should be a minimum of six times thicker than the diaphragm and the ratio of b/a should be greater than 0.15 for the boss to be effective [3]. The ratio of b/a is fundamental to the behavior of the diaphragm as shown in the following equations.

The characteristic equation of a bossed diaphragm under pressure is given by (6.26), where A_p is a stiffness coefficient calculated from (6.27), and B_p is the stiffness coefficient of the nonlinear term given by (6.28).

$$P = \frac{E h^3}{A_p a^4} (y) + B_p \frac{E h^3}{a^4} (y^3) \tag{6.26}$$

$$A_p = \frac{3(1-v^2)}{16} \left(1 - \frac{b^4}{a^4} - 4\frac{b^2}{a^2} \log \frac{a}{b} \right) \tag{6.27}$$

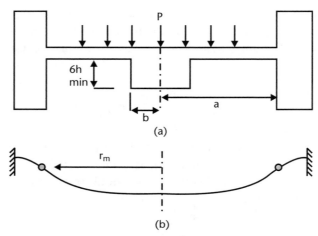

(a)

(b)

Figure 6.13 (a) Bossed diaphragm geometry and (b) its associated displacement under uniform pressure.

$$B_p = \frac{\dfrac{7-v}{3}\left(1+\dfrac{b^2}{a^2}+\dfrac{b^4}{a^4}\right)+\dfrac{(3-v)^2}{1+v}\dfrac{b^2}{a^2}}{(1-v)\left(1-\dfrac{b^4}{a^4}\right)\left(1-\dfrac{b^2}{a^2}\right)} \tag{6.28}$$

The deflection of a bossed diaphragm is shown in Figure 6.13(b). The maximum radial bending stress occurs at the outer perimeter where the diaphragm is clamped and the inner perimeter where the boss begins. The stress on the outer perimeter is equal and opposite to that occurring at the inner, and at the midpoint of the annulus [r_m on Figure 6.13(b)] the stress is zero. The radial stress is given by

$$\sigma_{r_{outer}} = \sigma_{r_{inner}} = \pm\frac{3P}{4h^2}\left(a^2 - b^2\right) \tag{6.29}$$

6.4.5   Corrugated Diaphragms

Corrugations in a diaphragm enable operation at larger displacements with improved linearity. The corrugations can have sinusoidal, triangular, rectangular, trapezoidal, and toroidal profiles. While this has a small influence on the behavior of the diaphragm, the depth of corrugation (H), material thickness (h), wavelength (l), and ratio of corrugations to diaphragm radius (l/a) are the main factors. The characteristic equation of corrugated diaphragms is shown in (6.30). A_p, B_p, and q are given by (6.31), (6.32), and (6.33), respectively.

$$\frac{Pa^4}{Eh^4} = A_p\frac{y}{h} + B_p\frac{y^3}{h^3} \tag{6.30}$$

$$A_p = \frac{2(3+q)(1+q)}{3\left[1-\left(\dfrac{v}{q}\right)^2\right]} \tag{6.31}$$

$$B_p = \frac{32}{q^2-9}\left[\frac{1}{6}-\frac{3-v}{(q-v)(q+3)}\right] \tag{6.32}$$

$$q = \left(\frac{s}{l}\left(1+1.5\frac{H^2}{h^2}\right)\right)^{\frac{1}{2}} \tag{6.33}$$

Rigid centers, or bosses, can be incorporated into corrugated diaphragms, and these will increase the stiffness of the diaphragm if sufficiently large.

6.4.6   Traditional Diaphragm Transduction Mechanisms

The generally small displacements associated with traditional metal diaphragms typically require electromechanical transduction techniques to sense their magnitude. The most common method employed is metal strain gauges located on the face of the diaphragm. These are positioned at the points of maximum strain in order to maximize the sensitivity of the gauge. Metal gauges can be incorporated onto the diaphragm face by bonding foil gauges or by depositing and patterning insulator and metal materials using thin-film techniques such as sputtering or CVD [5]. Another resistive approach is the use of screen printed thick-film strain gauge resistors. These can be printed on the top surface of a metal diaphragm, previously coated with a printed dielectric layer, and offer improved sensitivity compared with bonded strain gauges. Maximum resistive strain gauge sensitivity can be achieved by bonding a silicon strain gauge to the metal diaphragm. This approach utilizes the piezoresistive nature of silicon, which increases the output of the strain gauge for a given deflection. The relative merits of these resistive methods and their associated gauge factors are discussed in Chapter 5.

Other transduction techniques include capacitance, inductance, reluctance, and piezoelectric. The capacitive approach uses the diaphragm as one electrode of a parallel capacitor structure. Diaphragm displacement causes a change in capacitance between it and a fixed electrode. Inductance can be used to monitor the displacement of the diaphragm by mechanically linking it to the core of a linear variable differential transformer (LVDT). This consists of a symmetrical arrangement of a primary coil and two secondary coils. Movement of the magnetic core causes the mutual inductance of each secondary coil to vary relative to the primary. Variable reluctance transducers remove the mechanical link to the core and use the permeability of the diaphragm material itself to alter the inductance within two coils positioned on either side of the diaphragm. The coils are typically wired in an inductive half bridge, and a change in inductance alters the impedance of each coil unbalancing the bridge. Unbalances result in the ac drive signal being coupled across to the output, and the physical arrangement is suitable for differential pressure-sensing applications. Piezoelectric pressure sensors utilize a piezoelectric sensing element mechanically linked to the diaphragm. Movements in the diaphragm induce a strain in the piezoelectric and hence a charge is generated. These sensors are only suitable for measuring dynamic pressures and are not suitable for static applications because piezoelectric materials only respond to changing strains.

6.5   MEMS Technology Pressure Sensors

Research into solid-state pressure sensors began as far back as the 1960s [6–8]. Since then there have been many developments both in micromachining and sensing techniques, which have enabled MEMS pressure sensors to mature into a commercially successful solution for many sensing applications. The mechanical sensor element is typically (but not exclusively) a micromachined diaphragm. This section commences with a brief analysis of rectangular silicon diaphragms. The different sensing principles employed to date will be introduced and illustrated with both commercially available and research based devices. Finally, the state of the art in micromachined pressure sensor technology will be discussed.

6.5.1   Micromachined Silicon Diaphragms

MEMS pressure sensors typically employ a diaphragm as the sensor element. This is because of its compatibility with a range of bulk and surface silicon micromachining processes. The most common fabrication method is anisotropic wet silicon etching, which allows good control over diaphragm dimensions and is a batch process capable of producing hundreds of devices simultaneously across a group of wafers. When combined with a (100) wafer orientation, a wet potassium hydroxide (KOH) etch, for example, produces a rectangular diaphragm with sloping side walls that follow the (111) planes. A cross-section of a typical diaphragm is shown in Figure 6.14. Diaphragm thickness can be controlled by timing etch duration, or more precisely by using boron doping or electrochemical etch stops. Surface micromachining techniques are becoming increasing applied since they offer the opportunity for reduced device size and compatibility with integrated electronics.

When modeling complex micromachined structures, finite element (FE) packages such as those described in Chapter 3 are normally employed. Diaphragms represent one of the few MEMS structures that can be modeled analytically. Since the diaphragm is rectangular, the characteristic equations will differ from those describing the circular case above. The characterizing equations for a rectangular diaphragm, where a is the length of the shorter side, and with rigidly clamped edges and small deflections are given next.

$$y_0 = \alpha\left(\frac{Pa^4}{Eh^3}\right)(1 - v^2)$$
(6.34)

$$\sigma = \beta\left(\frac{Pa^2}{h^2}\right)$$
(6.35)

For a rectangular diaphragm, the coefficients α and β depend upon the ratio of the lengths of the diaphragm sides and the position of interest. Assuming a square diaphragm, α equals 0.0151, and β equals 0.378 for the maximum stress that occurs along the edge of the diaphragm and 0.1386 for the maximum stress at the center of the diaphragm.

Bossed diaphragms can also be fabricated using both anisotropic and isotropic etching. Such structures are typically modeled using FE techniques [9]; however, Sandmaier has presented a set of analytical equations enabling basic optimization of diaphragm design [10]. Corrugated silicon diaphragms have been discussed in the papers by van Mullem et al. [11] and Jerman [12]. The analytical equations presented in Section 6.4.5 provide an adequate approximation to the silicon case.

The dynamics of a micromachined diaphragm can be adequately characterized by linear plate theory. The undamped resonant frequency f_n of a clamped square diaphragm of uniform thickness and homogenous material is given by [13]

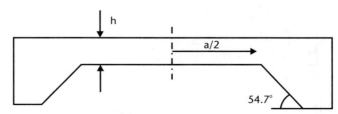

Figure 6.14 Anisotropically etched silicon diaphragm.

$$f_n = 1.654\left[\frac{E}{\rho}\left(1-v^2\right)\right]^{\frac{1}{2}} h/a^2 \qquad (6.36)$$

The amount of damping present will depend not only on the diaphragm design but also its packaging and surroundings. As a rough guide, resonant frequencies of typical diaphragms should range between ~80 kHz for a 1-bar device to 575 kHz for a 40-bar device [14]. Higher frequency devices have been developed; for example, the Entran EPIH Micro Miniature range high-frequency pressure sensor series offers a maximum resonant frequency of 1.7 MHz for the 20-bar device [15]. For this series, the pressurized media is in direct contact with the micromachined silicon structure, and therefore it is suitable only for dry gas or some noncorrosive fluid applications. The introduction of a stainless steel barrier diaphragm lowers the resonant frequency to 45 kHz for a 17-bar device [16].

6.5.2 Piezoresistive Pressure Sensors

The piezoresistive nature of silicon makes the use of diffused or implanted resistors an obvious and straightforward technique for measuring the strain in a micromachined silicon diaphragm. The piezoresistive effect of silicon was first exploited by bonding silicon strain gauges to metal diaphragms [7], but this is an unsatisfactory approach given the thermal mismatch between the metal, adhesive layer, and silicon. Diaphragms were first micromachined into the silicon itself by mechanical spark erosion and wet isotropic etching [8]. This was not a batch approach and therefore device costs were high. The use of anisotropic etching, anodic and fusion bonding, ion implanted strain gauges, and surface micromachining have since reduced the size and improved the accuracy of piezoresistive pressure sensors.

A cross-section and plan view of a typical anisotropically etched silicon piezoresistive pressure sensor is shown in Figure 6.15. The diaphragm is etched as described above and the resistors are located along the edge of the diaphragm, one on each side. The resistors are all orientated in the same direction, and therefore, two are in parallel with the maximum strain (R_l) and two are perpendicular (R_t). The change in resistance of each resistor is calculated from (5.10). The piezoresistive coefficients associated with these resistors will depend upon the orientation of the wafer and diaphragm, the type and amount of doping, and the temperature. Given a (100) wafer,

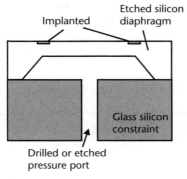
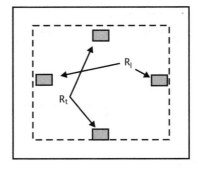

Figure 6.15   Cross-section and plan view of a typical bulk micromachined piezoresistive pressure sensor.

the edges of the diaphragm will be in the (110) directions. The piezoresistive coefficients of p- and n-type silicon are presented graphically by Kanda [17]. Assuming p-type doping, which produces the largest and most linear piezoresistive effect, π_l and π_t are equal and opposite at $+/-69$ m^2/N, respectively. From (5.10) it can be seen that the resistor orientation shown in Figure 6.15 will produce equal and opposite changes in the resistance of the two pairs of resistors. Placing the two pairs of resistors on opposite sides of a full bridge circuit will therefore maximize the sensitivity of the sensor to strains arising from pressure induced deflection of the diaphragm. The stress can be calculated from (6.35) and for a full bridge the fractional bridge output is given by (6.37). This is the most common resistor arrangement and has been modeled analytically extensively [18–20].

$$\left(\frac{\Delta V}{V}\right) = \frac{(\Delta R/R)_l - (\Delta R/R)_t}{2 + (\Delta R/R)_l + (\Delta R/R)_t} \tag{6.37}$$

Piezoresistive pressure sensors in the form described above have been commercially available for many years. Manifold absolute pressure sensors are an established application of these devices in the automotive industry. An example of such a device has been developed by Motorola and has been described in detail in [21]. Other, more recent automotive applications based upon piezoresistive sensing include diesel injection pressure [22] and exhaust gas recirculation systems [23]. Circular diaphragms are less common and have been analyzed by Matsuoka et al. [24]. Variations on the theme involve changes to the diaphragm structure (including bossed and ribbed diaphragms), temperature compensation techniques, and the use of alternative fabrication processes.

Modifications to the basic diaphragm structure have been investigated in order to improve the linearity and sensitivity of the sensors. Bossed diaphragms have been fabricated using anisotropic etching processes that incorporate the rigid center seen on traditional diaphragms [9, 25]. This approach enables a resistor layout shown in Figure 6.16, which enables equal and opposite strains to be experienced by the inner and outer resistor pairs. This arrangement improves the nonlinearity of the diaphragm in both directions, making it suitable for differential applications [26]. Another design uses a double boss at the diaphragm center [27] while researchers at Honeywell have used FE techniques to design a ribbed and bossed diaphragm [28]. The Honeywell device takes a standard diaphragm anisotropically etched from the

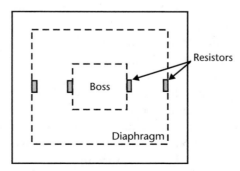

Figure 6.16   Resistor placement on a bossed diaphragm.

back and patterns the bosses and ribs on the front of the diaphragm. The resistors were positioned in the standard layout (Figure 6.15) and were located on the top surface of the rib, which served to magnify the stress by removing the resistor further from the neutral axis. The bosses were stiffened regions along each side of the diaphragm leaving the center unstiffened like a standard diaphragm.

Meandering resistors have also been applied to basic and bossed diaphragms [29]. The meander incorporates different levels of doping in each direction, which maximizes the strain sensitivity of the resistor. The meander pattern increases the length of the resistor, and this approach improves sensitivity compared with standard resistors.

The temperature cross-sensitivity is an obvious drawback of silicon piezoresistors. The change in resistance due to temperature will often exceed that arising from the change in the measurand. Several techniques are therefore employed to compensate for temperature. The first technique arises from the use of a full bridge with the resistors arranged as shown in Figure 6.15. In such an arrangement the change in temperature is a common mode effect acting on all resistors simultaneously, and therefore, the temperature effects should cancel out. Due to manufacturing tolerances, however, the temperature coefficients of each resistor will invariably be slightly different. The change in resistance due to temperature and its resulting effect on the output of the bridge can be expressed in the following equations [30]:

$$R(T) = R(0)(1 + \alpha T + \beta T^2)$$ (6.38)

$$\frac{\Delta V_0(T)}{V_A} = \frac{R_1(0)R_2(0)}{[R_1(0)+R_2(0)]^2} \times [(\alpha_1 - \alpha_2)T + (\beta_1 - \beta_2)T^2]$$
$$- \frac{R_1(0)R_2(0)}{[R_1(0)R_2(0)]^2} \times [(\alpha^1 - \alpha^2)T + (\beta_1 - \beta_2)T^2]$$ (6.39)

The incorporation of a temperature sensor onto the sensor chip can enable temperature compensation via a look-up table or algorithm. Such an approach, however, requires extensive temperature and pressure calibration, which is a time consuming and expensive operation. An alternative technique is to include a dummy bridge on the sensor chip in addition to the pressure sensitive bridge. The dummy resistors should be positioned at least 100 μm away from the edge of the diaphragm to ensure they do not experience any pressure-induced stresses [31]. This compensation technique has been applied with the dummy resistors arranged in either a full bridge [29] or a half-bridge [32]. The temperature limits of the implanted piezoresistive approach are approximately 120°C due to the limitations of the p-n junction. This temperature limit can be extended by using doped polysilicon resistors deposited on the top surface of the diaphragm. Polysilicon resistors are, however, less sensitive to applied stress (see Chapter 5).

Over the years, developments in materials and fabrication processes have also had an effect on piezoresistive pressure sensors. Silicon fusion bonding, for example, has enabled a reduction in chip size by enabling a diaphragm wafer to be bonded to the back of an anisotropically etched cavity as shown in Figure 6.17 [33]. The use of SOI wafers has improved performance in several ways. The buried oxide can act as an etch stop, facilitating fabrication [34] and precisely controlling the diaphragm

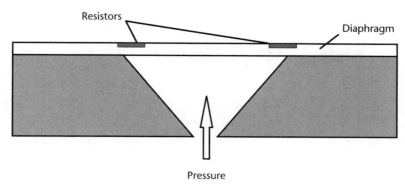

Figure 6.17 Fusion bonded piezoresistive pressure sensor.

thickness, or as an electrical insulator, enabling higher temperature operation [35–37]. Ultimate high-temperature operation of piezoresistive pressure sensors has been developed using micromachined silicon carbide [38]. The diaphragms are etched by a photoelectrochemical process in a diluted HF etchant. A prototype device has been demonstrated operating at 600°C [39] and in a dynamic sensing application on a gas turbine engine [40]. Finally, silicon nitride diaphragms have been realized by bulk wet anisotropic etching. The nitride membrane is formed by wet etching through the silicon entirely from the back of the wafer. The wet etch stops upon reaching the nitride, and the piezoresistors are protected due to the high-dose boron implant used to define them [41]. Nitride membranes are stronger than their silicon counterpart but may suffer from in-built stresses due to the deposition process.

The need to reduce the size of devices, and therefore the cost of production, has led to the use of surface micromachining to fabricate the mechanical sensing element and resistors [42]. In addition to reduced size, surface micromachining is more compatible with IC fabrication technology. It is a flexible fabrication approach enabling the diaphragm to be fabricated from a range of deposited materials such as polysilicon [43] and silicon nitride [44]. In both cases an underlying sacrificial layer is removed. For the polysilicon diaphragm the sacrificial material is silicon dioxide and a wet etch is used to remove it. The nitride membrane uses a polysilicon sacrificial material. In both cases the lateral dimensions of the membrane are defined by previous patterning of the oxide, or doping of the polysilicon, respectively. Both devices use polysilicon resistors to sense diaphragm deflections. Both are absolute pressure sensors since a CVD process is used to deposit nitride to seal sacrificial etch holes. The vacuum used in the CVD process is therefore trapped in the sealed volume under the diaphragm. A cross-section of each device is shown in Figure 6.18. Other examples of surface-micromachined piezoresistive pressure sensors include a cardiovascular pressure sensor for measurement of blood pressure inside coronary arteries [45]. This is based on a square polysilicon diaphragm with edge lengths of 103 μm with a vacuum-sealed cavity underneath. One polysilicon resistor is used to detect the deflection of the diaphragm, and a second dummy resistor is used for temperature compensation.

As discussed in the earlier analysis, the boundary conditions of the diaphragm will play an important role in the behavior of the diaphragm. With surface

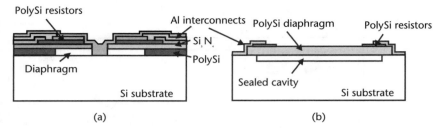

Figure 6.18   Surface-micromachined pressure sensors with (a) nitride and (b) polysilicon diaphragms.

micromachining there are more variations in the nature of the clamping at the edge of the diaphragm. Depending on the profile of the sacrificial layer, the diaphragm could be flat along its entire length [Figure 6.19(a)] or have a step at the edge from where the diaphragm material was deposited over the sacrificial layer [Figure 6.19(b)]. Flat membranes have been found to be preferable since the stepped structure exhibits inferior drift characteristics [46].

The extra flexibility offered by surface micromachining has also enabled more complex pressure-sensing structures to be realized. An example of this is a duel beam pressure sensor, which couples the diaphragm deflection to a cantilever beam. A polysilicon piezoresistive strain gauge is located on the top surface of the cantilever, as shown in Figure 6.20 [47]. The cantilever, and its attachment to the underside of the diaphragm, acts as a mechanical lever, amplifying the strain experienced by the piezoresistor compared to straightforward mounting on the diaphragm. For

Figure 6.19   Diaphragm edge conditions: (a) flat diaphragm, and (b) stepped diaphragm.

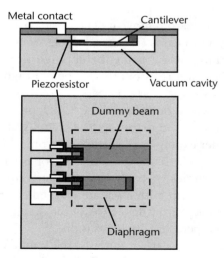

Figure 6.20   Dual beam pressure sensor configuration.

temperature compensation, a second beam with piezoresistor is positioned alongside but not coupled to the diaphragm. The diaphragm is a polysilicon layer that coats the entire chip surface (except bond pads), thereby physically, electrically, and thermally isolating the strain gauges and beams from the pressurized media.

6.5.3 Capacitive Pressure Sensors

Capacitive pressure sensors are typically based upon a parallel plate arrangement whereby one electrode is fixed and the other flexible. As the flexible electrode deflects under applied pressure, the gap between electrodes decreases and the capacitance increases. The principles of capacitive sensing have been described in Chapter 5. Capacitive pressure sensors were first developed in the late 1970s and early 1980s [18, 48]. An early device, shown in Figure 6.21, consists of an anisotropically etched silicon diaphragm with the fixed electrode being provided by a metallized Pyrex 7740 glass die [49]. The glass and silicon die were joined using anodic bonding at die level. This device demonstrated the main attractions of capacitive sensing, these being high sensitivity to pressure, low power consumption, and low temperature cross-sensitivity. The combination of materials and bonding mechanisms demonstrated remain a common choice for capacitive sensors [50, 51]. All silicon devices fabricated by silicon fusion bonding [52, 53] and glass frit bonding [54] have also been reported along with many surface-micromachined devices, which are discussed below. An example of an all-silicon fusion bonded device is a vacuum sensor developed by NASA [55]. This sensor uses a circular diaphragm and demonstrates a sensitivity of ~ 1 pF mbar^{-1}. Quartz has also been used to realize micromachined capacitive sensors [56]. This technology uses fused quartz components laser-welded together, and the fixed electrode is another diaphragm that is free to deflect but does not experience any pressure (see Figure 6.22). This means it is free to deflect under acceleration and will therefore move in the same manner as

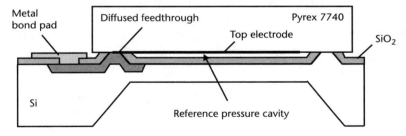

Figure 6.21 Early silicon/Pyrex capacitive pressure sensor.

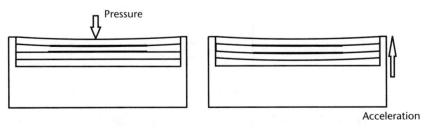

Figure 6.22 Acceleration compensated quartz capacitive pressure sensor.

the pressure-sensitive diaphragm. This technique greatly reduces the cross-sensitivity to accelerations.

The main drawbacks associated with the capacitive approach are the inherently nonlinear output of the sensor and the complexity of electronics (compared with the resistive bridge). Assuming parallel deflection in the flexible diaphragm, the change in capacitance is inversely proportional to the gap height. In addition to this, a basic diaphragm such as that shown in Figure 6.21 will bend as it deflects. The diaphragm will therefore no longer be parallel to the fixed electrode and this introduces a further nonlinearity in the sensor output. The use of bossed diaphragms will mitigate this effect to some degree [57, 58]. Another linearizing approach is to pattern the electrodes such that the sensing capacitance is measured from a particular part of the diaphragm. Maximum deflection occurs at the diaphragm center but this is also the location of maximum nonlinearity. By sensing the capacitance at an annulus removed a short distance from the diaphragm center, non-linearity is reduced but at the expense of sensitivity [59, 60]. Another approach, again at the expense of sensitivity, is to clamp the center of the diaphragm such that the pressure-sensitive structure becomes a ring shape. The sensitivity of such a structure is reported to be half that of an equivalent flat plate diaphragm, but nonlinearity falls to 0.7% FS [61]. The final approach commonly employed to improve linearity is to operate the sensor in touch mode, where the diaphragm touching the fixed electrode. The center of the diaphragm is bought into contact by a sufficient pressure, and as pressure increases an increasing area of the diaphragm touches the fixed electrode [62–64]. The output of such a sensor is more linear than that of a typical sensor operated in noncontact mode, as shown by the graph in Figure 6.23. One potential drawback of touch-mode devices is hysteresis arising from friction between the surfaces as they move together and apart, as well as the risk of stiction.

The increased circuit complexity associated with capacitive devices and the influence of parasitic capacitances on sensor performance has lead to the development of capacitive interface chips and further research into integrated sensor and circuit solutions. Capacitive interface chips have been designed by a number of manufacturers (including Microsensors Capacitive Readout IC MS3110, Analogue Microelectronics CAV414, Xemics XE2004, and Smartec's Universal Transducer Interface chip

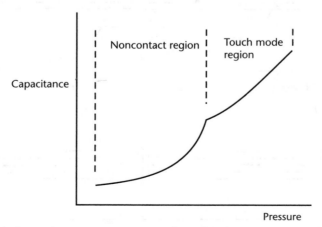

Figure 6.23 Typical capacitance versus pressure relationship for noncontact and touch-mode pressure sensors.

(UTI)). However, in order to reduce the effects of parasitic capacitance and achieve higher performance devices, the pressure sensor should ideally be integrated with electronics. This has been achieved by combining a bulk-etched device similar to that shown in Figure 6.21 with basic CMOS circuitry [65, 66], but the more common solution is to employ surface micromachining. Standard sacrificial surface micromachining processes have been combined with CMOS capacitance measurement circuitry in a number of devices [67–69]. A common theme with these sensors is the use of an array of sensing diaphragms to increase the measured capacitance signal. In some instances, diaphragms with different pressure sensitivities have been incorporated onto the same die in order to broaden the range of operation [70, 71]. A common application of capacitive pressure sensor arrays with integrated electronics is intravascular blood pressure measurement [72] and intracranial pressure [73]. This last device was coated in a silicon elastomer, NUSIL, for reasons of biocompatibility. A discussion of biocompatible coatings is included in Chapter 4.

Similar devices to the surface-micromachined pressure sensors have also been realized using SOI wafers [74]. These devices use the buried oxide as the sacrificial layer, and the hole to allow the undercutting etch is located at the center of the diaphragm. The hole is sealed afterwards by silicon nitride deposition, which results in a ring shaped diaphragm as described previously. The buried oxide also isolates the diaphragm from the surrounding silicon, thereby reducing parasitic capacitances. A cross-section of the device is shown in Figure 6.24.

Another more recent development is the integration of planar coils on the capacitive pressure sensor chip. The capacitor and coil form a resonant LC circuit the frequency of which varies with applied pressure. By integrating the coil on the sensor chip itself, it can also be used to inductively couple power into the sensor chip from an external coil. After energizing the sensor circuit, the external coil is used as an antenna to detect the resonant frequency. This approach is attractive for wireless sensing and can be used in applications where wire links are not suitable (e.g., harsh environments). Several devices have been reported in the literature from different research groups including two integrated devices using electroplated coils [75, 76] and a prototype microsystem on a ceramic substrate with a printed gold coil [64].

6.5.4   Resonant Pressure Sensors

Resonant pressure sensors typically use a resonating mechanical structure as a strain gauge to sense the deflection of the pressure-sensitive diaphragm. Resonant sensing has been discussed in Chapter 5. The resonant approach is more technically challenging for a number of reasons discussed below, but it does offer performance specifications beyond that achievable with piezoresistive and capacitive techniques.

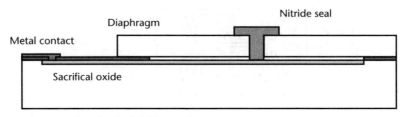

Figure 6.24   Cross-section through SOI capacitive pressure sensor.

Resonant pressure sensors have been successfully commercialized, and these shall be discussed in detail below to highlight the principles involved.

The technical challenges associated with resonant pressure sensors are as follows:

- Fabrication of mechanical resonator structure on top of pressure-sensing structure;
- In the case of silicon resonators, the incorporation of vibration excitation and detection mechanisms;
- The vacuum encapsulation of the resonator to negate gas-damping effects.

The earliest MEMS resonant pressure sensor was developed by Greenwood [77] and later commercialized by Druck [78]. A cross-section of the sensor is shown in Figure 6.25 along with a plan view of the resonator and its mode of vibration. The butterfly-shape resonator is attached via four arms to pillars that form part of the diaphragm. As the diaphragm deflects, the angle on the arms causes the resonator to be placed in tension and the resonant frequency to change. The two halves of the resonator are coupled together via a small physical link and the arms are positioned at node points in the optimum mode of operation. The resonator and diaphragm are fabricated using the boron etch stop technique and the resonator driven electrostatically and its vibrations detected capacitively via metal electrodes on the support chip. A vacuum is trapped around the resonator by mounting the support chip on a glass stem and sealing the end of the stem while in a vacuum. The assembly is then mounted by the stem, which provides some measure of isolation from packaging stresses (see Chapter 4). The resonator has a Q-factor of 40,000, and the sensor has a resolution of 10 ppm and total error of less than 100 ppm [79].

Another successfully commercialized device was developed by the Yokogawa Electric Corporation (DPharp, EJA series differential pressure sensor [80]). This consists of two resonators located on a diaphragm, the differential output of which provides the sensor reading [81]. The resonators are driven electromagnetically by placing the device in a magnetic field and running an alternating current through the structure. The pressure sensor arrangement is shown in Figure 6.26. The fabrication

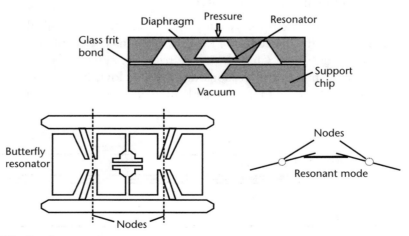

Figure 6.25 Druck resonant pressure sensor.

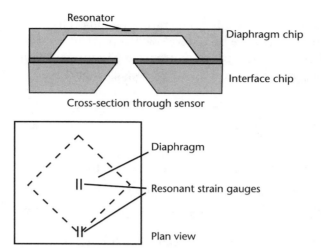

Figure 6.26   Yokogawa differential resonant pressure sensor.

process associated with this device is particularly impressive. The beams are vacuum encapsulated at wafer level using a series of epitaxial depositions, selective etches, and finally annealing in nitrogen, which drives the trapped gases left by the sealing process through the cavity walls or into the silicon. This leaves a final cavity pressure of below 1 mTorr, and the resonator possesses a Q-factor of more than 50,000 [82].

Other similar devices have been fabricated using a variety of techniques including silicon fusion bonding [83], surface-micromachined resonators on bulk etches diaphragms [84], and more recently using SOI wafer technology [85] and entirely surfaced-micromachined sensors [86]. Surface micromachining offers the opportunity for using comb-drive structures to excite and detect lateral resonances, but the polycrystalline materials used to fabricate the resonator are inferior to single crystal silicon. An alternative coupling mechanism to using the resonator as a strain gauge on the top surface of a diaphragm is to use a hollow structure open to the measurand. Changes in the applied pressure alter the shape of the resonator and hence the frequency shifts [87]. This approach means the media is in contact with the resonator and this introduces a cross-sensitivity to media density changes in which will shift resonant frequency in a manner indistinguishable from the pressure measurand. This device has also been used to demonstrate burst operation of the resonator, which involves exciting and detecting the vibrations at separate intervals [88]. Another pressure coupling mechanism has been demonstrated by Andrews et al. [89], where the measured pressure surrounds the resonator. Squeezed film damping effects, which vary with the pressure around the resonator, alter the resonant frequency. This device is designed as a vacuum sensor for use between 1 Pa and atmosphere.

Quartz is an attractive material for resonant applications given its piezoelectric properties and single crystal material properties. The piezoelectric nature of quartz simplifies the excitation and detection of resonant modes, and quartz is routinely used in high-stability time-based applications. The main drawback associated with quartz is the limited choice of micromachining options compared with silicon and the lack of suitability for integrating circuits. High-performance quartz resonant pressure sensors have been developed in particular for high-pressure applications in the oil and gas industry were accurate pressure measurement is essential [90, 91]. These devices are designed as capsules (see Figure 6.27), which place the resonating element at the center of the rigid structure designed to be squeezed hydrostatically. Their all-quartz construction makes them extremely stable. Resonating quartz diaphragms designed for lower pressures have also been developed [92].

6.5.5 Other MEMS Pressure Sensing Techniques

The main pressure sensing techniques have been discussed in some detail above, but there are other, less widely known approaches, which will be covered briefly here.

- Optical techniques typically employ a microsensor structure that deforms under pressure, this deformation producing a change in an optical signal. For example, diaphragm-based pressure sensors have been fabricated that incorporate optical waveguides on the top surface. Deflections in the diaphragm alter the phase of a light wave via the elasto-optic effect [93]. This is detected by having a reference waveguide unaffected by pressure and arranging the guides in a Mach-Zehnder interferometer [94, 95]. Another approach is to use Fabry-Pérot interferometers, which require a cavity. Micromachined diaphragms can be addressed by optical fibers with the gap between the fiber tip and silicon diaphragm forming the cavity [96]. An alternative approach involves actually fabricating the cavity on the end of the fiber itself [97]. Cavities can be etched into the end of the fiber and silicon diaphragms anodically bonded over the top.
- SAW resonators use surface acoustic waves excited on piezoelectric substrate (typically quartz) using interdigital (IDT) electrodes to detect pressure. A surface wave is excited at one end of the substrate and detected at the other. Applied strains can affect both the time of flight [98] and the frequency of the SAW [99]. By placing the SAW resonator on a pressure-sensitive structure, pressure can be measured. This approach is ideal for wireless, self-powered

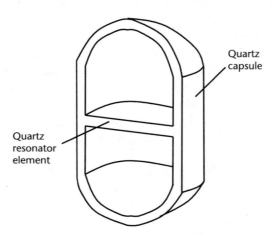

Figure 6.27 Cross-section of a quartz pressure-sensing capsule. (*After:* [90].)

sensing applications since they can be remotely operated by rf electromagnetic waves. This approach is being explored commercially for type pressure-sensing applications.

- MOS transistors can also utilize the piezoresistive effect to sense strain and therefore pressure [100]. The piezoresistive effect alters channel carrier mobility and therefore the characteristics of the transistor [101].

- Inductive coupling has also been used on a MEMS pressure sensing by micromachining two planar coils, one fixed beneath a diaphragm and the other located on top of the diaphragm. An ac current is applied through the primary coil on the diaphragm, and the induced current in the second coil varies with applied pressure [102].

- Force balance is an established sensing principle whereby and actuating force is applied to maintain the sensor structure in position during the application of the measurand. Electrostatic actuation has been applied to diaphragm structures for pressure-sensing applications. The actuating voltage required provides a measure of the applied pressure [103, 104]. This approach complicates the fabrication of the diaphragm since an actuation electrode is required in addition to the diaphragm deflection sensing mechanism. However, this approach can improve dynamic range and linearity [105].

6.6 Microphones

Microphones are a particular type of MEMS pressure sensor designed to transduce acoustic signals into electrical output. MEMS technology is an attractive approach for mass-producing miniature devices in, for example, hearing aid applications. Microphone diaphragms, or membranes, should be highly sensitive, exhibit suitable dynamic behavior, and be packaged so as to remain insensitive to static pressures [106]. Different membrane designs have been simulated and fabricated including corrugated [107] and even one based upon the ear of the parasitic fly *Ormia ochracea* [108]. This approach was adopted in order to mimic the directionality achieved by the fly's ear. Immunity to static pressures is typically achieved by ensuring both sides of the diaphragm are open to atmosphere, but only one side is subject to the incoming acoustic pressure waves. The microphone die is typically packaged within a chamber designed to tune the response of the diaphragm.

The distinction between types of microphone is typically based upon the sensing technology used to detect the membrane displacements. These can be summarized as follows.

- Capacitive microphones (also known as condenser microphones) are the most widely used form of MEMS device. They have demonstrated the highest achievable levels of sensitivity and very low noise levels [109]. These devices consist of a parallel plate-based capacitive pressure sensor with a flexible membrane positioned in close proximity to a fixed electrode. This fixed electrode, or backplate, is normally perforated with acoustic holes to minimize damping and ensure suitable dynamic characteristics. A schematic of a typical condenser microphone is shown in Figure 6.28 [110]. This device consists of a nitride/aluminum diaphragm and a boron-doped backplate with acoustic holes etched through. Membrane materials successfully used include polysilicon, both flat [111, 112] and corrugated [113, 114], nitride (as shown in Figure 6.28), and boron-doped silicon [115]. Other examples can be found in the literature [106, 109, 116] including differential devices [117], acoustic arrays of microphones [118], and hydrophones [119].

- Electret microphones are a form of capacitive microphone that utilizes a material that holds a permanent charge. This avoids the need to dc bias a capacitive device. The electret material is typically silicon dioxide, silicon nitride [107, 113], or Teflon [120]. Otherwise, the design and fabrication of these devices is very similar to those of the capacitive microphones.
- Piezoresistive microphones consist of thin diaphragms with four piezoresistors arranged as with standard piezoresistive pressure sensors described in Section 5.5.2 [121, 122]. These are not widely used due to their relatively low sensitivity.
- Piezoelectric microphones utilize a thin-film piezoelectric layer deposited on the top surface of a structure sensitive to acoustic pressures. As the structure deforms, charge is generated. Microphones are a dynamic sensing application and therefore well suited to piezoelectric sensing techniques. Example membrane-based devices include bulk etched silicon nitride membranes with thin-film ZnO and Al electrodes [123] or spin-coated P(VDF/TrFE) film [124], a boron-doped etch stop defined diaphragm coated with a sol-gel layer of PZT [125], and a nitride/parylene membrane with ZnO piezoelectric sensing elements [126]. This last device incorporates ZnO-coated cantilevers coated and integrated by a 1-μm-thick parylene layer that forms the membrane. Piezoelectric microphones based purely on cantilever structures have also been demonstrated. Cantilever structures are more compliant than membranes and are capable of larger displacements for a given acoustic pressure. The first device of this kind used a sputtered thin-film ZnO layer on a nitride cantilever [127], but later research on a similar structure demonstrated the improved sensitivity of PZT films [128]. Piezoelectric microphones are capable of comparable sensitivities to capacitive devices but suffer from higher noise levels.

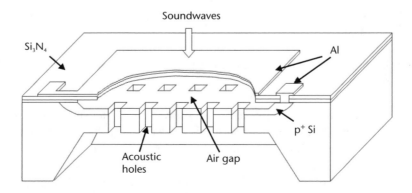

Figure 6.28 Typical condenser microphone. (*After:* [110].)

6.7 Conclusions

Pressure measurement is certainly one of the most mature applications of MEMS, and this chapter has described the many types of micromachined sensor that have been developed, both in industry and academia, over the years. The commercial success of this branch of MEMS serves as an excellent model for other potential MEMS applications. The market pull provided by the automotive industry—for example, for manifold air pressure sensors—has led to the development of successful devices and technologies that have benefited a wide range of other pressure sensing applications. This is made possible by the advantage of batch fabrication micromachining technologies capable of manufacturing sensors at very low unit cost. The importance of the material properties of silicon must also not be underestimated. Its inherent piezoresistive behavior facilitates resistive strain gauge pressure sensors, while, at the other end of the performance spectrum, its mechanical properties make it ideal for complex resonant-based pressure-sensing solutions. Micromachined pressure sensors are now an accepted, and in many instances, the preferred option in many pressure-sensing applications. As MEMS technology advances, fabrication processes become more capable, and a broader range of materials becomes available, micromachined pressure sensors will find many new opportunities.

References

[1]   Cook, J., "Fundamentals of Sensor Technology: Pressure," *Proc. Sensors Expo.*, Baltimore, MD, May 4–6, 1999, pp. 401–415.

[2]   Hagen, R., "Choosing the Right Low-Pressure Sensor," *Sensors*, September 1998, http://www.sensorsmag.com.

[3]   Di Giovanni, M., *Flat and Corrugated Diaphragm Design Handbook*, New York: Marcel Dekker, 1982.

[4]   Timoshenko, S., and S. Woinowsky-Krieger, *Theory of Plates and Shells*, New York: McGraw-Hill, 1959.

[5]   Van Vessem, P., and D. Williams, "Rediscovering the Strain Gauge Pressure Sensor," *Sensors Online*, Vol. 16, No. 4, April 1999.

[6]   Tufte, O. N., P. W. Chapman, and D. Long, "Silicon Diffused Element Piezoresistive Diaphragms," *J. Appli. Phys.*, Vol. 33, November 1962, pp. 3322–3327.

[7]   Sanchez, J. C., "Semi-Conductor Strain Gauge Pressure Sensors," *Instruments and Control Systems*, November 1963, pp. 117–120.

[8]   Peake, E. R., A. R. Zias, and J. V. Egan, "Solid-State Digital Pressure Devices," *IEEE Trans. on Electron Devices*, ED-16, No. 19, October 1969, pp. 870–876.

[9]   Mallon, J. R., et al., "Low-Pressure Sensors Employing Bossed Diaphragms and Precision Etch-Stopping," *Sensors and Actuators*, Vol. A21–23, 1990, pp. 89–95.

[10]   Sandmaier, H., "Non-Linear Analytical Modeling of Bossed Diaphragms for Pressure Sensors," *Sensors and Actuators*, Vol. A25–27, 1991, pp. 815–819.

[11]   van Mullem, C. J., K. J. Gabriel, and H. Fujita, "Large Deflection Performance of Surface Micromachined Corrugated Diaphragms," *Proc. Transducers '91, the 1991 Int. Conf. on Solid State Sensors and Actuators*, San Francisco, CA, June 24–27, 1991, pp. 1014–1017.

[12]   Jerman, J. H., "The Fabrication and Use of Micromachined Corrugated Silicon Diaphragms," *Sensors and Actuators*, Vol. A23, 1990, pp. 988–992.

[13]   Timoshenko, S. P., and S. Woinowski-Krieger, *Theory of Plates and Shells*, New York: McGraw-Hill, 1983.

[14] Stankevic, V., and C. Šimkevicius, "Use of a Shock Tube in Investigations of Micromachined Piezoresistive Pressure Sensors," *Sensors and Actuators*, Vol. 86, 2000, pp. 58–65.

[15] Entran EPIH series data sheet, http://www.entran.com.

[16] Entran EPO series data sheet, http://www.entran.com.

[17] Kanda, Y., "Graphical Representation of Piezoresistive Coefficients in Silicon," *IEEE Trans. Electron Devices*, ED-29, 1982, pp. 64–70.

[18] Clark, S. K., and K. D. Wise, "Pressure Sensitivity in Anisotropically Etched Thin-Diaphragm Pressure Sensors," *IEEE Trans. Electron Devices*, ED-26, 1979, pp. 1887–1896.

[19] Elgamel, H. E., "Closed-Form Expressions for the Relationships Between Stress, Diaphragm Deflection, and Resistance Change with Pressure in Silicon Piezoresistive Pressure Sensors," *Sensors and Actuators*, Vol. A50, 1995, pp. 17–22.

[20] Lin, L., H.-C. Chu, and Y.-W. Lu, "A Simulation Program for the Sensitivity and Linearity of Piezoresistive Pressure Sensors," *IEEE Journal of Microelectromechanical Systems*, Vol. 8, No. 4, December 1999, pp. 514–522.

[21] Goldman, K., et al., "A Vertically Integrated Media-Isolated Absolute Pressure Sensor," *Sensors and Actuators*, Vol. A66, 1998, pp. 155–159.

[22] Marek, J., and M. Illing, "Micromachined Sensors for Automotive Applications," *Proc. of IEEE Sensors 2002, 1st Int. Conf. on Sensors*, Vol. 2, Orlando, FL, June 12–14, 2002, pp. 1561–1564.

[23] Czarnocki, W. S., "Media-Isolated Sensor," *Sensors and Actuators*, Vol. A67, 1998, pp. 142–145.

[24] Matsuoka, Y., et al., "Low-Pressure Measurement Limits for Silicon Piezoresistive Circular Diaphragm Sensors," *J. Micromech. Microeng.*, Vol. 5, 1995, pp. 32–35.

[25] Sandmaier, H., and K. Kuhl, "A Square-Diaphragm Piezoresistive Pressure Sensor with a Rectangular Central Boss for Low Pressure Ranges," *IEEE Trans. on Electron Devices*, Vol. 40, No. 10, 1993, pp. 1754–1759.

[26] Kanda, Y., and A. Yasukawa, "Optimum Design Considerations for Silicon Piezoresistive Pressure Sensors," *Sensors and Actuators*, Vol. A62, 1997, pp. 539–542.

[27] Wu, X. P., "A New Pressure Sensor with Inner Compensation for Nonlinearity and Protection to Overpressure," *Sensors and Actuators*, Vol. 21, Issue 1–3, 1990, pp. 65–69.

[28] Johnson, R., et al., "A High-Sensitivity Ribbed and Bossed Pressure Transducer," *Sensors and Actuators*, Vol. A35, 1992, pp. 93–99.

[29] Dziuban, J., et al., "Self-Compensating Piezoresistive Pressure Sensor" *Sensors and Actuators*, Vol. A41–42, 1994, pp. 368–374.

[30] Boukabache, A., et al., "Characterization and Modelling of the Mismatch of TCRs and Their Effects on the Drift of the Offset Voltage of Piezoresistive Pressure Sensors," *Sensors and Actuators*, Vol. 84, 2000, pp. 292–296.

[31] Suzuki, K., et al., "Nonlinear Analysis of a CMOS Integrated Silicon Pressure Sensor," *IEEE Trans. Electron Devices*, ED-34, 1987, pp. 1360–1367.

[32] Akbar, M., and M. A. Shanblatt, "A Fully Integrated Temperature Compensation Technique for Piezoresistive Pressure Sensors," *IEEE Trans. Instrumentation and Measurement*, Vol. 42, No. 3, June 1993.

[33] Bryzek, J., et al., *Silicon Sensors and Microstrauctures*, Fremont, CA: Novasensor, 1990.

[34] Diem, B., et al., "SOI 'SIMOX': From Bulk to Surface Micromachining, a New Age for Silicon Sensors and Actuators," *Sensors and Actuators*, Vol. A46, 1995, pp. 8–16.

[35] Diem, B., et al., "'SIMOX': A Technology for High-Temperature Silicon Sensors," *Sensors and Actuators*, Vol. A23, 1990, pp. 1003–1006.

[36] von-Berg, J., et al., "A Piezoresistive Low-Pressure Sensor Fabricated Using Silicon-on-Insulator (SOI) for Harsh Environments," *TRANSDUCERS '01, EUROSENSORS XV, 11th International Conference on Solid-State Sensors and Actuators, Digest of Technical Papers*, Vol. 1, 2001, pp. 482–485.

[37] Kroetz, G. H., M. H. Eickhoff, and H. Moeller, "Silicon Compatible Materials for Harsh Environment Sensors," *Sensors and Actuators*, Vol. A74, 1999, pp. 182–189.

[38] Okojie, R. S., A. A. Ned, and A. D. Kurtz, "Operation of a (6H)-SiC Pressure Sensor at 500C," *Sensors and Actuators*, Vol. A66, 1998, pp. 200–204.

[39] Ned, A. A., R. S. Okojie, and A. D. Kurtz, "6H-SiC Pressure Sensor Operation at 600°C," *IEEE 4th Int. High Temperature Electronics Conference (HITEC)*, June 14–18, 1998, Albuquerque, NM, pp. 257–260.

[40] Ned, A. A., et al., "Dynamic Pressure Measurements Using Silicon Carbide Transducers," *19th IEEE Int. Congress on Instrumentation in Aerospace Simulation Facilities*, August 27–30, 2001, Cleveland, OH, pp. 240–245.

[41] Folkmer, B., P. Steiner, and W. Lang, "A Pressure Sensor Based on a Nitride Membrane Using Single-Crystalline Piezoresistors," *Sensors and Actuators*, Vol. A54, 1996, pp. 488–492.

[42] Lin, L., and W. Yun, "Design, Optimization and Fabrication of Surface Micromachined Pressure Sensors," *Mechatronics*, Vol. 8, 1998, pp. 505–519.

[43] Guckel, H., "Silicon Microsensors: Construction, Design and Performance," *Microelectronic Engineering*, Vol. 15, 1991, pp. 387–398.

[44] Sugiyama, S., et al., "Micro-Diaphragm Pressure Sensor," *Proc. IEEE IEDM*, Los Angeles, CA, 1986, pp. 184–187.

[45] Kälvesten, E., et al., "The First Micromachined Pressure Sensor for Cardio Vascular Pressure Measurement," *Proc. 11th IEEE Int. Workshop on MEMS*, June 25–29, 1998, Heidelberg, Germany, pp. 574–579.

[46] Lisec, T., M. Kreutzer, and B. Wagner, "Surface Micromachined Piezoresistive Sensors with Step-Type Bent and Flat Membrane Structures," *IEEE Trans. Electron Devices*, Vol. 43, No. 9, September 1996, pp. 1457–1552.

[47] Melväs, P., E. Kälvesten, and G. Stemme, "A Temperature Compensated Dual Beam Pressure Sensor," *Sensors and Actuators*, Vol. A100, 2002, pp. 46–53.

[48] Sander, C. S., J. W. Knutti, and J. D. Meindl, "A Monolithic Capacitive Pressure Sensor with Pulsed Period Output," *IEEE Trans. Electron Devices*, Vol. ED-17, May 1980, pp. 927–930.

[49] Lee, Y. S., and K. D. Wise, "A Batch Fabricated Silicon Capacitive Pressure Transducer with Low Temperature Sensitivity," *IEEE Trans. Electron Devices*, Vol. ED-29, No. 1, January 1982, pp. 42–48.

[50] Pons, P., and G. Blasquez, "Low-Cost High Sensitivity Integrated Pressure and Temperature Sensor," *Sensors and Actuators*, Vol. A41–42, 1994, pp. 398–401.

[51] Rogers, T., and J. Kowal, "Selection of Glass, Anodic Bonding Conditions and Material Compatibility for Silicon-Glass Capacitive Sensors," *Sensors and Actuators*, Vol. 46–47, 1995, pp. 113–120.

[52] Rudolf, F., and H. de Lambilly, "Low Cost Pressure Sensor Microsystems," *Proc. 4th Int. Conf. on Micro Opto Mechanical Systems*, Berlin, Germany, October 19–21, 1994, pp. 703–711.

[53] Goustouridis, D., et al., "A Miniature Self-Aligned Pressure Sensing Element," *J. Micromech. Microeng.*, Vol. 6, 1996, pp. 33–35.

[54] Hanneborg, A., et al., "A New Integrated Capacitive Pressure Sensor with Frequency Modulated Output," *IEEE Digest 1985 Int. Conf. on Solid State Sensors and Actuators (Transducers '85)*, June 11–14, 1985, Philadelphia, PA, pp. 186–188.

[55] Catling, D. C., "High-Sensitivity Silicon Capacitive Sensors for Measuring Medium-Vacuum Gas Pressures," *Sensors and Actuators*, Vol. A64, 1998, pp. 157–164.

[56] Mulkins, D. F., and Pogany, K., "Quartz Capacitive Pressure Sensor: Product Applications and Technical Description," *Sensors Expo West Proceedings*, 1989, pp. 206A/1–206A/12.

[57] Zhang, Y., and K. D. Wise, "An Ultra Sensitive Capacitive Pressure Sensor with Bossed Dielectric Diaphragm," *Technical Digest, 1994 Solid-State Sensor and Actuator Workshop*, Hilton Head, SC, 1994, pp. 205–208.

[58] Beeby, S. P., M. Stuttle, and N. M. White, "Design and Fabrication of a Low-Cost Microen-
      gineered Silicon Pressure Sensor with Linearized Output," *IEE Proc. Sci. Meas. Technol.*,
      Vol. 147, No. 3, May 2000, pp. 127–130.

[59] Pons, P., G. Blasquez, and R. Behocaray, "Feasibility of Capacitive Pressure Sensors without
      Compensation Circuits," *Sensors and Actuators*, Vol. A37–38, 1993, pp. 112–115.

[60] Hyeoncheol, K., Y.-G. Jeong, and K. Chun, "Improvement of the Linearity of Capacitive
      Pressure Sensor Using an Interdigitated Electrode Structure," *Sensors and Actuators*, Vol.
      A62, 1997, pp. 586–590.

[61] Omi, T., et al., "Capacitive Pressure Sensor with Center Clamped Diaphragm," *IEICE
      Trans. Electron.*, Vol. E80-C, No. 2, February 1997, pp. 263–268.

[62] Park, J. S., and Y. B. Gianchandani, "A Low Cost Batch Sealed Capacitive Pressure Sensor,"
      IEEE Tech Digest, 12th Intl. Conf. on MEMS, Orlando, FL, January 17–21, 1999,
      pp. 82–87.

[63] Wang, Q., and W. H. Ko, "Modeling of Touch Mode Capacitive Sensors and Diaphragms,"
      Sensors and Actuators, Vol. A75, 1999, pp. 230–241.

[64] Suster, M., D. J. Darrin, and W. H. Ko, "Micro-Power Wireless Transmitter for High Tem-
      perature MEMS Sensing and Communication Applications," *IEEE Tech Digest 15th Intl.
      Conf. on MEMS*, Las Vegas, NV, January 20–24, 2002, pp. 641–644.

[65] Matsumoto, Y., and M. Esasi, "An Integrated Capacitive Absolute Sensor," *Electronics and
      Communication in Japan*, Pt. 2, Vol. 76, No. 1, 1993, pp. 451–461.

[66] Kung, J. T., and H.-S., Lee, "An Integrated Air-Gap-Capacitor Pressure Sensor and Digital
      Readout with Sub-100 Attofarad Resolution," *IEEE J. MEMS*, Vol. 1, No. 3, September
      1992, pp. 121–129.

[67] Scheiter, T., et al., "Full Integration of a Pressure Sensor into a Standard BiCMOS Process,"
      Sensors and Actuators, Vol. A67, 1998, pp. 211–214.

[68] Paschen, U., et al., "A Novel Tactile Sensor System for Heavy-Load Applications Based on
      an Integrated Capacitive Pressure Sensor," *Sensors and Actuators*, Vol. A68, 1998,
      pp. 294–298.

[69] Trieu, H. K., N. Kordas, and W. Mokwa, "Fully CMOS Compatible Capacitive Differential
      Pressure Sensors with On-Chip Programmabilities and Temperature Compensation," *Proc.
      First IEEE Intl. Conf. on Sensors*, Orlando, FL, June 12–14, 2002, Vol. 2, pp. 1451–1455.

[70] Dudaicevs, H., et al., "Surface Micromachined Pressure Sensors with Integrated CMOS
      Read-Out Electronics," *Sensors and Actuators*, Vol. A43, Issues 1–3, May 1994,
      pp. 157–163.

[71] Chavan, A. V., and K. D. Wise, "A Monolithic Fully Integrated Vacuum Sealed CMOS Pres-
      sure Sensor," *IEEE Trans. Electron Devices*, Vol. 49, No. 1, January 2002, pp. 164–169.

[72] Kandler, M., et al., "A Miniature Single-Chip Pressure and Temperature Sensor," *J. Micro-
      mech. Microeng.*, Vol. 2, 1992, pp. 199–201.

[73] Hierold, C., et al., "Implantable Low Power Integrated Pressure Sensor System for Minimal
      Invasive Telemetric Patient Monitoring," *Proc. 11th IEEE Intl. Workshop on MEMS*,
      Heidelberg, Germany, June 25–29, 1998, pp. 568–573.

[74] Renard, S., et al., "Miniature Pressure Acquisition Microsystem for Wireless In Vivo
      Measurements," *Proc. 1st Annual Intl. IEEE-EMBS Special Topic Conf. on Microtechnolo-
      gies in Medicine and Biology*, Lyon, France, October 12–14, 2000, pp. 175–179.

[75] Akar, O., T. Akin, and K. Najafi, "A Wireless Batch Sealed Absolute Capacitive Pressure
      Sensor," *Sensors and Actuators*, Vol. A95, 2001, pp. 29–38.

[76] deHennis, A., and K. D. Wise, "A Double Sided Single Chip Wireless Pressure Sensor,"
      IEEE Tech Digest, 15th Intl. Conf. on MEMS, Las Vegas, NV, January 20–24, 2002, pp.
      252–255.

[77] Greenwood, J. C., "Etched Silicon Vibrating Sensor," *J. Phy. E. Sci. Instrum.*, Vol. 17,
      1984, pp. 650–652.

[78] Druck RPT (Resonant Pressure Transducer) Series Datasheet.

[79] Greeenwood, J., and T. Wray, "High Accuracy Pressure Measurement with a Silicon Resonant Sensor," *Sensors and Actuators*, Vol. A37–38, 1993, pp. 82–85.

[80] Harada, K., et al., "Various Applications of Resonant Pressure Sensor Chip Based on 3-D Micromachining," *Sensors and Actuators*, Vol. A73, 1999, pp. 261–266.

[81] Ikeda, K., et al., "Silicon Pressure Sensor Integrates Resonant Strain Gauge on Diaphragm," *Sensors and Actuators*, Vol. A21–23, 1990, pp. 146–150.

[82] Ikeda, K., et al., "Three-Dimensional Micromachining of Silicon Pressure Sensor Integrating Resonant Strain Gauge on Diaphragm," *Sensors and Actuators*, Vol. A21–23, 1990, pp. 1007–1010.

[83] Petersen, K., et al., "Resonant Beam Pressure Sensor Fabricated with Silicon Fusion Bonding," *Proc. 6th Intl. Conf. on Solid State Sensors and Actuators (Transducers '91)*, San Francisco, CA, June 1991, pp. 664–667.

[84] Welham, C. J., J. W. Gardner, and J. Greewood, "A Laterally Driven Micromachined Resonant Pressure Sensor," *Proc. 8th Int. Conf. on Solid State Sensors and Actuators (Transducers '95) and Eurosensors IX*, Stockholm, Sweden, June 25–29, 1995, pp. 586–589.

[85] Beeby S. P., et al., "Micromachined Silicon Resonant Stain Gauges Fabricated Using SOI Wafer Technology," *IEEE J. Microelectromechanical Systems*, Vol. 9, No. 1, March 2000, pp. 104–111.

[86] Melvås, P., E. Kälvesten, and G. Stemme, "A Surface Micromachined Resonant Beam Pressure Sensor," *IEEE J. Microelectromechanical Systems*, Vol. 10, No. 4, December 2001, pp. 498–502.

[87] Stemme, E., and G. Stemme, "A Balanced Dual-Diaphragm Resonant Pressure Sensor in Silicon," *IEEE Trans. Electron Devices*, Vol. 37, No. 3, March 1990, pp. 648–653.

[88] Melin, J., et al., "A Low-Pressure Encapsulated Deep Reactive Ion Etched Resonant Pressure Sensor Electrically Excited and Detected Using Burst Technology," *J. Micromech. Microeng.*, Vol. 10, 2000, pp. 209–217.

[89] Andrews, M. K., et al., "A Miniature Resonating Pressure Sensor," *Microelectronics Journal*, Vol. 24, 1993, pp. 831–835.

[90] Karrer, H. E., and J. Leach, "A Quartz Resonator Pressure Transducer," *IEEE Trans. Industrial Electronics and Control Instrumentation*, Vol. IECI-16, No. 1, July 1969, pp. 44–50.

[91] Besson, R. J., et al., "A Dual-Mode Thickness Shear Quartz Pressure Sensor," *IEEE Trans. Ultrasonics, Ferroelectrics and Frequency Control*, Vol. 40, No. 5, September 1993, pp. 584–591.

[92] Wagner, H.-J., W. Hartig, and S. Büttgenbach, "Design and Fabrication of Resonating AT-Quartz Diaphragms as Pressure Sensors," *Sensors and Actuators*, Vol. 41–42, 1994, pp. 389–393.

[93] Yamada, A., et al., "Relationship Between Sensitivity and Waveguide Position on Diaphragm for Silicon Based Integrated Optic Pressure Sensor," *IEEE Tech Digest 4th Pacific Rim Conf. on Lasers and Electro Optics*, Chiba, Japan, July 15–19, 2001, Vol. 1, pp. 1-420–1-421.

[94] Fischer, K., et al., "Elasto-optical Properties of SiON Layers in an Integrated Optical Interferometer Used as a Pressure Sensor," *IEEE J. Lightwave Tech.*, Vol. 12, No. 1, 1994, pp. 163–169.

[95] Benaissa, K., and A. Nathan, "IC Compatible Optomechancial Pressure Sensors Using Mach-Zehnder Interferometry," *IEEE Trans. Electron Devices*, Vol. 43, No. 9, September 1996, pp. 1571–1582.

[96] Kim, Y., and D. P. Neikirk, "Micromachined Fabry-Pérot Cavity Pressure Transducer," *IEEE Photonics Technol. Lett.*, Vol. 7, No. 12, December 1995, pp. 1471–1473.

[97] Abeysinghe, D. C., et al., "Novel MEMS Pressure and Temperature Sensor Fabricated on Optical Fibers," *J. Micromech. Micromeng.*, Vol. 12, 2002, pp. 229–235.

[98] Reindl, L., et al., "A Wireless AQP Pressure Sensor Using Chirped SAW Delay Line Structures," *Proc. IEEE 1998 Ultrasonics Symp.*, Sendai, Japan, October 5–8, 1998, Vol. 1, pp. 355–358.

[99] Buff, W., et al., "Universal Pressure and Temperature SAW Sensor for Wireless Applications," *Proc. 1997 Ultrasonics Symp.*, Toronto, Canada, October 5–8, 1997, Vol. 1, pp. 359–362.

[100] Canali, C., et al., "Piezoresistivity Effects in MOS-FET Useful for Pressure Transducers," *J. Phys. D:Appl. Phys.*, Vol. 12, 1979, pp. 1973–1983.

[101] Alcántara, S., et al., "MOS Transistor Pressure Sensor," *Proc. IEEE Int. Conf. on Devices, Circuits & Systems (ICCDCS '98)*, Venezuela, 1998, pp. 381–285.

[102] Okojie, R. S., and N. Carr, "An Inductively Coupled High Temperature Silicon Pressure Sensor," *Proc. 6th IOP Conf. on Sensors and Their Applications*, Manchester, England, September 12–15, 1993, pp. 135–140.

[103] Wang, Y., and M. Esashi, "The Structures for Electrostatic Servo Capacitive Vacuum Sensors," *Sensors and Actuators*, Vol. A66, 1998, pp. 213–217.

[104] Gogoi, B. P., and C. H. Mastrangelo, "A Low-Voltage Force Balanced Pressure Sensor with Hermetically Sealed Servomechanism," *Proc. 12th IEEE Intl. Workshop on MEMS*, Orlando, FL, January 17–21, 1999, pp. 493–498.

[105] Park, J.-S., and Y. B. Gianchandani, "A Servo-Controlled Capacitive Pressure Sensor Using a Capped-Cylinder Structure Microfabricated by a Three-Mask Process," *J. MEMS*, Vol. 12, No. 2, April 2003, pp. 209–220.

[106] Scheeper, P. R., et al., "A Review of Silicon Microphones," *Sensors and Actuators*, Vol. A44, 1994, pp. 1–11.

[107] Kressman, R., K. Klaiber, and G. Hess, "Silicon Condenser Microphones with Corrugated Silicon Oxide/Nitride Electret Membranes," *Sensors and Actuators*, Vol. A100, 2002, pp. 301–309.

[108] Yoo, K., et al., "Fabrication of Biomimetic 3-D Structured Diaphragms," *Sensors and Actuators*, Vol. A97–98, 2002, pp. 448–456.

[109] Miao, J., et al., "Design Considerations in Micromachined Silicon Microphones," *Microelectronics Journal*, Vol. 33, 2002, pp. 21–28.

[110] Kronast, W., et al., "Single-Chip Condenser Microphone Using Porous Silicon as a Sacrificial Layer for the Air Gap," *Sensors and Actuators*, Vol. A87, No. 3, 2001, pp. 188–93.

[111] Hsu, P.-C., C. H. Mastrangelo, and K. D. Wise, "A High Sensitivity Polysilicon Condenser Microphone," *Proc. 11th IEEE Intl. Workshop on MEMS*, Heidelberg, Germany, June 25–29, 1998, pp. 580–585.

[112] Brauer, M., et al., "Silicon Microphone Based on Surface and Bulk Micromachining," *J. Micromech. Microeng.*, Vol. 11, 2001, pp. 319–322.

[113] Zou, Q. B., et al., "Design and Fabrication of a Novel Integrated Floating-Electrode-'Electret'-Microphone (FEEM)," *Proc. 11th IEEE Intl. Workshop on MEMS*, Heidelberg, Germany, June 25–29, 1998, pp. 586–590.

[114] Li, X., et al., "Sensitivity Improved Silicon Condenser Microphone with a Novel Single Deeply Corrugated Diaphragm," *Sensors and Actuators*, Vol. A92, 2001, pp. 257–262.

[115] Kabir, A. E., et al., "High Sensitivity Acoustic Transducers with Thin p+ Membranes and Gold Back Plate," *Sensors and Actuators*, Vol. A78, 1999, pp. 138–142.

[116] Scheeper, P. R., et al., "Fabrication of Silicon Condenser Microphones Using Single Wafer Technology," *J. Microelectromechanical Systems*, Vol. 1, No. 3, September 1993, pp. 147–203.

[117] Rombach, P., et al., "The First Low Voltage, Low Noise Differential Silicon Microphone, Technology Development and Measurement Results," *Sensors and Actuators*, Vol. 95, 2002, pp. 196–201.

[118] Chowdhury, S., et al., "A MEMS Implementation of an Acoustical Sensor Array," *Proc. IEEE Int. Symp. on Circuits and Systems (ISCAS 2001)*, Sydney, Australia, May 6–9, 2001, Vol. 2, pp. 273–276.

[119] Bernstein, J., et al., "Advanced Micromachined Condenser Microphone," *Proc. 1994 Solid State Sensor and Actuator Workshop*, Hilton Head, SC, June 13–16, 1994, pp. 73–77.

[120] Hsieh, W., T.-Y. Hsu, and Y.-C. Tai, "A Micromachined Thin-Film Teflon Electret Microphone," *Proc. Intl. Conf. on Solid State Sensors and Actuators (Transducers '97)*, Chicago, IL, June 16–19, 1997, pp. 425–428.

[121] Kälvesten, E., L. Löfdahl, and G. Stemme, "Small Piezoresistive Silicon Microphones Specially Designed for the Characterization of Turbulent Gas Flows," *Sensors and Actuators*, Vol. A46–47, 1995, pp. 151–155.

[122] Huang, C., et al., "A Silicon Micromachined Microphone for Fluid Mechanics Research," *J. Mircomech. Microeng.*, Vol. 12, 2002, pp. 767–774.

[123] Ko, S. C., et al., "Micromachined Piezoelectric Membrane Acoustic Device," *Sensors and Actuators*, Vol. A103, 2003, pp. 130–134.

[124] Schellin, R., et al., "Micromachined Silicon Subminiature Microphones with Piezoelectric P(VDF/TrFE) Payers and Silicon Nitride Membranes," *Proc. IEE 8th Int. Symp. on Electrets (ISE 8)*, Paris, France, September 7–9, 1994, pp. 1004–1009.

[125] Bernstein, J., et al., "Micromachined Ferroelectric Transducers for Acoustic Imaging," *Proc. Intl. Conf. on Solid State Sensors and Actuators (Transducers '97)*, Chicago, IL, June 16–19, 1997, pp. 421–424.

[126] Han, C.-H., and E. S. Kim, "Fabrication of Piezoelectric Acoustic Transducers Built on Cantilever-Like Diaphragm," *Tech Digest, 14th IEEE Intl. Conf. on MEMS*, Interlaken, Sweden, June 21–25, 2001, pp. 110–113.

[127] Lee, S. S., and R. M. White, "Piezoelectric Cantilever Acoustic Transducer," *J. Micromech. Microeng.*, Vol. 8, 1998, pp. 230–238.

[128] Tian-Ling, R., et al., "Design Optimization of Beam-Like Ferroelectric-Silicon Microphone and Microspeaker," *IEEE Trans. Ultrasonics Ferroelectrics and Frequency Control*, Vol. 49, No. 2, February 2002, pp. 266–270.

CHAPTER 7

Force and Torque Sensors

Professor Barry E. Jones and Dr. Tinghu Yan

7.1 Introduction

In a highly mechanized world, force and torque are among the most important of all measured quantities [1–4]. They play a significant role in products from weighing machines and load cells used in industrial and retailing applications, to automotive and aerospace engines, screw caps on medicine bottles, and nut and bolt fasteners. Forces and torques can range from greater than 10 kN to less than 1 μN, and from 50 kNm to below 1 Nm, respectively. Measurement accuracy levels required can vary widely from, say, 5% to better than 0.01% of full scale ranges, depending on the application. Hysteresis and nonlinear effects in the mechanical structures of measuring devices need to be small, and measurement resolutions need to be high. Measurement devices need to be robust to withstand changing environmental influences such as temperature, vibration, and humidity, and they must also provide reliable measurement over long periods of time. Mechanical interfacing of the devices can be difficult and can influence final measurement. The forces and torques may change rapidly, and so the devices must have adequate frequency and transient responses.

There are several methods to measure forces and torques. Often, the force to be measured is converted into a change in length of a spring element. The change in dimensions is subsequently measured by a sensor, for example, a piezoresistive, a capacitive or a resonant sensor.

It is not so surprising, therefore, that most force and torque measurement devices utilize the long and well-established resistance strain gauge technology. Unfortunately, the metallic resistance strain gauge is relatively insensitive such that in use it is normal to obtain only several millivolts of analog voltage before amplification, and the gauges must not be significantly overstrained. The rangeability and overloading capabilities are seriously restricted. Also, the gauges consume relatively high electrical power (e.g., 250 mW).

In general, measurement instrumentation now needs smaller sensing devices of lower power consumption and with greater rangeability and overload capabilities. Greater compatibility with digital microelectronics is highly desirable. Noncontact and wireless operation is sometimes needed, and in some cases batteryless devices are desirable. Production of measurement devices using metallic resistance strain

153

gauges can be relatively labor intensive and skilled, and may require relatively inefficient calibration procedures.

In recent years some instrument manufacturers of force and torque measurement devices have moved away from using resistance strain gauges. Already, one leading manufacturer of weighing machines for retail and industrial applications now uses metallic and quartz resonant tuning fork technologies, and smaller companies have established niche markets using surface acoustic wave (SAW) technology, optical technology, and magnetoelastic technology.

Further commercial developments are taking place to enhance device manufacturability and improve device sensitivity and robustness in operation. Measurement on stiffer structures at much lower strain levels is now possible. The worldwide sensor research base is very active in exploring MEMS for sensing force and torque, and the rest of this chapter will review the current situation and future prospects.

7.2 Silicon-Based Devices

Strain gauges based on semiconductor materials such as silicon have been used for a long time, and although they are rather more expensive and more difficult to apply to a surface than metal strain gauges, their big advantage is a very high gauge factor of about ± 130, allowing measurement of small strain (e.g., 0.01 microstrain). It should be noted that the same factor for metal strain gauges is about 2. In semiconductor gauges most of the resistance change comes from the piezoresistance effect [5]. This gauge is rather nonlinear at comparatively high strain levels—that is, the gauge factor varies with strain. For example, if the gauge factor is 130 at 0.2% of strain, then it is about 112 at 0.4% of strain, which is the elastic limit of the gauge. Also, the gauge factor varies significantly with temperature about −0.15%/°C, which is more than 10 times worse than the metal gauges. This temperature sensitivity can be substantially reduced by using two gauges, each consisting of two pieces of semiconductor material having almost equal but opposite sign gauge factors. The two gauges are mounted with their axes at right angles on the member to be strained by a force and the four resistances are connected in the bridge as shown in Figure 7.1 [6], all these resistances have very similar temperature coefficients of resistance. The bridge output is proportional to strain, but little unbalance occurs due to

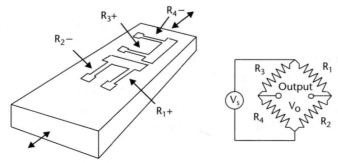

Figure 7.1 Temperature-compensated semiconductor strain gauges (the plus and minus signs indicate positive and negative gauge factors). (*From:* [6]. © 1977 B. E. Jones, Inc. Reprinted with permission.)

temperature change. Other gauge arrangements are also used. Semiconductor strip strain gauges can be very small, ranging in length from 0.7 to 7 mm, and having width typically a tenth of the element length; thus, they are useful in the measurement of highly localized strains.

In a diffused semiconductor strain gauge (Figure 7.2), an n-Si base has a p-Si diffused layer, and this layer works as a stress-sensitive conductor when its resistance is measured between leads attached to deposited metallizations. A cantilever with four C-shaped diffused gauges is stretched and compressed at its upper and lower surfaces, respectively, when the cantilever undergoes bending deformation under force F. All the gauges are identical since they are made on the same die and during the same technological cycle.

MEMS technology makes use of silicon as a mechanical structural material because of its excellent mechanical properties and the relative ease of fabricating in high volumes small mechanical devices by the process of micromachining [7, 8].

Silicon is an excellent piezoresistive material, with good mechanical properties. Amorphous silicon can be deposited directly on a mechanical part, for example, glass or plastics. The basic structure of such a sensor is shown in Figure 7.3 [9]. A thin amorphous silicon layer (n-, p-, or micro-compensated) acts as the sensitive area, with size $300 \times 300\,\mu$m, and four metallic contacts. Two of these contacts are used to apply a fixed current to the sensing element, while the other two, orthogonal to the previous ones, provide as output a voltage proportional to the mechanical stress. When a mechanical stress is applied, an anisotropic modification of resistivity occurs.

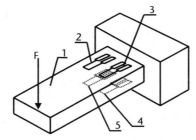

Figure 7.2 Cantilever integrated strain gauge element. F = force, 1 = cantilever, 2–5 = C-shaped strain gauge.

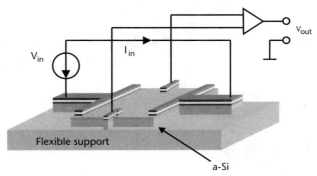

Figure 7.3 Structure of the sensing element. (*From:* [9]. © 2003 IEEE. Reprinted with permission.)

A silicon piezoresistive force sensor has been used in a tonometric transducer [10]. A plunger is positioned with silicon gel-like glue to press onto the force sensor. The other end of the plunger has a disposable protecting latex cap to touch the eyeball cornea.

The simultaneous use of silicon bulk-machined components and miniaturized high precision mechanical structures in a hybrid configuration can solve industrial measurement problems elegantly. As one example, a micro-torque sensor based on differential force for use in the watch industry has been developed [11] with a resolution better than 0.5 μNm over the range –200 to 200 μNm; it has a volume $3 \times 3 \times 1$ cm. The torque sensor is schematically represented in Figure 7.4. It consists of two piezoresistive force sensors. A 100-μm-thick spring blade made of copper beryllium and mounted perpendicular to the torque axis converts the torque to a force acting on the two force sensors. The force sensors are micromachined silicon cantilevers. A perpendicular bar mounted on the torque axis acts on the spring blade by way of two adjustable screws. The spring blade acts through two points on the two cantilever force sensors. A torque applied on the axis will increase the pressure on one force sensor and decrease the pressure on the other.

Load cells are force sensors that are used in weighing equipment [3]. In most conventional load cells the spring element is made from steel or aluminum, and metal resistance strain gauges are used as the sensor elements. Silicon does not suffer from hysteresis and creep, and therefore, a load cell made from silicon might be a good alternative to traditional load cells made from steel. Bending beam structures may be used for loads up to 150 kg, but for high loads, certainly above 1,000 kg, a load cell has to be based on the compression of silicon as shown in Figure 7.5 [7]. This sensor consists of two bonded silicon wafers. The edge of the sensor chip is compressed under the load, and the amount of compression can be measured by measuring the change in capacitance between two capacitor plates located in the center. An improved design to apply the load homogeneously will be discussed in Section 7.5. Another design of silicon load cell for loads up to 1,000 kg has been reported [12]. Besides large forces/torques, very small quantities can be sensed; a micro-torque sensor based on differential force measurement was reported more than 10 years ago [11].

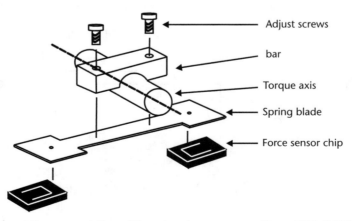

Figure 7.4 Schematic representation of the micro-torque sensor. (*From:* [11]. © 2003 IEEE. Reprinted with permission.)

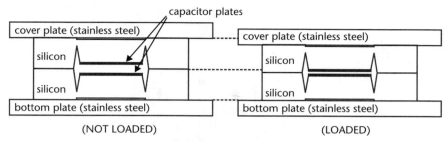

Figure 7.5   Principle of a load cell based on compression of silicon. (*From:* [7]. © 2001 Springer-Verlag Berlin Heidelberg. Reprinted with permission.)

7.3   Resonant and SAW Devices

Sensors utilizing a frequency shift as an output are highly attractive. They can be extremely sensitive and possess a wide dynamic range. The nature of the output signal makes these devices easy to integrate into digital systems and provides a reasonable immunity to noise. For these reasons, metallic and quartz tuning fork resonators have been successfully applied in industry [13–17], and sensors using bulk silicon technologies have also been demonstrated [18–21].

Recently, metallic digital strain gauges have been developed [22]. The metallic triple-beam resonator with thick-film piezoelectric elements to drive and detect vibrations is shown in Figure 7.6. The resonator substrate was fabricated by a double-sided photochemical-etching technique, and the thick-film piezoelectric elements were deposited by a standard screen-printing process. The resonator, 15.5 mm long and 7 mm wide, has a favored mode at 6.2 kHz and a Q-factor of 3,100, and load sensitivity about 13 Hz/N. Other means of resonator drive and detection are possible, for example, the use of an optical fiber to reflect light from a beam edge, and an electromagnetic drive [23].

A surface-micromachined force sensor using tuning forks as resonant transducers has been successfully demonstrated [24]. Figure 7.7 shows the basic design of a micromachined DETF. One end of the structure is anchored to the substrate and the other is left free for the application of an axial force. The dimensional design of the DETF determines the desired operating frequency and sensitivity [25]. In the center of each of the lines is an electrostatic transducer, such as a comb or parallel plate drive. When this tuning fork is used as an oscillator (lateral balanced mode), the

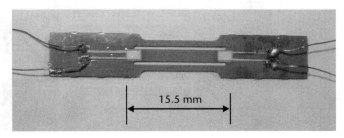

15.5 mm

Figure 7.6   Photograph of metallic resonator. (*From:* [22]. © 2003 IEE. Reprinted with permission.)

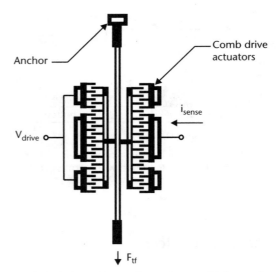

Figure 7.7 A basic tuning fork design using surface micromachining technology. (*From:* [24]. © 1995 ASME. Reprinted with permission.)

resulting frequency is a function of the applied force. The change in this frequency is the output of the device. The force sensor constructed used two tuning forks in a differential or push-pull structure, such that the output of the device was a shift in the frequency difference between them. This arrangement cancelled out temperature effects and allowed the force being measured to be amplified by mechanical leverage to the connection point of the two forks. In vacuum with closed loop feedback the fork frequencies were each close to 228 kHz and sensor sensitivity was about 4,300 Hz/μN.

A fully integrated silicon force sensor for static load measurement under high temperature has been demonstrated [26]. In this case load coupling, the excitation and detection of the vibration of the microresonator were integrated in one and the same single crystal silicon package. The complete single crystal design together with a single-mode optical fiber on-chip detection method should allow measurement to high temperatures well over 100°C. A perforated mass was suspended on two beams of 25-μm thickness and 0.5-mm length (Figure 7.8). Tests in a vacuum showed the

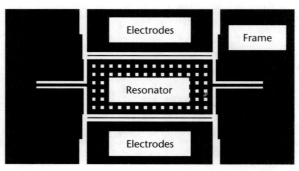

Figure 7.8 Resonant structure: perforated mass suspended on two beams. (*From:* [26]. © 2000 SPIE. Reprinted with permission.)

resonant structure vibrated with an amplitude of 100 nm in resonance at about 104 kHz with a Q of 30,000. Load sensitivity was about 4,000 Hz/N.

Relatively small SAW resonators can be used for noncontact torque measurement [27–35]. The sensitivity of SAW devices to strain is sufficient to perform measurements on a shaft that has not been weakened. Usually two SAW devices are used in one sensor, as shown in Figure 7.9 [30], and differential measurement of either phase delay or resonant frequency is performed in order to achieve temperature compensation and eliminate sensitivity to shaft bending. Both types of SAW sensors rely on the fact that the torque M applied to the shaft creates two principal components of strain, $s_{xx} = -s_{yy} = s$. As a result, one of the SAW devices is under tension and the other one is under compression, causing the opposite change of phase delay or resonant frequency in the devices. The resonators have the same or better performance for the same size of substrate and are less demanding in terms of the receiver bandwidth and sensitivity. Resonator Q factors are about 10,000. The torque sensor interrogation system can employ continuous frequency tracking of reflected frequencies from the two SAW resonators, having slightly different frequencies, for example, 200 and 201 MHz. For torque of ±10 Nm, and using ST-X quartz SAW resonators, device sensitivity to torque at room temperature has been measured as 4.65 kHz/Nm. This torque sensitivity has a temperature coefficient of 0.2%/°C. Therefore the sensor needs to measure both torque and temperature to allow for the temperature compensation of the measured results. SAW devices can break if the strain in the substrate is more than approximately 1,500 microstrain. If the sensor has to withstand a 30-fold overload, then the nominal strain can be equal to 50 microstrain. As a consequence, interrogation error gives torque measurement error of about 1%.

7.4   Optical Devices

Measurement of torque has always been an important challenge for numerous industries like aerospace and automotive. In particular there is increasing interest in electric power-assisted steering (EPAS) systems among vehicle manufacturers and component suppliers [36–39]. One of the key components of an EPAS system is a torque sensor with a basic specification as follows: torque measuring range of around ±10 Nm, an overload torque capability (nonmeasuring) of about ±110 Nm, and maximum rotational speed of around 90 rev/min. The sensor must meet the appropriate environmental and electromagnetic compatibility (EMC)

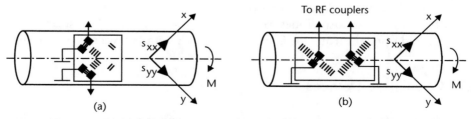

Figure 7.9   Torque sensing element based on (a) SAW reflective delay lines and (b) SAW resonators. (*From:* [30]. © 2003 IEEE. Reprinted with permission.)

specifications, and optical methods are largely immune to such interference effects. Also, a noncontacting sensor is desirable to improve reliability.

One commercial system [40, 41] employs an optical sensor to measure the relative angular movement between the ends of a torsion bar as torque is applied to the shaft. The torsion bar is a compliant portion of shaft designed to increase the angular displacement for a given torque—for example, approximately 2.5° of twist for 10 Nm of applied torque. Two rotating discs are fitted to the shaft; one mechanically links to each end of the torsion bar. Each rotating disc has a pattern of slots forming two tracks of alternating transparent and opaque zones. The optical arrangement is such that the light intensity received relates to torque, the direction of torque can be determined, and the variation in light source intensity is eliminated from the torque sensor output signal. The device has a 5% accuracy level, which is adequate for this application.

A sensitive optical torque sensor based on the optical measure of the torsion angle by using a sensitive polarimetric method has been demonstrated [42]; the torsion angle has been measured with 0.001° accuracy. A birefringent torque sensor [43] uses a photoelastic polymer illuminated by polarized light and experiencing torsional strain. A two-dimensional fringe pattern is viewed through an optical polarizer. The strain that causes this observed image pattern is a complex function of the torque applied to the shaft. A neural network can be trained with the fringe patterns to activate the applied torque for both static shafts (accuracy level 1%) and rotating shafts up to 1,500 rev/min (accuracy level 4%).

An optical torque sensor has been demonstrated that uses a modified moiré fringe method for measuring torque in a rotating shaft [44]. The system utilizes two circular gratings mounted concentrically at either end of the torsion shaft; one grating is ruled radially, and the other has rulings that are tangential to a small central generating circle. Circular fringes are formed that move radially as the angular displacement between the gratings charges; the fringe movement may be read out using a CCD array.

Two 11-bit absolute position optical encoders have been used for torque measurement with the torsion bar [45]. Seven additional bits provide interpolation (relative) position determination between adjacent codes. An ASIC computes the two optically encoded positions for speeds up to 1,600 rev/min. A toolholder torque sensor (spindle-mounted dynamometer) using an optical method has been developed [46]. Torque is detected by monitoring the change of clearance of a V-notch between two flanges, with a focused light ray from the side of the cluck.

A miniaturized optoelectronic torque sensor with hysteresis and nonlinearity less than 1% for maximum torque measurement of 0.15 Ncm in both directions has been developed [47]. The torsion of a cross-spring-bearing moves a precision slit in front of twin-photo diodes, transforming the torque into an electric signal. An integrated optical torque measurement microsystem has been reported [48].

7.5   Capacitive Devices

Noncontact torque measurement on a rotating shaft can be achieved by the use of a capacitive sensor, although susceptibility of the telemetry circuits to radio frequency and electromagnetic interference can be a problem [49, 50]. Torque can be measured by the use of a set of electrodes on one end of a torsion bar connected to a shaft and a second set of electrodes on the outside of a thin tube of dielectric material [51]. This tube is fixed to the shaft at the other end of the torsion bar. One set of electrodes moves with respect to the second set when torque is applied such that there is capacitance variation between the two sets of electrodes. The capacitance is part of a resonance circuit inductively linked to a coil on the stationary part. Each end of a torsion bar can have a noncontact capacitive angular displacement sensor, and the torque twist is monitored by electrical phase change.

Silicon micromachining has been used to realize a differential capacitive force sensor [52]. The principle of the variable gap force sensor is shown in Figure 7.10. The capacitors are made out of two electrically isolated thin plates (electrodes) with a very small distance between them. If one capacitance increases, the other decreases. A differential reading of the capacitance ensures better linearity and higher sensitivity. A gap of $10\,\mu$m gives a capacitance of 1 pF. Force measurements in the range 0.01N to 10N can be made, and the sensing element could be used as a tactile sensor or in the field of nanorobotic technology.

A micromachined silicon load cell has been developed for measuring loads up to 1,000 kg [53]. The sensing surface contains a matrix (or array) of capacitive sensing elements to make the load cell insensitive to nonhomogeneous load distributions. A schematic diagram of the load cell is shown in Figure 7.11. The design is realized in two wafers that are bonded on top of each other. The load is now applied to an area of 1 cm^2 in the center of the chip. In this area, the top wafer contains an array of poles that bear the load. The bottom wafer contains an electrode pattern that forms an array of capacitors with the top wafer as a common electrode. On application of a load, the poles will be compressed and the distance between the metal electrodes and the top wafer at the position of the capacitors will decrease, thereby increasing the capacitance. Each capacitance is measured individually and the total capacitance is given as the sum of the reciprocal values of the individual capacitors. In this case the total load, even if not distributed homogeneously, is obtained. There are 25

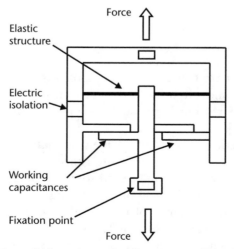

Figure 7.10 Principle of the variable gap capacitor force sensor. (*From:* [52]. © 1993 IOP Publishing Ltd. Reprinted with permission.)

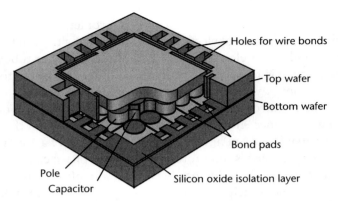

Figure 7.11 Distributed capacitive load cell. (*From:* [53]. © 2003 IEEE. Reprinted with permission.)

poles each of diameter 2 mm, height 200 μm, resulting in a change in height of 0.2 μm at a load of 1,000 kg. The distance between capacitor plates is 1 μm, and capacitance values are of the order of a few picofarads. Repeatability better than 0.05% has been achieved with a design having a larger number of smaller capacitors. The design can be easily adapted for higher loads by increasing the chip area or using multiple chips in a single package.

Three-dimensional microfabrication for a multidegree-of-freedom capacitive force sensor using optical fiber-to-chip coupling has been reported [54]. The sensor has been designed to operate in the 0- to 500-μN force range and the 0- to 10-μNm torque range. The intended application of this sensor is to obtain force-feedback during micromanipulation of large egg cells or during sperm injection. An elastically suspended rigid body is used, which is capable of moving in all six degrees of freedom when coupled to a glass fiber. Nonsymmetric comb capacitors allow for decoupling between displacements in the x and y directions. The z direction can be sensed through planar electrodes under the chip.

7.6 Magnetic Devices

Torque sensors are generally big components. In most cases, the shaft where the torque is to be measured has to be cut to install the torque sensor in between the resulting two parts. Furthermore, the signal is transmitted by slip rings or a coaxial transformer [1].

The magnetic head type of torque sensor allows the shaft to remain as one part and to receive the signal without slip rings [55–60]. This principle is based on the strong magnetostrictive properties of some ferromagnetic materials like amorphous alloy CoSiB ribbons. The ribbon has strong magnetoelastic properties and transforms torque into a change of permeability μ. A schematic of the arrangement is shown in Figure 7.12 [56]. Installed above the ribbon, a sensor head made of a ferromagnetic yoke with exciting and induction coils allows detection of the change in permeability caused by mechanical stress, without contacting the ribbon. A problem

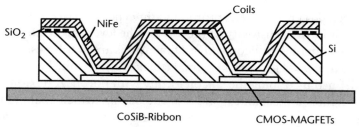

Figure 7.12 Micromachined sensor head. (*From:* [56]. © 1995 IOP Publishing Ltd. Reprinted with permission.)

arises with the variation of the air gap between the sensor head and the amorphous ribbon attached to the shaft. These air gap variations are typically $\pm 10 \ \mu m$, and when torque is applied, there is modulation of the signal of the induction coil.

A very small micromachined sensor head has been designed to eliminate this modulation of the signal by air gap variation, by means of a differential measuring method [57]. The magnetic flux density is directly measured in the air gap under the pole pieces and in the stray field with CMOS-MAGFETs because the signal of induction coils is too small. This device consists of CMOS-MAGFETs at the front side of a silicon wafer, and planar exciting coils, combined with a ferromagnetic yoke at the rear side of the wafer. To measure torque, two sensor heads are repositioned with an angle $\pm 45°$ to the shaft axis. These directions show the maxima of tensile and compressive stress on the shaft. Torque is proportional to the tensile stress.

Improvements have been made by replacing MAGFETs with magnetic field resistors (MAGRES), which have lower noise levels [60]. This torque sensor system is fabricated in silicon—planar coils, NiFe yoke, magnetic sensors—all integrated on one chip, thus avoiding common problems like telemetry or bulky designs.

Microfabricated ultrasensitive piezoresistive cantilevers for torque magnetometry have been investigated [61]. The basic arrangement is shown in Figure 7.13 and

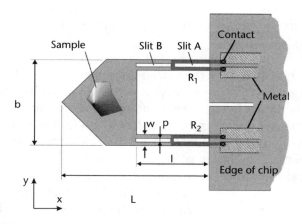

Figure 7.13 Schematic diagram of piezoelectric cantilever for torque magnetometry: two-leg cantilever with two piezoresistors for measuring bending and torsion independently. (*From:* [61]. © 1999 Elsevier Science. Reprinted with permission.)

has been designed to detect torque with respect to two axes, x and y. The legs have slits to improve sensitivity of the levers. The levers protrude from a carrier chip that is etched out of the silicon wafer. Dimensions are as follows: $L = 200\,\mu m$, $l = 100\,\mu m$, $w = 10\,\mu m$, $p = 4\,\mu m$, $b = 117\,\mu m$, t (thickness) $= 5\,\mu m$. Two piezoresistors R_1 and R_2 are defined by doping the silicon locally with boron. A flexion and/or torsion of the lever creates a mechanical stress in the beams, which changes the resistance of the piezoresistors by ΔR owing to the piezoresistive effect. The torque about the x-axis can be extracted by measuring $\Delta(R_1 - R_2)$. The torsion creates stress with the opposite sign symmetrically around a location at the middle of the lever. Sensitivity ($\Delta R/R$) to torsion has a value 1.5×10^{-6} per pNm. Piezoresistance values are typically 2 to 3 kΩ and resonance frequency is about 78 kHz, and so the device has a short response time. Sensitivity is high (up to $\approx 10^{-14}$ Nm). An external magnetic field applied to the sample having a magnetic moment generates a torque on it and to the cantilever.

7.7 Atomic Force Microscope and Scanning Probes

There is a growing need to measure and characterize finer and finer surfaces. This requirement imposes considerable demands on the instruments that measure and characterize these surfaces. The scanning force microscope (SFM), which includes the atomic force microscope (AFM), has become a well-established technique for the analysis of surfaces. Basically, a cantilever either dynamically in vibration scans across a sample surface or scans across the surface in a static contacting mode. The cyclic contact SFM may not damage the surface of soft samples as does the contact SFM. Miniaturized standalone SFMs are needed for use in wafer inspection, ultra-high vacuum SFM, and liquid environments. The cantilever deflection sensing and alignment maintenance arrangements during scanning need to be small. Force-sensing cantilevers for miniaturized SFMs include the following: the piezoresistive type, the piezoelectric type, and the capacitive type. The piezoelectric cantilever can perform the actuation of z-axis tip-sample spacing by a superimposed dc voltage, when the cantilever executes the self-force sensing at the same time. For the miniaturized dynamic SFMs, the use of the piezoelectric cantilever enables the necessary components to become just one piezoelectric microcantilever and an x-y axes scanner [62].

A schematic diagram of a cyclic contact SFM with a PZT force sensor is shown in Figure 7.14 [62], and the silicon micromachined PZT force sensor is shown as well ($200\,\mu m$ long, $50\,\mu m$ wide, thickness $4\,\mu m$, PZT thin-film layer $1\,\mu m$ thick). For resonant vibration in air, the viscous and acoustic damping is predominant and the micro-cantilever quality factor Q is 200 with a resonance frequency of 60 kHz. Cantilever sensitivity has a value 0.6 fC/nm and a cantilever spring constant 5 N/m. Vertical amplitude resolution of such a microcantilever system is about 0.2nm (2Å). Vibrational amplitude range is about 100 nm.

The magnetic force microscope (MFM) is widely used as a simple technique for the investigation of stray fields at the surface of magnetic samples with submicron resolution. A complementary technique, which is both noninvasive and quantitative, is the scanning hall probe microscope (SHPM), which is capable of magnetic

Micromachined
PZT force sensor

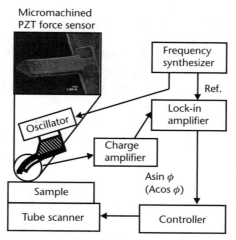

Figure 7.14 Schematic diagram of a cyclic contact SFM with a PZT force sensor using the piezoelectric charge detection method. (*From:* [62]. © 1997 American Institute of Physics. Reprinted with permission.)

imaging at spatial resolutions down to 150 nm. A new type of SHPM is based on the piezoresistive AFM [63]. Piezoresistive AFM cantilevers are commonly fabricated from p-type Si. The new SHPM cantilever is shown in Figure 7.15. The plan view shows the two primary sensors required for the dual magnetic and topographic imaging. The first sensor, a Hall cross-situated near the end of the cantilever, is electrically contacted via the four gold leads at either side of the cantilever. The piezoresistor is placed at the base of the cantilever where bending stresses are at a maximum. At the very end of the cantilever is a sharp (<100-nm diameter) AFM tip, which, by inclining the cantilever, is used to map the sample surface. With a

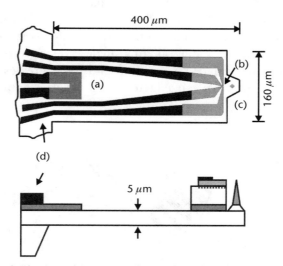

Figure 7.15 Plan and side view of SHPM cantilever (vertical axis not to scale). (a) The piezoresistor is fabricated at the cantilever base. (b) The Hall probe and (c) the tip are fabricated at the very end of the cantilever. (d) The Hall probe and the piezoresistor are electrically contacted via Au/Ge alloyed contacts. (*From:* [63]. © 2003 IOP Publishing Ltd. Reprinted with permission.)

drive amplitude of 1 nm, the resonant frequency of the cantilever was 20 kHz with a Q-factor of 400 in air.

Chemical imaging as well as topographical information of solid surfaces can now be undertaken using SFMs [64]. A micromachined integrated sensor for combined AFM and near-field scanning optical microscopy (NSOM) has been reported [65, 66]. This sensor consists of a microfabricated cantilever with an integrated waveguide and a transparent near-field aperture tip.

7.8   Tactile Sensors

The intensified miniaturization of devices requires an appropriate handling of microparts during fabrication and assembly. Indeed, investigation at the atomic scale level needs more accurate sample manipulation by means of a "nanorobot" having, say, resolution of 10 nm and a 1-cm^3 working space. Micromachined grippers are required, and when the gripper clamps or touches an object, force sensing would be a great advantage. A tactile microgripper with both actuation and sensing integrated has been developed [67]. A thermal bimorph actuator and piezoresistive force sensor are used. A 6-μm-thick, 250-μm-wide silicon beam finger has a 300-μm stroke and time constant of 11 ms. Gripping force is about 250 μN, and this is sensed by diffused boron piezoresistors in a Wheatstone bridge.

A silicon micromachined piezoelectric tactile sensor has been integrated on to the tip of an endoscopic grasper used by a surgeon to manipulate tissue [68]. The grasper has the usual rigid tooth-like surface (Figure 7.16). It consists of upper silicon, a perspex substrate, and a patterned polyvinylidene fluoride (PVDF) film that is sandwiched between the two layers. Force dynamic range is 0.1N to 2N with a resolution of 0.1N and bandwidth from near dc to several megahertz.

The silicon substrate used for micromachined tactile sensors is rigid and mechanically brittle, and therefore not capable of sustaining large deformation and sudden impact. Recently a two-dimensional tactile sensor array based solely on polymer (polyimide) micromachining and thin-film metal resistors has been demonstrated [69]. A schematic diagram of a single taxel is given in Figure 7.17. The magnitude of in-plane surface stresses is found to be greatest at the periphery of the membrane. The effective gauge factor of the taxels is approximately 1.3.

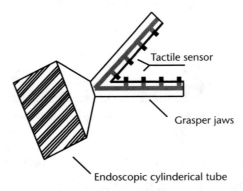

Figure 7.16   Endoscopic grasper with integrated tooth-like tactile sensor. (*From:* [68]. © 2003 IEEE. Reprinted with permission.)

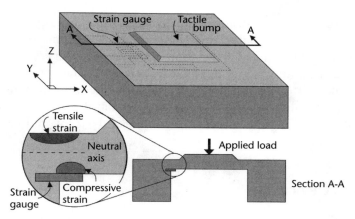

Figure 7.17 Schematic diagram of a single taxel. (*From:* [69]. © 2003 IOP Publishing Ltd. Reprinted with permission.)

A 5 × 5 silicon micromachined tactile sensor array for the detection of extremely small force (micro-Newton range) has also been demonstrated recently [70]. The central contacting pads are trampoline-shaped suspended structures with a piezoresistive layer of polysilicon embedded in each of the four sensor beams to detect the displacement of the suspended contacting pad. Each square tactile has dimension of 200 × 200 μm with 250-μm center-to-center spacing. The entire sensor area is 1.25 × 1.25 mm. Each of the sensor beams has dimension 90 μm long and 10 μm wide, while the central square contacting plate is 40 × 40 μm. Linear sensitivity of the 10-kΩ sensors is about 0.02 mV/μN at the output of a Wheatstone bridge with a single sensor and 10-V excitation.

The precise and inexpensive measurement of multiple-axis displacements and forces is an important concern for microsystems, which include very small mechanical structures that execute complex motions. A six-axis (three translations and three rotations) tactile sensor has been demonstrated employing microfield-emitting and detecting elements on separate chips [71]. The basic structure is shown in Figure 7.18. The field emitter is a cross-shaped conductor driven with alternating current. The detector array consists of eight open-gated FETs, each of

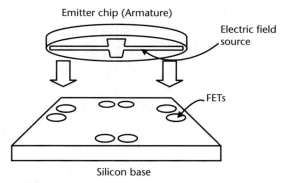

Figure 7.18 Schematic diagram of a multiaxial strain transducer. (*From:* [71]. © 1990 ASME. Reprinted with permission.)

which produces a drain current proportional to the charge induced in its gate by the local electric field. Based upon the pattern of outputs of the field-detecting elements, the position and orientation of the field emitters relative to the field detector array can be inferred, with some redundancy. The compliance properties of the elastomer material separating the two chips determine the sensitivities to forces and torques. Resolutions of 75 nm translational in the x-y plane and of $0.01°$ rotation about the z-axis seem possible. A packaged device would have dimensions of the order $5 \times 5 \times 5$ mm.

7.9 Future Devices

It will be evident from the review undertaken in this chapter that MEMS technologies are already playing a significant role in force and torque measurements.

Besides industrial weighing, power-assisted steering, shaft power, force/torque drives, and fastener fixing, sensors are used in a number of advanced applications such as force microscopy, dexterous and dynamic manipulation of fragile or irregular objects with microgrippers [67] and robotic hands [69]. "Smart" structures in aerospace [72] need low-cost integrated force/torque sensor systems, and microsystems technologies will be at the forefront of new sensor developments. New approaches [73] and new applications are being established all the time [74].

Miniaturization and integration of several technologies, such as silicon micromaching, microelectronics, optical fibers, and thin/thick-films, will contribute to these developments. MEMS actuation will be combined with the MEMS sensors to develop miniature MEMS feedback control systems to control small forces and torques [75]. Arrays of devices will provide two- and three-dimensional capabilities. Resonant force/torque sensors requiring only very low levels of power will continue to be developed for noncontact/remote applications, and batteryless/wireless/autonomous devices should become available. Small electrical energy generators operating on available local vibration will also be used.

References

[1]   Schicker, R., and G. Wegener, *Measuring Torque Correctly*, Darmstadt, Germany: Hottinger Baldwin Messtechnik (HBM) GmbH, 2002.

[2]   Garshelis, I. J., "Torque and Power Measurement," in *The Mechatronics Handbook*, R. H. Bishop, (ed.), Boca Raton, FL: CRC Press, 2002, pp. 19.48–19.61.

[3]   *Guide to the Measurement of Force*, London, England: The Institute of Measurement and Control, 1998.

[4]   *Resistance Strain Gauge Load Cells NMP7*, Watford, England: Training Publications Ltd., 2002.

[5]   Kanda, Y., "Graphic Representation of the Piezoresistance Coefficients in Silicon," *IEEE Trans. on Electron Devices*, Vol. 29, No. 1, 1982, pp. 64–70.

[6]   Jones, B. E., *Instrumentation, Measurement, and Feedback*, New York: McGraw-Hill, 1977.

[7]   Elwenspoek, M., and R. Wiegerink, *Mechanical Microsensors*, Berlin, Germany: Springer-Verlag, 2001, pp. 97–106.

[8]   Sharpe, W. N., "Mechanical Properties of MEMS Materials," in *The MEMS Handbook*, M. Gad-el-Hak, (ed.), Boca Raton, FL: CRC Press, 2002, pp. 3.6–3.16.

[9]   Biagiotti, L., et al., "A New Stress Sensor for Force/Torque Measurements," *Proc. IEEE Int. Conf. on Robotics and Automation*, Vol. 2, Washington, D.C., May 11–15, 2002, pp. 1655–1660.

[10]   Hachol, A., and R. Dzik, "Metrological Properties Study of a Planatic Tonometric Transducer Operating with Micromachined Force Sensor," *Conference Optoelectronic and Electronic Sensors II*, Szczyrk, Poland, May 13–16, 1996, *SPIE*, Vol. 3054, 1996, pp. 104–110.

[11]   Gass, V., et al., "Micro-Torque Sensor Based on Differential Force Measurement," *Proc. IEEE Workshop on Micro Electro Mechanical Systems*, Oiso, Japan, January 25–28, 1994, pp 241–244.

[12]   Wiegerink, R., et al., "Quasi-Monolithic Silicon Load Cell for Loads up to 1000 kg with Insensitivity to Non-Homogeneous Load Distributions," *Sensors and Actuators*, Vol. A80, No. 2, 2000, pp. 189–196.

[13]   Shinko Denshi, "Vibra," http://www.atimco.dk/shinko_denshi/hvorforstemmegaffel.htm, November 2003.

[14]   Weigh-Tronix, "Quartzell Technology," http://www.wt-nci.com/innovations.html, November 2003.

[15]   Whitehead, N., B. E. Jones, and D. Rees, "Non-Contact Torque Measurement on a Rotating Shaft Incorporating a Mechanical Resonator," in *Sensors and Their Applications X*, N. M. White and A. T. Augousti, (eds.), Bristol, England: Institute of Physics Publishing, 1999.

[16]   Hauptmann, P., "Resonant Sensors and Applications," *Sensors and Actuators*, Vol. A26, No. 1–3, 1991, pp. 371–377.

[17]   Bailleu, A., and W. Thelen, "Doppelplattenresonanzsensor zur Kraftmessung," *Conf. SENSOR 97*, Nuremberg, Germany, May 13–15, 1997, Vol. A5.1, pp. 7–12.

[18]   Greenwood, J. C., "Silicon in Mechanical Sensors," *J. Phys. E: Sci. Instrum.*, Vol. 21, 1988, pp. 1114–1128.

[19]   Greenwood, J. C., "Miniature Silicon Resonant Pressure Sensor," *IEE Proc.*, Pt. D, Vol. 135, No. 5, 1988, pp. 369–372.

[20]   Tilmans, H. A. C., M. Elwenspoek, and J. H. J. Fluitman, "Micro Resonant Force Gauges," *Sensors and Actuators*, Vol. A30, No. 1–2, 1992, pp. 35–53.

[21]   Esashi, M., "Resonant Sensors by Silicon Micromachining," *Proc. IEEE Int. Frequency Control Symposium*, Honolulu, HI, June 5–7, 1996, pp. 609–614.

[22]   Yan, T., et al., "Thick-Film PZT-Metallic Triple Beam Resonator," *Electronics Letters*, Vol. 39, No. 13, 2003, pp. 982–983.

[23]   Spooncer, R. C., B. E. Jones, and G. S. Philp, "Hybrid and Resonant Sensors and Systems with Optical Fiber Links," *J. Institution of Electronic and Radio Engineers*, Vol. 58, No. 5, 1988, pp. S85–S91.

[24]   Roessig, T., A. P. Pisano, and R. T. Howe, "Surface-Micromachined Resonant Force Sensor," *Proc. ASME International Mechanical Engineering Congress and Exposition*, Part 2 (of 2), San Francisco, CA, November 12–17, 1995, *ASME Dyn. Syst. Control Div. Publ. DSC*, Vol. 57, No. 2, 1995, pp. 871–876.

[25]   Cheshmehdoost, A., and B. E. Jones, "Design and Performance Characteristics of an Integrated High-Capacity DETF-Based Force Sensor," *Sensors and Actuators*, Vol. A52, No. 1–3, 1996, pp. 99–102.

[26]   Haueis, M., et al., "Packaged Bulk Micromachined Resonant Force Sensor for High Temperature Applications," *Conf. Design, Test, Integration, and Packaging of MEMS/MOEMS*, Paris, France, May 9–11, 2000, *SPIE*, Vol. 4019, 2000, pp. 379–388.

[27]   Morten, B., G. De Cicco, and M. Prudenziati, "A Novel Torque Sensor Based on Elastic Waves Generated and Detected by Piezoelectric Thick Films," *Sensors and Actuators*, Vol. A41, No. 1–3, 1994, pp. 33–38.

[28] Wolff, U., et al., "Radio Accessible SAW Sensors for Non-Contact Measurement of Torque and Temperature," *Proc. IEEE Ultrasonics Symposium*, Vol. 1, San Antonio, TX, November 3–6, 1996, pp. 359–362.

[29] Reindl, L., et al., "SAW Devices as Wireless Passive Sensors," *Proc. IEEE Ultrasonics Symposium*, Vol. 1, San Antonio, TX, November 3–6, 1996, pp. 363–367.

[30] Beckley, J., et al., "Non-Contact Torque Sensors Based on SAW Arrays," *Proc. IEEE Int. Frequency Control Symp.*, New Orleans, LA, May 29–31, 2002, pp. 202–213.

[31] Jerems, F., et al., "A New Generation Non-Torsion Bar Torque Sensors for Electromechanical Power Steering Applications," *Proc. Intl. Congress on Electronic Systems for Vehicles*, Baden Baden, Germany, September 27–28, 2001, pp. 202–212.

[32] Londsdale, A., "Dynamic Rotary Torque Measurement Using Surface Acoustic Waves," *Sensors Express*, Vol. 18, No. 10, 2001, pp. 51–55.

[33] Sachs, T., et al., "Remote Sensing Using Quartz Sensors," *Smart Structures and Materials: Smart Sensing, Processing, and Instrumentation Conference*, San Diego, CA, February 26–28, 1996, *SPIE*, Vol. 2718, 1996, pp. 47–58.

[34] Pohl, A., G. Ostermayer, and F. Seifert, "Wireless Sensing Using Oscillator Circuits Locked to Remote High-Q SAW Resonators," *IEEE Trans. on Ultrasonics, Ferroelectrics and Frequency Control*, Vol. 45, No. 5, 1998, pp. 1161–1168.

[35] Grossmann, R., et al., "Measurement of Mechanical Quantities Using Quartz Sensors," *Proc. 10th European Frequency Time Forum*, Brighton, England, March 5–7, 1996, IEE Conf. Publ. No. 418, pp. 376–381.

[36] Nakayama, T., and E. Suda, "The Present and Future of Electric Power Steering," *Intl. J. Vehicle Design*, Vol. 15, No. 3–5, 1994, pp. 243–254.

[37] Zabler, E., et al., "A Non-Contact Strain-Gage Torque Sensor for Automotive Servo-Driven Steering Systems," *Sensors and Actuators*, Vol. A41, No. 1–3, 1994, pp. 39–46.

[38] Fleming, W. J., "Automotive Torque Measurement: A Summary of Seven Different Methods," *IEEE Trans. on Vehicular Technology*, Vol. VT-31, No. 3, 1982, pp. 117–124.

[39] Mörbe, M., and C. von Hörsten, "Force and Torque Sensors," in *Sensors for Automotive Technology*, J. Marek, et al. (eds.), Weinheim, Germany: Wiley-VCH Verlag GmbH, 2003, pp. 450–462.

[40] Hazelden, R. J., "High Integrity Optical Torque Sensor for Electric Power Steering Systems," *Proc. of Int. Conf. Advanced Measurement Technique and Sensory Systems for Automotive Applications*, Ancona, Italy, June 29–30, 1995, pp. 329–337.

[41] Hazelden, R. J., "Optical Torque Sensor for Automotive Steering Systems," *Sensors and Actuators*, Vol. A37–38, 1993, pp. 193–197.

[42] Javahiraly, N., et al., "An Ultra Sensitive Polarimetric Torque Sensor," *Thermosense XXIV Conf.*, Orlando, FL, April 1–4, 2002, *SPIE*, Vol. 4710, 2002, pp. 721–727.

[43] Chung D., et al., "A Birefringent Torque Sensor for Motors," *Proc. Three-Dimensional Imaging, Optical Metrology, and Inspection IV*, Boston, MA, November 2–3, 1998, *SPIE*, Vol. 3520, 1998, pp. 254–261.

[44] Spooncer, R. C., R. Heger, and B. E. Jones, "Non-Contacting Torque Measurement by a Modified Moiré Fringe Method," *Sensors and Actuators*, Vol. A31, No. 1–3, 1992, pp. 178–181.

[45] Mortara, A., et al., "An Opto-Electronic 18b/Revolution Absolute Angle and Torque Sensor for Automotive Steering Applications," *IEEE Intl. Conf. Solid-State Circuits*, San Francisco, CA, February 7–9, 2000, pp. 182–183, 445.

[46] Hatamura, Y., T. Nagao, and Y. Watanabe, "Development of a Tool-Holder Type Torque Sensor," *Proc. 5th Intl. Conf. on Production Engineering*, Tokyo, Japan, July 1984, pp. 124–129.

[47] Halmai, A., and A. Huba, "Miniaturised Optoelectronic Torque Sensor," *Proc. Joint Hungarian British Int. Mechatronics Conference*, Budapest, Hungary, September 21–23, 1994, pp. 273–278.

[48] Ebi, G., et al., "Integrated Optical Noncontact Torque Measurement Microsystem," *Opt. Eng.*, Vol. 38, No. 2, 1999, pp. 240–245.

[49] Meckes, A., et al., "Capacitive Silicon Microsensor for Force and Torque Measurement," *Proc. TRANSDUCERS '01 – Eurosensors XV*, Munich, Germany, June 10–14, 2001, pp. 498–501.

[50] Cermak, S., et al., "Capacitive Sensor for Torque Measurement," *Proc. Intl. Measurement Confederation XVI IMEKO World Congress*, Vol. III, Vienna, Austria, September 25–28, 2000, pp. 25–28.

[51] Turner, J. D., and L. Austin, "Sensors for Automotive Telematics," *Measurement Science and Technology*, Vol. 11, No. 2, 2000, pp. R58–R79.

[52] Despont, M., et al., "New Design of Micromachined Capacitive Force Sensor," *J. Micromech. Microeng.*, Vol. 3, No. 4, 1993, pp. 239–242.

[53] Wiegerink, R., et al., "Quasi-Monolithic Silicon Load Cell for Loads Up to 1000 kg with Insensitivity to Non-Homogeneous Load Distributions," *Proc. 12th IEEE Intl. Conf. Micro Electro Mechanical Systems*, Orlando, FL, January 17–21, 1999, pp. 558–563.

[54] Enikov, E. T., and B. J. Nelson, "Three-Dimensional Microfabrication for a Multi-Degree-of-Freedom Capacitive Force Sensor Using Fiber-Chip Coupling," *J. Micromech. Microeng.*, Vol. 10, No. 4, 2000, pp. 492–497.

[55] Sasada, I., S. Uramoto, and K. Harada, "Noncontact Torque Sensors Using Magnetic Heads and a Magnetostrictive Layer on the Shaft Surface-Application of Plasma Jet Spraying Process," *IEEE Trans. on Magnetics*, Vol. 22, No. 5, 1986, pp. 406–408.

[56] Rombach, P., H. Steiger, and W. Langheinrich, "Planar Coils with Ferromagnetic Yoke for a Micromachined Torque Sensor," *J. Micromech. Microeng.*, Vol. 5, No. 2, 1995, pp. 136–138.

[57] Rombach, P., and W. Langheinrich, "Modeling of a Micromachined Torque Sensor," *Sensors and Actuators*, Vol. A46, No. 1–3, 1995, pp. 294–297.

[58] Romback, P., and W. Langheinrich, "An Integrated Sensor Head in Silicon for Contactless Detection of Torque and Force," *Sensors and Actuators*, Vol. A42, No. 1–3, 1994, pp. 410–416.

[59] Sahashi, M., et al., "A New Contact Amorphous Torque Sensor with Wide Dynamic Range and Quick Response," *IEEE Trans. on Magnetics*, Vol. 23, No. 5, 1987, pp. 2194–2196.

[60] Umbach, F., et al., "Contactless Measurement of Torque," *Mechatronics*, Vol. 12, No. 8, 2002, pp. 1023–1033.

[61] Brugger, J., et al., "Microfabricated Ultrasensitive Piezoresistive Cantilevers for Torque Magnetometry," *Sensors and Actuators*, Vol. 73, No. 3, 1999, pp. 235–242.

[62] Lee, C., et al., "Characterization of Micromachined Piezoelectric PZT Force Sensors for Dynamic Scanning Force Microscopy," *Rev. Sci. Instrum.*, Vol. 68, No. 5, 1997, pp. 2091–2099.

[63] Brook, A. J., et al., "Micromachined III-V Cantilevers for AFM-Tracking Scanning Hall Probe Microscopy," *J. Micromech. Microeng.*, Vol. 13, No. 1, 2003, pp. 124–128.

[64] Lee, D. W., et al., "Switchable Cantilever Fabrication for a Novel Time-of-Flight Scanning Force Microscope," *Microelectric Engineering*, Vol. 67–68, 2003, pp. 635–643.

[65] Drews, D., et al., "Micromachined Aperture Probe for Combined Atomic Force and Near-Field Scanning Optical Microscopy (AFM/NSOM)," *Proc. Conf. on Materials and Device Characterization in Micromachining*, Santa Clara, CA, September 21–22, 1998, *SPIE*, Vol. 3512, 1998, pp. 76–83.

[66] Abraham, M., et al., "Micromachined Aperture Probe Tip for Multifunctional Scanning Probe Microscopy," *Utramicroscopy*, Vol. 71, No. 1–4, 1998, pp. 93–98.

[67] Greitmann, G., and R. A. Buser, "Tactile Microgripper for Automated Handling of Micro-parts," *Sensors and Actuators*, Vol. A53, No. 1–3, 1996, pp. 410–415.

[68] Dargahi, J., M. Parameswaran, and S. Payandeh, "A Micromachined Piezoelectric Tactile Sensor for an Endoscopic Grasper—Theory, Fabrication and Experiments," *J. of Microelectromechanical Systems*, Vol. 9, No. 3, 2000, pp. 329–335.

[69] Engel, J., J. Chen, and C. Liu, "Development of Polyamide Flexible Tactile Sensor Skin," *J. Micromech. Microeng.*, Vol. 13, No. 3, 2003, pp. 359–366.

[70] Lomas, T., A. Tuantranont, and F. Cheevasuvit, "Micromachined Piezoresistive Tactile Sensor Array Fabricated by Bulk-Etched Mumps Process," *Proc. IEEE Intl. Symp. Circuits and Systems*, Bangkok, Thailand, May 25–28, 2003, pp. 856–859.

[71] Jeglinski, S. A., S. C. Jacobsen, and J. E. Wood, "Six-Axis Field-Based Transducer for Measuring Displacements and Loads," *Winter Annual Meeting of the American Society of Mechanical Engineers*, Dallas, TX, November 25–30, 1990, *ASME Dyn. Syst. Control Div. Publ. DSC*, Vol. 19, 1990, pp. 83–98.

[72] Lyshevski, S. E., "Distributed Control of MEMS-Based Smart Flight Surfaces," *Proc. American Control Conf.*, Arlington, VA, June 25–27, 2001, pp. 2351–2356.

[73] Lee, C., T. Itoh, and T. Suga, "Sol-Gel Derived PNNZT Thin Films for Micromachined Piezoelectric Force Sensors," *Thin Solid Films*, Vol. 299, No. 1–3, 1997, pp. 88–93.

[74] Svedin, N., E. Stemme, and G. Stemme, "A Static Turbine Flow Meter with a Micromachined Silicon Torque Sensor," *Proc. 14th IEEE Conf. Micro Electro Mechanical Systems*, Interlaken, Switzerland, January 21–25, 2001, pp. 208–211.

[75] McKenzie, J. S., K. F. Hale, and B. E. Jones, "Optical Actuators," in *Advances in Actuators*, A. P. Dorey and J. H. Moore, (eds.), Bristol, England: Institute of Physics Publishing, 1995, pp. 82–111.

CHAPTER 8

Inertial Sensors

8.1 Introduction

Micromachined inertial sensors are a very versatile group of sensors with applications in many areas. They measure either linear acceleration (along one or several axes) or angular motion about one or several axes. The former is usually referred to as an accelerometer, the latter as a gyroscope. Until recently, medium to high performance inertial sensors were restricted to applications in which the cost of these sensors was not of crucial concern, such as military and aerospace systems. The dawn of micromachining has generated the possibility of producing precision inertial sensors at a price that allows their usage in cost-sensitive consumer applications. A variety of such applications already exists, mainly in the automotive industry for safety systems such as airbag release, seat belt control, active suspension, and traction control. Inertial sensors are used for military applications such as inertial guidance and smart ammunition. Medical applications include patient monitoring, for example, for Parkinson's disease. Many products, however, are currently in their early design and commercialization stage, and only one's imagination limits the range of applications. A few examples are:

- Antijitter platform stabilization for video cameras;
- Virtual reality applications with head-mounted displays and data gloves;
- GPS backup systems;
- Shock-monitoring during the shipment of sensitive goods;
- Novel computer input devices;
- Electronic toys.

Clearly, micromachined sensors are a highly enabling technology with a huge commercial potential. The requirements for many of the above applications are that these sensors be cheap, can fit into a small volume, and their power consumption must be suitable for battery-operated devices. Micromachined devices can fulfill these requirements since they can be batch-fabricated and they benefit from similar advantages as standard integrated circuits.

Tables 8.1 and 8.2 give an overview of some existing and future applications for accelerometers and gyroscopes, respectively. Typical values for required bandwidth, resolution, and dynamic range are quoted (these are provided for approximate guidance only).

As can be seen from the tables, the typical performance requirements for each application are considerably different. This implies that it is highly unlikely that

Table 8.1 Typical Applications for Micromachined Accelerometers

Application	Bandwidth	Resolution	Dynamic Range
Automotive			
Airbag release	0–0.5 kHz	<500 mG	±100G
Stability and active	0–0.5 kHz	<10 mG	±2G
control systems	dc–1 kHz	<10 mG	100G
Active suspension			
Inertial navigation	0–100 Hz	<5 μG	±1G
Seismic activity			
Shipping of fragile goods	0–1 kHz	<100 mG	±1 kG
Space microgravity	0–10 Hz	<1 μG	±1 G
measurements			
Medical applications	0–100 Hz	<10 mG	±100G
(patient monitoring)			
Vibration monitoring	1–100 kHz	<100 mG	±10 kG
Virtual reality (head-mounted	0–100 Hz	<1 mG	±10G
displays and data gloves)			
Smart ammunition	10 Hz to 100 kHz	1 G	±100 kG

there will be a single inertial sensor capable of being used for all applications areas; rather, all inertial sensors are application specific, which explains the great variety of sensor types.

For any given application the inertial sensor is part of a larger control system, whereas the mere information about acceleration or angular motion of a body of interest is usually of little interest. For example, a gyroscope detects the angular motion of a car and if this is above a critical level, the safety system will actively control the steering angle and the brakes at each wheel to prevent the vehicle from overturning.

Micromachined inertial sensors have been the subject of intensive research for over two decades since Roylance et al. [1] reported the first micromachined accelerometer in 1979. Since then many authors have published work about various types of MEMS accelerometer. The development of gyroscopes based on micromachined silicon sensing elements lags behind by about one decade: the first real MEMS gyroscope was reported by Draper Labs in 1991 [2].

Table 8.2 Typical Applications for Micromachined Gyroscopes

Application	Bandwidth	Resolution	Dynamic Range
Automotive			
Rollover protection	0–100Hz	<1°/sec	±100°/sec
Stability and active	0–100Hz	<0.1°/sec	±100°/sec
control systems			
Inertial navigation	0–10 Hz	<10⁻⁴°/sec	±10°/sec
Platform stabilization	0–100 Hz	<0.1°/sec	±100°/sec
(e.g., for video camera)			
Virtual reality (head-mounted	dc–10 Hz	<0.1°/sec	±100 °/sec
displays and data gloves)			
Pointing devices for	dc–10 Hz	<0.1°/sec	±100°/sec
computer control			
Robotics	dc–100 Hz	<0.01°/sec	±10°/sec

This chapter will introduce the fundamental principles and describe in more detail some of the most important research prototype and commercial devices. Furthermore, it will provide an outlook about the developments in this field to be expected in the near future.

8.2   Micromachined Accelerometer

8.2.1   Principle of Operation

8.2.1.1   Mechanical Sensing Element

Many types of micromachined accelerometers have been developed and are reported in the literature; however, the vast majority has in common that their mechanical sensing element consists of a proof mass that is attached by a mechanical suspension system to a reference frame, as shown in Figure 8.1.

Any inertial force due to acceleration will deflect the proof mass according to Newton's second law. Ideally, such a system can be described mathematically in the Laplace domain by

$$\frac{x(s)}{a(s)} = \frac{1}{s^2 + \dfrac{b}{m} + s\dfrac{k}{m}} \tag{8.1}$$

where x is the displacement of the proof mass from its rest position with respect to a reference frame, a is the acceleration to be measured, b is the damping coefficient, m is the mass of the proof mass, k is the mechanical spring constant of the suspension system, and s is the Laplace operator. The natural resonant frequency[1] of this system is given by

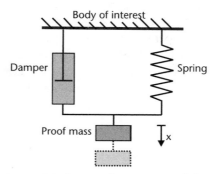

Figure 8.1   Lumped parameter model of an accelerometer consisting of a proof (or seismic) mass, a spring, and a damping element.

1.   Sometimes it is preferred to write the transfer function in terms of the natural frequency and the quality factor Q:

$$\frac{x(s)}{a(s)} = \frac{1}{s^2 + \dfrac{\omega_n}{Q}s + \omega_n^2} \quad \text{with } Q = \frac{\omega_n m}{b} = \frac{\sqrt{mk}}{b}$$

$$\omega_n = \sqrt{\frac{k}{m}}$$

and the sensitivity (for an open sensor) by

$$S = \frac{m}{k}$$

As an accelerometer can typically be used at a frequency below its resonant frequency, an important design trade-off becomes apparent here since sensitivity and resonant frequency increase and decrease with m/k, respectively. This trade-off can be partly overcome by including the sensing element in a closed loop, force-feedback control system, as will be described later.

For the dynamic performance of an accelerometer, the damping factor is crucial. For maximum bandwidth the sensing element should be critically damped; it can be shown that for $b = 2m\omega_n$ this is the case. It should be noted here that in micromachined accelerometers the damping originates from the movement of the proof mass in a viscous medium. Depending on the mechanical design, however, the damping coefficient cannot be assumed to be constant; rather, it increases with the deflection of the proof mass and also with the frequency of movement of the proof mass—this phenomenon is called squeeze film damping. This is a complex fluid dynamic problem and goes beyond of the scope of this book. For further reading on this topic, the interested reader is referred to the literature [3–6].

A common factor for all micromachined accelerometers is that the displacement of the proof mass has to be measured by a position-measuring interface circuit, and it is then converted into an electrical signal. Many types of sensing mechanisms have been reported, such as capacitive, piezoresistive, piezoelectric, optical, and tunneling current. Each of these has distinct advantages and drawbacks (as described in Chapter 5). The first three sensing mechanisms are the most commonly used. The characteristic and performance of any accelerometer is greatly influenced by the position measurement interface, and the main requirements are low noise, high linearity, good dynamic response, and low power consumption. Ideally, the interface circuit should be represented by an ideal gain block, relating the displacement of the proof mass to an electrical signal.

8.2.1.2   Open Loop Accelerometer

If the electrical output signal of the position measurement interface circuit is directly used as the output signal of the accelerometer, this is called an open loop accelerometer, as conceptually shown in Figure 8.2.

Most commercial micromachined accelerometers are open loop in that they are the most simple devices possible and are thus low cost. The dynamics of the mechanical sensing element are mainly to determine the characteristics of the sensor. This can be problematic as the mass and spring constant are usually subject to considerable manufacturing tolerances (depending on the fabrication process, this could be up to $\pm 20\%$). Furthermore, second order effects for larger proof mass deflection

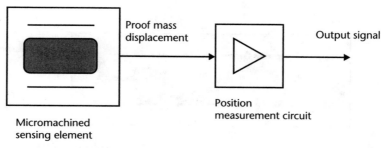

Figure 8.2 Open loop accelerometer.

introduce nonlinear effects; squeeze film damping was mentioned earlier. Another effect is that any silicon suspension system will have nonlinear behavior, such as a spring stiffening effect, for larger deflections, or cross-axes sensitivity. Nevertheless, for most automotive and other low-cost applications the achievable performance is still acceptable.

8.2.1.3 Closed Loop Accelerometer

The output signal of the position measurement circuit can be used, together with a suitable controller, to steer an actuation mechanism that forces the proof mass back to its rest position. The electrical signal proportional to this feedback force provides a measure of the input acceleration. This is usually referred to as a closed loop or force balanced accelerometer. This approach has several advantages:

1. The deflection of the proof mass is reduced considerably; hence, nonlinear effects from squeeze film damping and the mechanical suspension system are reduced considerably.
2. The sensitivity is now mainly determined by the control system; hence, the trade-off between the sensitivity and bandwidth can be overcome.
3. The dynamics of the sensor can be tailored to the application by choosing a suitable controller (i.e., the bandwidth, dynamic range, and sensitivity can be increased compared with the open loop case).

The drawback of a closed loop accelerometer is mainly the added complexity in interface and control electronics.

There is a range of possible actuation mechanisms to keep the proof mass at its rest position, such as electrostatic, magnetic, and thermal. Electrostatic forces are by far the most commonly used type since for small gap sizes these forces are relatively large, allowing typical supply voltages of between 5V and 15V. If capacitive position sensing is used, the same electrodes can be used for sensing and actuation. Care has to be taken, however, to ensure that the sense and actuation signal do not interact. One major problem of electrostatic forces is that they are always attractive and non-linear because they are proportional to the voltage squared and inversely to the gap squared. Consequently, it is difficult to produce a linear, negative feedback signal.

Analog Force-Feedback   Consider the simple sensing element in Figure 8.3: a proof mass between two electrodes forms an upper and lower capacitor.

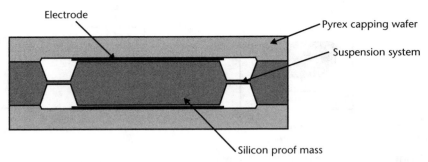

Figure 8.3   Typical bulk-micromachined capacitive sensing element.

This can be incorporated in a closed loop, force-feedback system, which is diagrammatically shown in Figure 8.4.

Assuming the proof mass is at zero potential, any voltage on the top or bottom electrode will produce an electrostatic force on the proof mass. To achieve linear, negative feedback, it is necessary to superimpose a feedback voltage, V_F, on a bias voltage on both electrodes, V_B, which results in a net electrostatic force on the mass, given by

$$F = F_1 - F_2 = \frac{1}{2}\varepsilon A\left[\frac{(V_B + V_F)^2}{(d_0 - x)^2} - \frac{(V_B - V_F)^2}{(d_0 + x)^2}\right] \tag{8.2}$$

Under closed loop control, the proof mass deflection will be small; hence, it can be assumed that $d^2 << x^2$. Using this assumption and rearranging yields

$$F = F_1 - F_2 = 2\varepsilon A\left[\frac{d_0 x(V_B^2 + V_F^2) - V_B V_F d_0^2}{d_0^4}\right] \tag{8.3}$$

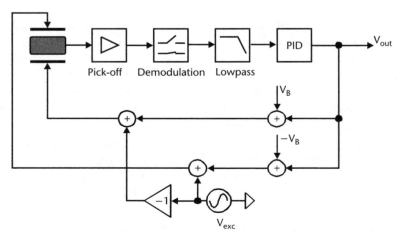

Figure 8.4   Capacitive accelerometer incorporated in an analog force-feedback loop.

If the limit $x \to 0$ is taken, (8.3) yields

$$F = F_1 - F_2 \xrightarrow{\lim x = 0} -2\varepsilon A \left[\frac{V_B}{d_0^2} V_F \right] \tag{8.4}$$

which is a linear, negative feedback relationship.

If we further assume the simplest form of controller, a pure proportional controller, the feedback voltage can be expressed as $V_F = k_p x$ with k_p as the proportional gain constant. This can be substituted into (8.2) and (8.4) to plot the resulting electrostatic force on the proof mass for the exact and linearized solution, respectively. Figure 8.5 shows the electrostatic feedback force for different bias voltages as a function of proof mass deflection.

It can be seen that the proof mass is pulled back to its nominal position by the feedback force, as long as the deflection is assumed small, which is the case under normal operating conditions. However, if the proof mass is deflected further from its nominal position, the feedback force first becomes nonlinear and eventually even changes polarity. This would result in a latch-up or electrostatic pull-in situation and hence the instability of the sensor. Larger deflections can be caused by an acceleration on the sensor that exceeds the nominal dynamic range of a sensor (e.g., a car driving into a pothole). This potential instability is a major drawback of this form of analog feedback. A potential solution is to include mechanical stoppers to prevent the proof mass from being deflected close enough to the electrodes to cause electrostatic pull-in.

Digital Feedback   Another form of electrostatic feedback is to incorporate the sensing in a sigma-delta type control system, which is schematically shown in Figure 8.6.

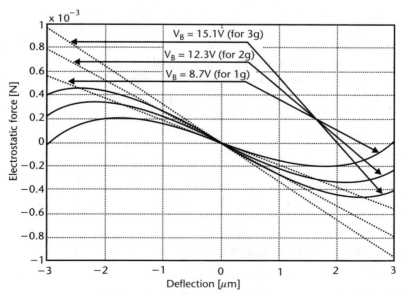

Figure 8.5   Net electrostatic force on the proof mass with analog force-feedback. The solid line is according to (8.2); the dashed line shows the linearized solution of (8.4). Only for small proof mass deflections is the feedback force negative and linear; for larger deflections it becomes nonlinear and eventually changes polarity, which can lead to electrostatic pull-in.

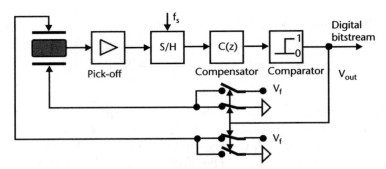

Figure 8.6   Closed loop accelerometer with analog force-feedback.

In this approach the information about the deflection of the proof mass is fed into a comparator. Depending on the dynamics of the proof mass, a compensator may be required to stabilize the loop. The comparator controls a range of switches, which applies a feedback voltage to the electrode that the proof mass is further away from in this moment; the other electrode is being grounded. This is done for a fixed time interval, which is locked to the sampling frequency of the comparator. As with their electronic counterpart, this electromechanical sigma-delta modulator is an oversampling system; hence, the clock frequency has to be many times higher than the bandwidth of the sensor. This approach has a number of advantages over analog force-feedback:

1. No electrostatic pull-in is possible as an electrostatic feedback force is only produced in one direction (i.e., pulling the proof mass to its nominal position).
2. The output signal, taken from the comparator, is a direct digital signal in the form of a pulse-density modulation (i.e., the number of high-bits in a given interval is a measure of the input acceleration). The output signal can interface directly to a digital signal processor (DSP), which can perform the necessary lowpass filtering and further signal processing if required.
3. As with any sigma-delta modulator, such a sensor will produce a self-sustained oscillation at the output even if no input acceleration is present. As a constant signal at the output would indicate a sensor failure, this can be used as a simple form of functionality test.

This approach has gained much popularity in recent years, with a number of researchers reporting accelerometers with such a closed loop control system [7–10].

8.2.2   Research Prototype Micromachined Accelerometers

Many prototype micromachined accelerometers have been reported during the last two decades. In the following sections, an overview of key devices will be given and a few examples of interesting and representative devices will be described in more detail. The classification used here is mainly based on the position sensing mechanism.

8.2.2.1   Piezoresistive Accelerometers

The first micromachined, batch-fabricated accelerometer was reported by Roylance and Angell [1] at Stanford University in 1979. It used a bulk-micromachined sensing element consisting of a central silicon wafer that forms the proof mass and a cantilever as its suspension system. The silicon wafer is bonded between two glass wafers into which cavities are etched to allow the mass to move as a response to acceleration. The glass wafers also protect the proof mass and act as a shock stopper. The motion of the proof mass was detected with piezoresistors, which were fabricated by implanting Boron directly in the beams of the silicon suspension system of the proof mass. In general, early devices tended to use a piezoresistive position measurement interface, as these are easy to fabricate in silicon and the read-out circuit is relatively simple; they provide a low-impedance output signal and a conventional resistive bridge circuit can be used. Furthermore, early piezoresistive accelerometers were directly based on the expertise gained through the development of micromachined pressure sensors. A serious drawback, however, is that the output signal tends to have a strong temperature dependency because the piezoresistors inherently produce thermal noise and the output signal is relatively small [11]. Typical performance figures for these devices show a sensitivity of 1 to 3 mV/g, 5g to 50g dynamic range, and an uncompensated temperature coefficient of 0.2%/C. Examples of early devices are described in [12–14]. They typically consist of a multiwafer assembly with the central wafer comprising the bulk-micromachined proof mass and suspension system and either silicon or Pyrex glass wafers on top and bottom to provide over-range protection and near critical damping due to squeeze film effects.

The disadvantages of piezoresistive signal pick-off can be partially overcome by integrating the read-out electronics on the same chip. A good example is the accelerometer presented by Seidel et al. [15]. The sensing element consists of a bulk-micromachined proof mass, which is attached to the substrate by three cantilever beams. On the main cantilever four piezoresistors are implanted and form a full Wheatstone bridge. A cross-section of the sensor is shown in Figure 8.7.

The sensing element is encapsulated by top and bottom wafers, which are bonded to the middle layer at wafer level. Small air gaps were formed into the cap-wafers by dry-etching in order to provide near-critical damping. The electronic read-out circuitry is integrated onto the same chip and was fabricated in a standard 3-μm CMOS process. The remaining processing steps for the fabrication of the mechanical sensing element were done after the CMOS process. They mainly included a wet-etch step of the device wafer to form the sensing element, for which the n-well was used as an electrochemical etch-stop and the implantation of the

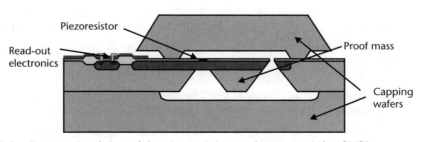

Figure 8.7   Cross-sectional view of the piezoresistive accelerometer. (*After:* [15].)

piezoresistors. A reference structure was used for offset and temperature drift cancellation. The electronic circuitry was operated at a supply voltage of 5V and a supply current of 22 mA. This not only provided filtering, amplification, and buffering of the output signal from the Wheatstone bridge, but also active offset and temperature drift compensation by subtracting the output signal of the reference structure and the sensing element. The reported performance of this device was a full-scale measurement range up to ±20G, a resonance frequency of 1.2 kHz, a sensitivity of 0.4 mV/V/G with a sensitivity drift of −1.8‰/K and an offset drift of 8 μV/V/K.

Another example of an integrated piezoresistive accelerometer is described by Reithmueller et al. [16], who developed a similar fabrication technique and experimented with different device designs that resulted in comparable performance.

8.2.2.2 Capacitive Accelerometers

Measuring the displacement of the proof mass capacitively has some inherent advantages over piezoresistive signal pick-off. It provides a large output signal, good steady-state response, and better sensitivity due to low noise performance. The main drawback is that capacitive sensors are susceptible to electromagnetic fields from their surroundings; hence, they have to be shielded carefully. It is also unavoidable that parasitic capacitances at the input to the interface amplifiers will degrade the signal. Usually, a differential change in capacitance is detected. As the proof mass moves away from an electrode, the capacitance decreases, and as it moves towards the electrodes, the capacitance increases. Neglecting fringe field effects, the change in capacitance is given by

$$\Delta C = \varepsilon A \left(\frac{1}{d_0 - x} - \frac{1}{d_0 + x} \right) \xrightarrow{x^2 << d_0^2} 2\varepsilon A \frac{x}{d_0^2} \qquad (8.5)$$

which is proportional to the deflection caused by the input acceleration only if the assumption of small deflections is made. For precision accelerometers this assumption may be not justifiable, and hence, closed loop control can be used to keep the proof mass deflections small.

Early types of capacitive sensors were typically also fabricated by bulk micromachining and the assembling of several wafers by bonding techniques. Most devices had the axis of sensitivity perpendicular to the wafer plane, with the cap wafers on the top and bottom, which, in addition to providing the damping, form the electrodes for capacitive detection. A typical example is shown Figure 8.8 [17].

An early, high-precision accelerometer, which can be operated in open and closed loop mode, was reported by Rudolf et al. [18]. The sensor consisted of a sandwich structure made up from Si-Glass-Si-Glass-Si and is shown schematically in Figure 8.9.

The chip size was 8.3 × 5.9 × 1.9 mm with the proof mass size of 4 × 4 × 0.37 mm and a mass of 14.7 × 10^{-6} kg. The distance of the mass to either electrode at the rest position was 7 μm, which is relatively large; hence, for closed loop operation a voltage of 15V was required. Three <100> silicon wafers were processed, the middle one containing the proof mass and suspension system. These are formed by

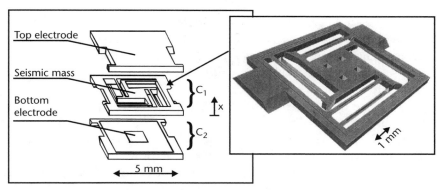

Figure 8.8   A bulk-micromachined accelerometer with capacitive signal pick-off.

time-controlled etching in KOH with silicon dioxide as a mask. The same etch is performed from the front and back sides of the wafer, resulting a highly symmetrical design. The upper and lower wafers are anodically bonded to a glass wafer onto which a thin layer of aluminum is deposited and patterned to form the electrodes. Over-range stoppers restrict the movement of proof mass and prevent it from touching the electrodes, which could lead to an electrostatic latch-up. The performance of the sensor depends on whether it is operated in open loop or closed loop mode, the latter principally based on an analog force-feedback as described in Section 2.1.3.1. For open loop operation the performance is well suited for general purpose and automotive applications, whereas in closed loop operation sub-μg resolution was reported to have made the device suitable for inertial navigation and guidance. The resolution was below 1 μg/vHz in a bandwidth up to 100 Hz with a temperature coefficient of offset and sensitivity of 30 μg/°C and 150 ppm/°C, respectively.

When capacitive sensors are operated in open loop mode, there exists one problem compared to piezoresistive devices in that the proof mass should move in parallel to the electrodes, like a piston, rather than rotating around an axis, as with a cantilever-type suspension system, which would introduce a nonlinearity for larger deflections. Although several other cantilever capacitive accelerometer prototypes

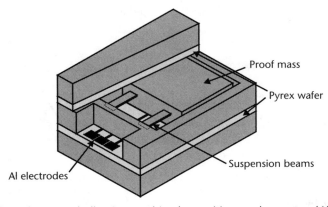

Figure 8.9   High-performance bulk-micromachined capacitive accelerometer. (*After:* [14].)

were presented [19, 20], more sophisticated suspension systems had to be designed where the proof mass was connected to the substrate by several tethers and/or folded beams. The design must be as symmetrical as possible in order to minimize cross-axis sensitivity (i.e., acceleration along an axis other than the sense-axis should not cause any change in capacitance) [21].

A range of high-performance devices has been reported, which were incorporated in a force-feedback sigma-delta modulator structure [7–10], as outlined in Section 2.1.3.2. Henrion et al. [7] achieves a dynamic range of 120-dB resolution. This, however, requires a high Q mechanical transfer function in order to achieve the appropriate noise shaping for the sigma-delta modulator. This implies that the sensing element has to be packaged in a vacuum. De Coulon et al. [8] used the sensing element described in [18] and demonstrated that the digital control loop is suitable to improve the performance. The bandwidth, in particular, has been improved considerably from 3 Hz in the open loop case to about 100 Hz for closed loop operation.

In the early to mid-1990s, the automotive market demanded cheap, reliable, and medium-performance accelerometers. Initially, bulk-micromachined accelerometers were used for these applications [14, 22], but this demand also led to a range of surface-micromachined sensors to be developed with the sensing element and electronics integrated on the same chip. Of particular interested are the accelerometers produced by Analog Devices [23–25] (described in more detail in Section 2.3). For these sensors, the axis of sensitivity is typically in the wafer plane. The proof mass is an order of magnitude smaller than that used in a bulk-micromachined device, and hence, the sensitivity is less, which is partly compensated by integrating the pick-off electronics on the same chip. The sensing element is typically formed by a 2-μm layer of deposited polysilicon on top of a sacrificial silicon dioxide layer.

A typical design for a surface-micromachined sensing element is shown in Figure 8.10 [26].

A range of tethers is connected to the proof mass, each one forming a capacitor to the fixed electrodes on each side. As this capacitor has a value of only a few femtofarads, many of them are required in parallel to give a total capacitance in the range of 100 fF. The minimum resolution of these sensors lies, nevertheless, in the milliG range or even below.

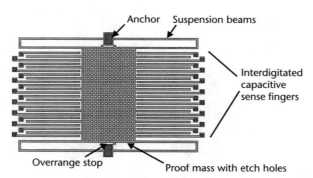

Figure 8.10 Typical design for an in-plane, capacitive surface-micromachined accelerometer. The interdigitated comb fingers can be used for capacitive sensing, and also for electrostatic forcing the proof mass in a closed loop configuration. (*After:* [25].)

Assuming typical values for such a sensor of a proof mass $m = 0.1 \times 10^{-9}$ kg, a resonant frequency of $f_R = 10$ kHz, and a nominal capacitance of 100 fF, the resulting static displacement for 1 mG is only 0.025Å and the resulting differential capacitance is about 10 attofarads. Measuring such tiny deflections and capacitances can only be achieved with reasonable performance by on-chip electronics.

These sensors have typical performance figures of a resolution below 0.1 mG in a bandwidth of about 100 Hz. Their performance is primarily limited by the Brownian noise from the proof mass as it is usually an order of magnitude smaller than that of bulk-micromachined devices. If the sensing element is packaged at a lower pressure, it is possible to reduce the Brownian noise floor considerably, at the expense of a more complex fabrication and packaging processes. The choice of control system is exactly the same as for bulk-micromachined sensors, open loop operation, or closed loop force-feedback. Examples of open loop devices are described in [27, 28], and examples of sensors using an analog force-feedback system are given in [29, 30]. Digital closed loop sensors are reported mainly by researchers from the University of California at Berkeley [31] with an excellent overview given in [32]. A more detailed example of such an accelerometer is given in Section 2.2.6.

One of the highest performance capacitive accelerometers created was developed by Yazdi and Najafi [33]. It uses a combination of bulk and surface micromachining that allows the fabrication of the sensing element on a single wafer, thereby avoiding the need to bond several wafers together, but nevertheless having the advantage of a wafer-thick proof mass. The latter is compliant to acceleration in the z-direction and moves between electrodes fabricated from polysilicon, which was deposited on a thin sacrificial silicon dioxide layer on the top and bottom wafer surface. These polysilicon electrodes are very thin (2 to 3 μm) but have an area of several square millimeters, and hence needed to be stiffened. This was achieved by etching 25- to 35-μm-wide vertical trenches into the wafers, which were refilled with polysilicon. The holes in the polysilicon electrodes lower the squeeze film damping effect, so that a design with critical damping is possible. Low cross-axis sensitivity of the sensor was achieved by a fully symmetrical suspension system consisting of eight beams, two on each side of the proof mass. The sensing element is shown in Figure 8.11.

This results in a high-precision accelerometer with a measured sensitivity of 2–19.4 pF/G for a proof mass area of 2×1 mm and 4×1 mm, respectively. The reported noise floor was around 0.2 μG/√Hz. The sensor was again incorporated in a sigma-delta modulator control system to electrostatically force-balance the proof mass.

8.2.2.3   Piezoelectric Accelerometers

Macroscopic accelerometers quite commonly use piezoelectric materials for the detection of the proof mass. There has been a range of micromachined accelerometers reported that are based on this principle. The advantage is the higher bandwidth of these sensors, which can easily reach several tens of kilohertz. The major drawback, however, is that they do not respond to static and low-frequency acceleration signals because of unavoidable charge leakage. An early device was reported by Chen et al. [34], which consisted of a cantilever beam onto which the piezoelectric material, ZnO, was sputtered. Interestingly, this sensor has integrated, simple

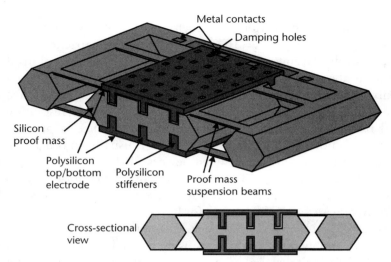

Figure 8.11 High-performance capacitive accelerometer using a combination of surface and bulk-micromachining techniques. The polysilicon electrodes include stiffening ribs. (*After:* [33].)

read-out electronics on the same chip. These comprise a MOS transistor in close proximity to the piezoelectric sensing element to minimize charge leakage. The accelerometer has a sensitivity of 1.5 mV/G with a flat frequency response of 3 Hz to 3 kHz.

Lead zirconate titanate (PZT) is another piezoelectric material often used for accelerometers. It can be sputtered at temperatures around 550°C to form thin-films of approximately 1 μm. Nemirovsky et al. [35] describe an accelerometer based on this technique, which resulted in a sensitivity of 320 mV/G and a very wide bandwidth from 1 Hz to 200 kHz. A more recent device is presented by Beeby et al. [36], which also uses PZT as piezoelectric material but employs a thick-film screen-printing technique to deposit layers of up to 60-μm thickness. The design of the sensing element and a SEM photograph are shown in Figure 8.12.

The fabrication process is simple and the yield was shown to be very high. The sensitivity was given as 16 pC/m/s^2, which was considerably higher than the devices using thin, sputtered zinc oxide (ZnO) layers.

8.2.2.4 Tunneling Accelerometers

The tunneling current from a sharp tip to an electrode is an exponential function of the tip-electrode distance and hence can be used for position measurement of a proof mass. The tunneling current is given by

$$I = I_0 \exp\left(-\beta\sqrt{\phi z}\right) \tag{8.6}$$

where I_0 is a scaling current dependent on material and tip shape (a typical value is 1.4 10^{-6}A), β is a conversion factor with a typical value of 10.25 eV$^{-1/2}$/nm, ϕ is the tunnel barrier height with a typical value of 0.5 eV, and z is the tip/electrode distance. The distance between the tunneling tip and the electrode has to be precisely

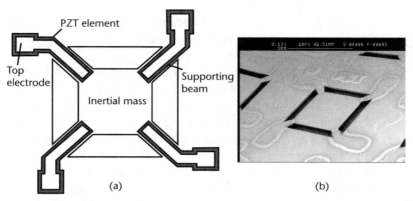

Figure 8.12 (a) Design of a piezoelectric accelerometer using thick-film printed PZT. (b) SEM photograph of the sensing element.

controlled; hence, these sensors have to be used in closed loop operation. Electrostatic force-feedback is employed for the majority of research devices and this keeps the separation distance approximately constant. The acceleration can then be inferred from the voltage required to produce the necessary electrostatic force. A typical sensing element is shown in Figure 8.13 [37]. The proof mass deflection electrode is used to pull the proof mass, by the electrostatic force, into close proximity so that a tunneling current begins to flow. The cantilever deflection electrode is used for closed loop control to maintain the distance between the tip and the cantilever constant.

Theoretically, this is the most sensitive detection mechanism. Several other accelerometers based on this principle have been reported, but no commercial device has been developed. One unresolved problem is the long-term drift of the tunneling current as material from the tip is removed by the high electric fields.

8.2.2.5   Resonant Accelerometers

Resonant accelerometers consist of a proof mass that changes the strain in an attached resonator, hence changing its resonant frequency, similar to tuning a guitar

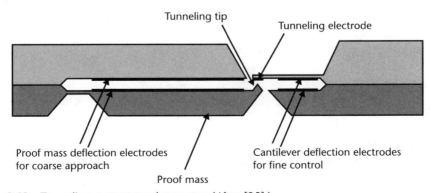

Figure 8.13 Tunneling current accelerometer. (*After:* [35].)

string. The main advantage of this approach is that a frequency output can be converted easily into a digital format by using a frequency counter and is, in general, more immune to noise.

A high resonant frequency is desirable to achieve a good sensitivity, which is in terms of change of frequency per acceleration ($\Delta f/G$). Consequently the resonator should have a high nominal frequency and hence be made of beams with small geometries, which lends itself to fabricating them in surface-micromachining technology. Furthermore, to achieve a high quality factor, the resonator should ideally be sealed in vacuum.

A resonant silicon accelerometer combining bulk and surface micromachining was presented by Burns et al. [38]. It consists of three wafers bonded together. The middle wafer contains the proof mass, which has the thickness of the full wafer. It is formed, together with the flexures, by a wet etching process. Prior to these bulk-micromachining process steps, the resonators are fabricated by surface micromachining. They consist of two beams of 200-μm length, 40-μm length, and 2-μm thickness. The beams are electrostatically excited to vibrate out of the wafer plane and have a base frequency of 500 kHz. They are located on the flexures at points where the highest stress occurs when the proof mass moves. As they are inside a vacuum enclosure, the air-damping is minimized, thereby resulting in a quality factor in excess of 20,000. Thus, ac voltages, in the range of only a few millivolts, are required to sustain the resonance. An additional dc bias voltage of 5V is required. Implanted piezoresistors are used to sense the resonance frequency. Two resonators are placed in such a way that the resonance frequency increases for one of them under applied acceleration, whereas the frequency of the other decreases, resulting in a differential output signal, which rejects common mode errors. A third resonator is used for temperature sensing, which can be used for compensating temperature drift effects. Accelerometers for ranges of $\pm 10G$, $\pm 20G$, and $\pm 50G$ have been fabricated and tested. The scale factor of the $\pm 20G$ device was as high as 743 Hz/G with a temperature frequency shift of about 45 ppm/°C.

A fully integrated, surface-micromachined resonant accelerometer was reported by Roessig et al. [39]. The nominal frequency of the double-ended tuning fork resonator was 68 kHz, and the scale factor of the sensor was measured to be 45 Hz/G. The resonator beams had comb drives attached to sense their motion via a capacitance change and to excite them into resonance using electrostatic forces. This is achieved by incorporating them into an oscillation loop. The coupling of the mechanical force caused by motion of the proof mass into the resonators was achieved by a novel mechanical leverage system that amplifies the force.

A range of other resonant devices has been reported in the literature. For further information, the reader is referred to [40, 41].

8.2.2.6 Multiaxis Accelerometers

A relatively recent innovation for micromachined accelerometers is sensors that are capable of measuring acceleration along two or three axes simultaneously. This is of interest for many applications, for example, inertial sensing, virtual reality, and medical applications. Although it is possible to mount three single-axis devices perpendicular to each other, an integrated version has advantages in cost, size, and alignment to the sense axes, as the sensing elements are defined by highly accurate photolithographic methods. Three-axis sensors with piezoresistive, piezoelectric, and capacitive position sensing mechanisms have been reported. Two approaches are possible: a single proof mass that is compliant to move along two or even three axes, or several proof masses integrated on one chip for the different sense axes. Cross-axis sensitivity is a major issue with multiaxis accelerometers.

An interesting prototype has been reported by Lemkin et al. [42]. It uses a surface-micromachined single proof mass, which is compliant to movement along all three axes, as shown in Figure 8.14. The 2.3-μm-thick sensing element has interdigitated comb fingers on all four sides so that acceleration can be sensed in the two in-plane axes. The sensing element is designed with a common centroid layout; hence, both translational and rotational off-axis accelerations become a common mode signal, which is rejected to the first order by the differential signal pick-off. It is also compliant to movements out of plane, which changes the air-gap of a capacitor formed by an electrode under the proof mass and the center section of the proof mass. The pick-off circuit in this direction is quasidifferential as it is referenced to a fixed capacitor, which is formed by a separate mechanical structure.

Figure 8.15 shows the front-end of the pick-off circuit. The proof mass, which acts as the common node center node to all three capacitive half-bridges, is driven with a step voltage, and the adjacent fixed electrodes are connected to a differential charge amplifier. The output voltage is proportional to the differential change in capacitance. Subsequently, this voltage is sampled and held, as the same capacitors are then used as electrostatic actuators. Feedback voltage pulses are applied to the electrode further away from the proof mass, forcing it to the nominal center position. The force-balancing system relies again on the incorporation of the proof mass in a sigma-delta modulator type control system, one for each sensing axes. The reported noise floor is about 0.73 mG/√Hz for all three axes. Especially impressive is the low cross-axis sensitivity of below 1%, which is better than many commercial grade single-axis devices.

The same authors developed a prototype with three separate proof masses on the same chip but using the same technology and interface and control strategy [43]. The performance for this sensor was improved, compared to that of the single-mass sensor, showing a measured noise floor of 160 mG/√Hz for x- and y-axes and 990 mG/√Hz for z-axis. This corresponds to a dynamic range of 84, 81, and 70 dB for

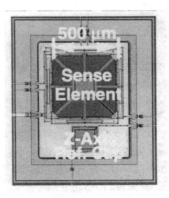

Figure 8.14   Sensing element of a three-axis capacitive accelerometer.

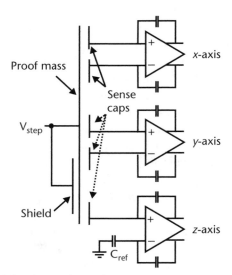

Figure 8.15 Pick-off circuit for three-axis accelerometer.

x-, y-, and z-axes, respectively. The improved performance is mainly attributed to the lower resonant frequencies and the larger sense capacitance compared to the single proof mass device.

Another three-axis capacitive accelerometer using bulk-micromachining technique was presented by Mineta et al. [44]. It uses a proof mass made from glass on which planar electrodes are sputtered. The mass is bonded to a silicon support structure, which is attached only from a central pillar to a lower Pyrex glass plate, as shown in Figure 8.16. This raises the center of gravity of the proof mass above the

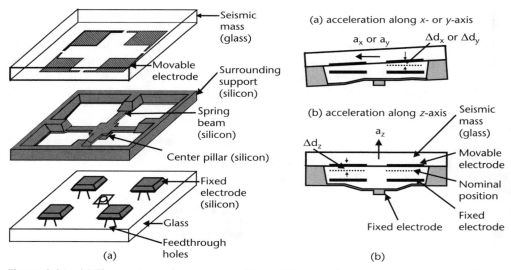

Figure 8.16 (a) Three-axis accelerometer consisting of three wafers: the top wafer contains the Pyrex proof mass, the middle wafer contains the silicon suspension system and the center pillar, and the bottom wafer comprises fixed silicon electrodes on a Pyrex wafer. (b) Acceleration along the x- and y-axes result in a tilt of the proof mass, whereas z-axis acceleration causes the proof mass to move out of plane.

suspending beams. In-plane accelerations cause the proof mass to tilt, and out-of-plane acceleration moves the proof mass perpendicular to the wafer plane; this is illustrated in Figure 8.16(b).

The effective spring constants for all three axes were designed to be the same, and also the rate of change for the differential shift in capacitance of acceleration along all three axes was equal; hence, uniform sensitivity was achieved for all axes. The sensor suffered from relatively high cross-axes sensitivity from z-axis to x-axis (10%) due to asymmetries in the beams of the suspension system. However, this could be removed by an arithmetic operation, yielding a cross-axis sensitivity below 0.8%. The signal pick-off electronics are off-chip, and hence, the commercial device based on this design would be a two-chip solution.

An example of a three-axis accelerometer with a modified piezoresistive pick-off is described by Takao et al. [45, 46]. A bulk-micromachined proof mass is suspended by four beams onto which sensing p-MOSFETs are integrated. They can be used directly as piezoresistive stress-sensing elements because the carrier mobility in the inversion layer of the transistor changes linearly with the induced stress. The same devices are used as input transistors to a CMOS differential amplifier. The modal response of the proof to acceleration along three axes is similar to the capacitive device described above. Optimizing the placement of the sensing MOSFETs results only in a differential output voltage for acceleration along one particular axis; cross-axis accelerations are common mode signals and are cancelled out.

Three axial accelerometers with a single proof mass are still in the prototype stage and have not been commercialized; however, this is expected to happen in the near future. Analog Devices offers a commercial dual-axis accelerometer, which is described later.

8.2.2.7   Other Position Measuring Methods

A range of other position measuring methods have been reported, but none of them has gained major importance so far. Optical means of detecting the proof mass position have the advantage of being insensitive to electromagnetic interference and not requiring electrical power directly at the proof mass. A drawback is that an optical fiber has to be brought into close proximity of the proof mass, which requires hand assembly, thereby negating the advantage of batch-fabrication. Schröpfer et al. [47] reports on an accelerometer with optical read-out; the optical fiber and the vertical sidewall of the sensing element, from which the light is reflected, form a simple Fabry-Perot interferometer with an optical cavity size between 45 and 135 μm. Any in-plane movement of the proof mass results in a wavelength shift that modulates the spectrum; the highest reported sensitivity, in terms of wavelength change per acceleration, was 462 nm/G.

Other researchers use a simple red LED and a PIN photodetector to measure the motion of the proof mass [48]. The proof mass consists of a grid structure with a pitch of 40 μm, 22-μm-wide beams, and 18-μm-wide slots. It acts as an optical shutter that modulates the flux of incident light from the LED to the detector, resulting in a proportional change of photodiode current.

The only class of accelerometer that does not rely on the displacement measurement of a mechanical proof mass is that of thermal devices. They work by heating up a small volume of air, which responds to acceleration. The temperature distribution under acceleration of the heated air bubble becomes asymmetric with respect to the heater and can be measured by temperature sensors placed symmetrically around the heater. Simple piezoresistors can be used both for heating and temperature sensing [49]. The sensor has a relatively low bandwidth from dc to 20 Hz. The authors claim, however, that with design modifications this can be extended to several hundred hertz and sensitivities in the microG range are possible.

Finally, there are sensors that sense the motion of the proof mass by electromagnetic means. Abbaspour-Sani et al. [50] designed an accelerometer with two 12-turn coils, one located on the proof mass, the other one on the substrate. Acceleration causes changes in the distance between the two coils, which results in a change of the mutual inductance. They achieved a sensitivity 0.175 V/G with a dynamic range of 0G to 50G. An advantage of this approach is the simple read-out electronics.

8.2.3 Commercial Micromachined Accelerometer

In this section, a selective overview of commercially available micromachined accelerometers is given. Often, detailed information about the design and fabrication process is not readily available, as this is often considered proprietary.

One of the most successful ranges of micromachined accelerometer was introduced by Analog Devices and is termed the ADXL range. These devices are primarily aimed at the automotive market; the first commercial device was the ADXL50, released in 1991. It is based on a surface micromachined technology with the sensing electronics integrated on the same chip. It is operated in an analog force-balancing closed loop control system and has a ±50G dynamic range with a 6.6-mG/√Hz noise floor, a bandwidth of 6 kHz, and a shock survivability of more than 2,000G, making it suitable for airbag deployment. The nominal sense capacitance is 100 fF and the sensitivity is 19 mV/G. A simplified control system block diagram is shown in Figure 8.17.

The sensor's fixed electrodes are excited differentially with a 1-MHz square wave, which are equal in amplitude but 180° out of phase. If the proof is not deflected, the two capacitors are matched and the resulting output voltage of the buffer is zero. If the proof is displaced from the center, the amplitude of the buffer voltage is proportional to the mismatch in capacitance. The buffer voltage is demodulated and amplified by an instrumentation amplifier referenced to 1.8V; this signal is fed back to the proof mass through a 3 MΩ isolation resistor. This results in an electrostatic force that maintains the proof mass virtually motionless over the dynamic range. The output signal for 0G is +1.8V with an output swing of ±0.95V for ±50G acceleration; with an internal buffer and level shifter this can be amplified to an output range from 0.25V to 4.75V. The sensor additionally has a self-test capability where a transistor-transistor logic (TTL) "high" signal is applied to one of the pins, which results in an electrostatic force approximately equal to a −50G inertial force. If the sensor operates correctly, a −1-V output signal is produced. The sensor is available in a standard 10-pin TO100 metal package.

Subsequently, Analog Devices has introduced a range of other micromachined accelerometers. The ADXL05 works in the same way as the ADXL50 but has a

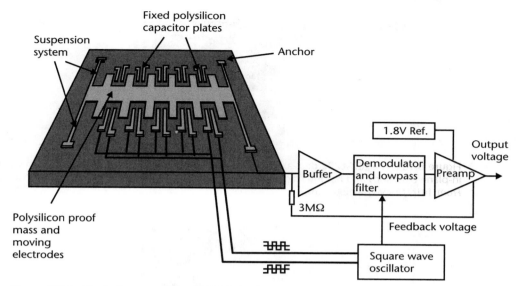

Figure 8.17   Block diagram of the ADXL50 accelerometer.

dynamic range that can be set with external resistors from ±1G to ±5G, resulting in a sensitivity between 200 mV/G and 1 V/G. The noise floor is 0.5 mG/√Hz, which is 12 times lower than for the ADXL50. The main difference to the ADXL50 is that the suspension system has a lower mechanical spring constant, which is achieved by a folded beam structure. This results in a higher compliance to inertial forces and hence to increased sensitivity.

The next generation (ADXL105 and ADXL150) was introduced in 1999 and showed an order of magnitude increase in performance. The ADXL105, with a dynamic range between ±1G and ±5G, has a 225 μG/√Hz noise floor, a 10-kHz bandwidth, and an on-chip temperature sensor, which can be used for calibration against temperature effects. A prototype of this sensor has been developed, based on a 3-μm-thick polysilicon structural layer, which increases the sense capacitance, which results in a lower noise floor of 65 μG/√Hz. The fabrication process and mechanical design of the sensing element are very similar to the previous models. A major difference is that the proof mass is operated in open loop mode, resulting in less complex interface electronics. This is mainly for economical reasons, as the chip size can be reduced by nearly a factor of two. The ADXL150 has a dynamic range of ±100G and is a popular choice for airbag release applications. Both sensors are packaged in a standard 16-pin surface mount package.

More recently, multiaxis accelerometers have been introduced by Analog Devices: a commercial dual-axis device is the ADXL202, which measures acceleration along the two in-plane axes. The proof mass is attached to four pairs of serpentine polysilicon springs affixed to the substrate by four anchor points. It is free to move in the two in-plane directions under the influence of static or dynamic acceleration. The proof mass has movable fingers extending radially on all four sides. These are interdigitated with the stationary fingers to form differential capacitors for x- and y-axes position measurement. A picture of the proof mass is shown in Figure 8.18 and the suspension system is depicted in Figure 8.19.

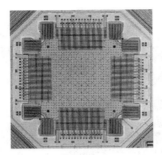

Figure 8.18 The ADXL202 dual-axis accelerometer. The proof mass is compliant to move in both in-plane directions and has interdigitated fingers on all four sides. (Courtesy Analog Devices, Inc. *From:* http://www/analog.com.)

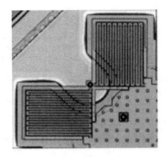

Figure 8.19 The suspension system of the ADXL202. (Courtesy Analog Devices, Inc. *From:* http://www.analog.com.)

The bandwidth of the ADXL202 may be set from 0.01 Hz to 6 kHz via external capacitors. The typical noise floor is 500 μg√Hz, allowing signals below 5 mg to be resolved for bandwidths below 60 Hz.

The latest model, introduced in January 2003, is the ADXL311, which is priced at only $2.50 in quantities greater than 10,000 units. It is also a dual-axis sensor and the working principle is very similar to the previous models. Improved fabrication tolerance controls have allowed improved performance. The main differences are that the noise floor has dropped to 300 μg√Hz and the sensor can now be operated from a single 3V power supply.

Two other companies offer commercial surface-micromachined accelerometers: Motorola and Bosch. The latter have only recently started selling their sensors separately. Previously they were only available embedded in complete automotive safety systems (e.g., for airbag release). Little more information is available other than that given on the datasheets.

Motorola's MMA1201P is a single-axis, surface-micromachined MEMS accelerometer rated for ±40G and is packed in a plastic 16-lead DIP package. The operating temperature range is –40°C to +85°C with a storage temperature range of –40°C to +105°C. The sensing element can sustain accelerations up to 2,000G from any axis and unpowered and powered accelerations up to 500G. The main components of the MMA1201P consist of a surface-micromachined capacitive sensing cell (g-cell) and a CMOS signal conditioning ASIC. The g-cell's mechanical structure is composed of three consecutive semiconductor plates, defining sensitivity along the

z-axis (orthogonal to flat plane of the chip). When the accelerometer system is sub-jected to accelerations with components parallel to the sensitive axis of the g-cell, the center plate moves relative to the outer stationary plates, causing two shifts in capacitance, one for each outer plate, proportional to the magnitude of force applied. The shifts in capacitance are then processed by the CMOS ASIC, which determines the acceleration of the system (using switched capacitor techniques), conditions and filters the signal, and returns a ratiometric high voltage output.

Many companies offer commercial bulk-micromachined accelerometers. For example, the Swiss company Colibrys produces high-performance sensors suitable for inertial guidance and navigation. The MS7000 and MS8000 devices (available from ±1G to ±100G) are their most recent and advanced range. Their devices excel, having high stability, low noise, low temperature drift, and high shock toler-ance. The typical long-term stability is less than 0.1% of the full-scale dynamic range, the bias temperature coefficient is less than 200 mG/°C, and the scale factor temperature coefficient is less than 200 ppm/°C. They use, contrary to Analog Devices, a hybrid approach, where the sensing element and the interface electronics are implemented on separate chips but packaged in a common, standard TO8 or LCC housing. The sensing element together with the ASIC is shown in Figure 8.20.

Table 8.3 gives an overview of a range of companies producing micromachined accelerometers with their most important features.

8.3 Micromachined Gyroscopes

8.3.1 Principle of Operation

Virtually all micromachined gyroscopes rely on a mechanical structure that is driven into resonance and excites a secondary oscillation in either the same structure or in a second one, due to the Coriolis force. The amplitude of this secondary oscillation is directly proportional to the angular rate signal to be measured. The Coriolis force is a virtual force that depends on the inertial frame of the observer. Imagine a person on a spinning disk, rolling a ball radially away from himself, with a velocity v_r. The person in the rotating frame will observe a curved trajectory of the ball. This is due to the Coriolis acceleration that gives rise to a Coriolis force acting perpendicularly to the radial component of the velocity vector of the ball. A way of explaining the origin of this acceleration is to think of the current angular velocity of the ball on its way from the center of the disk to its edge, as shown in Figure 8.21. The angular

Figure 8.20   Commercial bulk-micromachined accelerometer from Colibrys.

Table 8.3 Companies and Their Micromachined Accelerometers

Company	Sensor	Features	Comments
Analog Devices (http://www.analog.com)	Single axis (1.5G, 5G, 50G, 100G)	Analog output; bandwidth dc to 10 kHz; noise floor from 150 μG/√Hz (1.5G) to 4 mG/√Hz (100G); resolution from 1 mG (1.5G) to 40 mG (100G); 5V supply voltage; surfacemicromachined sensing element	Largest provider of commercial accelerometers. They were the first company to integrate a surface micromachined sensing element with the readout and interface electronics on one chip. (Appr. cost: $10 to $200)
	Dual axis (2G, 10G, 50G)	2G, 10G have a duty cycle output	
Applied MEMS (http://www.appliedmems.com)	Single axis (3G)	Analog output; bandwidth dc to 1,500 Hz; noise floor 300 nG/√Hz, 6V to 15V supply voltage; bulk-micromachined sensing element	dc coupled analog force-feedback
	Single axis (200 mG)	Digital output; bandwidth 1 kHz; noise floor 30 nG/√Hz	ASIC with fifth-order sigma delta modulator
	Triaxial (2.5G and 3G)	Analog output; bandwidth 1,500 Hz; noise floor 150 nG/√Hz (3G), 1 μG/√Hz (2.5G); 6V to 15V supply voltage	
Colibrys (http://www.colibrys.com)	Single axis (2G, 10G)	Ratiometric analog output; bandwidth 800 Hz (2G), 600 Hz (10G); output noise floor <18 μG/√Hz; resolution <100 μG (2G), <500 μG (10G); supply voltage 2V to 5V, bulk-micromachined sensing element	Custom design devices from 1G to 100G available
Bosch (http://www.bosch.com)	High-G sensors, single and dual axis (20G, 35G, 50G, 70G, 100G, 140G, 200G)	Analog and ratiometric output; bandwidth 400 Hz, bulk-micromachined sensing element	
	Low-G sensors (0.4G to 3.4G)	Surface-micromachined sensing element	
Endevco (http://www.endevco.com)	Single-axis piezoresistive devices (from 20G to 200,000G)	Analog output; bandwidth typically from tens of hertz to several kilohertz; sensitivity from 1 μV/G (200,000G) to 25 mV/G (20G); supply voltage 10V; bulk-micromachined sensing element	For applications ranging from biodynamics measurements and flutter testing to high shock measurements
	Single-axis capacitive devices (2G, 10G, 30G, 50G, 100G)	Analog output; bandwidth from 15 Hz (2G) to 1 kHz (50G, 100G); sensitivity from 20 mV/G (100G) to 1 V/G (2G); supply voltage 8.5V to 30V; bulk-micromachined sensing element	
	Triaxial (from 500G to 2,000G)	Analog output; bandwidth from tens of hertz to several kilohertz; sensitivity from 0.2 mV/G (2,000G) to 0.8 mV/G (500G); supply voltage 10V; bulk-micromachined device	
Honeywell (http://www.inertialsensor.com)	Single axis (20G, 30G, 60G, 90G)	Analog output; bandwidth 300 Hz; noise floor 0.6 G/vHz, resolution 1G (highest grade 60G device); noise floor 70 nG/vHz, resolution 10G (low grade 30G device); supply voltage 13V to 18V; etched quartz flexure sensing element	Quartz flexure accelerometer for applications ranging from aerospace, energy exploration, and industrial applications; resonating beam accelerometer
	Triaxial	Frequency output; resolution 1G, bandwidth 400 Hz	Assembly of three single-axis accelerometers to provide three-axis sensing
MEMSIC (http://www.memsic.com)	Dual axis (1G, 2G, 5G, 10G)	Analog absolute, analog ratiometric and digital output; bandwidth 17 to 160 Hz (depending on device grade); noise floor 0.2 to 0.75 mG/√Hz; resolution 2 mG; sensitivity for analog absolute from 500 mV/G for 1G to 50 mV/G for 10G, for ratiometric 1,000 mV/G for 1G, 50mV/G for 10G, for digital 20% duty cycle/G for 1G, 2% cycle/G for 10G; supply voltage 2.7V to 5.25V	Integrated MEMS sensors and mixed signal processing circuitry on single chip using standard CMOS process. Operation is based on heat transfer by convection of air. (Appr. cost: $12)
Kionix (http://kionix.com)	Single and dual axis (2G, 5G, 10G)	Analog output; bandwidth 250 Hz; noise floor 60 G/√Hz; resolution 0.1 to 0.3 mG; sensitivity from 200 mV/G (10G) to 1,000 V/G (2G); supply voltage 5V	
Kistler (http://kisler.com)	Single axis and triaxial K-Beam range (2G, 10G, 25G)	Analog output; bandwidth 0 to 300 Hz (2G), 0 to 180 Hz (10G), 0 to 100 Hz (25G); noise floor 38, 200, 570 μG/√Hz; resolution 540 G, 2.8 mG, 8 mG; sensitivity 1 V/G, 200 mV/G, 100 mV/G; supply voltage 3.8V to 16V, bulk-micromachined sensing element	Accelerometers for low-frequency applications. Device assembly provides triaxial sensing.

Table 8.3 (Continued)

	Single axis (20G, 50G), K-Beam range	Analog output; bandwidth 0 to 700 Hz; noise floor 7 $\mu G/\sqrt{Hz}$ (20G), 12 $\mu G/\sqrt{Hz}$ (50G); resolution 100, 170 μG; sensitivity 100, 60 mV/G, supply voltage 15V to 28V, bulk-micromachined sensing element	
	Single axis (2G), ServoK-Beam	Analog output; bandwidth 0 to 2 kHz; noise floor 0.8 $\mu G/\sqrt{Hz}$; resolution 2.5G; sensitivity 1.5 V/G; supply voltage 6V to 15V; bulk-micromachined sensing element	Employs analog electrostatic feedback.
Motorola (http://www.motorola.com)	Single axis (1.5G to 250G)	Ratiometric output; bandwidth from 50 to 400 Hz; noise floor 110 G/vHz; sensitivity from 1.2 V/G (1.5G) to 8 mV/G (250G); supply voltage 5V; surface-micromachined sensing element	Appr. cost: $8
	Dual axis (38G)	Bandwidth 400 Hz; sensitivity 50 mV/G	
Sensornor (http://sensornor.com)	Single axis (50G, 100G, 250G)	Ratiometric analog output; bandwidth 400 Hz; sensitivity 20 mV/G; supply voltage 5V to 11V	Piezoresistive detection, for airbag applications
	Dual axis (50G)	Ratiometric analog output; bandwidth 400 Hz; resolution 0.02G; sensitivity 40 mV/G; supply voltage 5V; bulk-micromachined sensing element	
STMicroelectronics (http://st.com)	Dual axis (2G, 6G)	Analog output; bandwidth 0 to 4 kHz; noise floor 50 $\mu G/\sqrt{Hz}$; sensitivity 1 V/G; supply voltage 5V	For handheld gamepad devices

velocity v_{ang} increases with the distance of the ball from the center ($v_{ang} = r\Omega$), but any change in velocity inevitably gives rise to acceleration in the same direction.

This acceleration is given by the cross product of the angular velocity Ω of the disk and the radial velocity v_r of the ball:

$$\text{Coriolis acceleration:} \; \vec{a}_c = 2\vec{\Omega} \times \vec{v}_r; \quad \text{Coriolis force:} \; \vec{F}_c = 2m\vec{\Omega} \times \vec{v}_r$$

Macroscopic mechanical gyroscopes typically use a flywheel that has a high mass and spin speed and hence a large angular momentum which counteracts all external torque and creates an inertial reference frame that keeps the orientation of the spin axis constant. This approach is not very suitable for a micromachined

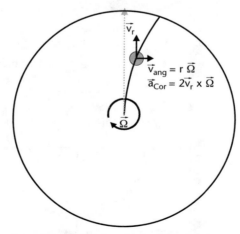

Figure 8.21 A ball rolling from the center of a spinning disk is subjected to Coriolis acceleration and hence shows a curved trajectory.

sensor since the scaling laws are unfavorable where friction is concerned, and hence, there are no high-quality micromachined bearings. Consequently, nearly all MEMS gyroscopes use a vibrating structure that couples energy from a primary, forced oscillation mode into a secondary, sense oscillation mode. In Figure 8.22, a lumped model of a simple gyroscope suitable for a micromachined implementation is shown. The proof mass is excited to oscillate along the x-axis with a constant amplitude and frequency. Rotation about the z-axis couples energy into an oscillation along the y-axis whose amplitude is proportional to the rotational velocity. Similar to closed loop micromachined accelerometers, it is possible to incorporate the sense mode in a force-feedback loop. Any motion along the sense axis is measured and a force is applied to counterbalance this sense motion. The magnitude of the required force is then a measure of the angular rate signal.

One problem is the relatively small amplitude of the Coriolis force compared to the driving force. Assuming a sinusoidal drive vibration given by $x(t) = x_0\sin(\omega_d t)$, where x_0 is the amplitude of the oscillation and ω_d is the drive frequency, the Coriolis acceleration is given by $a_c = 2v(t) \times \Omega = 2\Omega x_0 \omega_d \cos(\omega_d t)$. Using typical values of $x_0 = 1\ \mu$m, $\Omega = 1°$/s, and $\omega_d = 2\pi 20$ kHz, the Coriolis acceleration is only 4.4 mm/s². If the sensing element along the sense axis is considered as a second order mass-spring-damper system with a $Q = 1$, the resulting displacement amplitude is only 0.0003 nm [51]. One way to increase the displacement is to fabricate sensing elements with a high Q structure and then tune the drive frequency to the resonant frequency of the sense mode. Very high Q structures, however, require vacuum packaging, making the fabrication process much more demanding. Furthermore, the bandwidth of the gyroscopes is proportional to ω_d/Q; hence, if a quality factor of 10,000 or more is achieved in vacuum, the bandwidth of the sensor is reduced to only a few hertz. Lastly, it is difficult to design structures for an exact resonance frequency, due to manufacturing tolerances. A solution is to design the sense mode for a higher resonant frequency than the drive mode and then decrease the resonant frequency of the sense mode by tuning the mechanical spring constant using electrostatic forces [52].

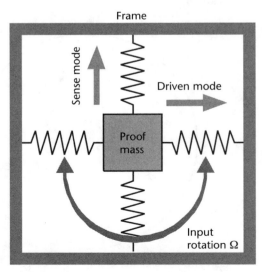

Figure 8.22 Lumped model of a vibratory rate gyroscope.

An acceptable compromise between bandwidth and sensitivity is to tune the resonant frequency of the sense mode close to the drive frequency (within 5% to 10%).

A second fundamental problem with vibratory rate micromachined gyroscopes is due to so-called quadrature error. This type of error originates from manufacturing tolerances manifesting themselves as a misalignment of the axis of the driven oscillation from the nominal drive axis. As a result, a small proportion of the driven motion will be along the sense axis. Even though the misalignment angle is very small, due to the minute Coriolis acceleration, the resulting motion along the sense axis may be much larger than the motion caused by the Coriolis acceleration.

8.3.2 Research Prototypes

8.3.2.1 Single-Axis Gyroscopes

Early micromachined gyroscopes were based on double-ended tuning forks. Two tines, which are joined at a junction bar, are excited to resonate in antiphase along one axis. Rotation causes the tines to resonate along the perpendicular axis. Different actuation mechanisms can be used to excite the primary or driven oscillation mode. Examples of electromagnetic actuation are given in [53–56] and have the advantage that large oscillation amplitudes are easily achievable. A severe disadvantage, however, is that it requires a permanent magnet to be mounted in close proximity to the sensing element, thereby making the fabrication process not completely compatible with that of batch processing. Piezoelectric excitation has also been reported, for example, by Voss et al. [57], who realized a double-ended tuning fork structure with the oscillation direction perpendicular to the wafer surface using bulk micromachining. The prevailing approach for prototype gyroscopes, however, is to use electrostatic forces to excite the primary oscillation.

For detecting the secondary or sense oscillation, different position measurement techniques have been used such as piezoresistive [56, 57], tunneling current [58], optical [59], and capacitive, the latter being by far the predominant method.

Greiff et al. [2], from the Charles Stark Draper Laboratories, presented a tuning fork sensor that can be regarded as one of the first micromachined gyroscopes suitable for batch-processing. The bulk-micromachined sensing element is shown in Figure 8.23. It is a two-gimbal structure supported by torsional flexures. The outer gimbal structure is driven into oscillatory motion at 3 kHz out of the wafer plane by

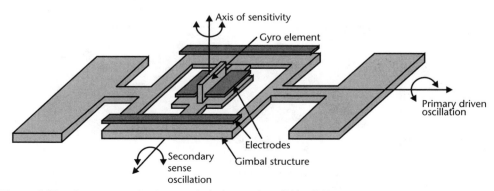

Figure 8.23 Gyroscope using a two-gimbal structure. (*After:* [2].)

electrostatic forces. An automatic gain control (AGC) control loop ensures that the oscillation amplitude is constant. In the presence of a rotation about the axis normal to the sensing element plane, energy is transferred to the inner gimbal structure, which starts vibrating at the same frequency at an amplitude proportional to the angular spin rate. Maximum sensitivity is achieved when the drive frequency of the outer structure is equal to the resonant frequency of the inner gimbal. The sensing element could be operated in a force-balance mode. Electrostatic forces generated by voltages on the feedback electrodes counterbalance the movement of the inner gimbal. The fixed electrodes above the inner and outer gimbal structure were fabricated by an EDP wet-etch that removes sacrificial silicon dioxide. The lower electrodes underneath the structure were implemented as p-type buried electrodes and are electrically isolated by a reverse biased p-n junction from the substrate. The gap between the fixed electrodes and the movable on the resonators is between 8 and 10 μm. To increase the mass of the inner resonator, an inertial mass made from gold, of 25-μm height, was electroformed.

The first polysilicon surface-micromachined vibratory rate gyroscope was presented in 1996 by Clark and Howe [51]. It is a direct implementation of the lumped model presented in Figure 8.22. Standard comb drive actuators were used to excite the structure to oscillate along one in-plane axis (x-axis), which allows relatively large drive amplitudes. Any angular rate signal about the out-of-plane axis (z-axis) excites a secondary motion along the other in-plane axis (y-axis). The sensing element is shown in Figure 8.24 and consists of a 2-μm-thick polysilicon structure. In this reference quadrature error is discussed in detailed and it is shown that a misalignment of the primary oscillation axis with the ideal x-axis of only one part in 3.6 million will result in a quadrature error equal to the signal of a 1°/sec rotation about the z-axis. No fabrication process can be accurate to such a degree, and hence, electrostatic tuning is used to alleviate this problem. The quadrature error is proportional to the position of the primary oscillation, whereas the Coriolis acceleration is proportional to the velocity of the primary oscillation; hence, the resulting forces are 90° out of phase (this explains the term *quadrature error*). The inner interdigitated

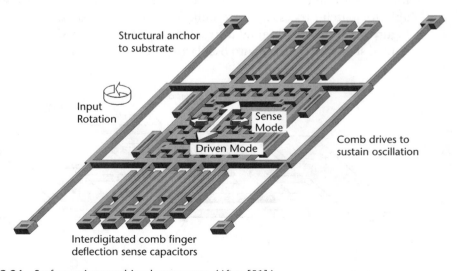

Figure 8.24 Surface-micromachined gyroscope. (*After:* [51].)

electrodes of the mechanical structure are used to exert an electrostatic force, which is proportional to the position of the primary oscillation. Applying a biasing voltage, together with a small differential voltage, results in an electrostatic force that allows counterbalancing of the unwanted motion of the proof mass of the primary oscillation due to quadrature error. The paper also discusses the required interface and control electronics for sustaining a constant amplitude and primary frequency oscillation. For the latter, a phase-locked loop is chosen; for the former an automatic gain control circuit is used. Furthermore, it is possible to tune the resonant frequencies of the primary and secondary oscillation modes by applying electrostatic negative springs. As a good compromise between bandwidth and sensitivity, a mismatch of about 5% to 10% is suggested.

Another surface-micromachined gyroscope was presented by Geiger et al. [60, 61]. It was manufactured using the Bosch foundry process [62], which features a polycrystalline structural layer with a thickness of 10.3 μm. This relatively large thickness for a surface-micromachined process is achieved by epitaxial deposition of silicon. Under the freestanding structures a second thinner layer of polycrystalline silicon is used for electrodes and as interconnects. The sensing element, shown in Figure 8.25, has two decoupled rotary oscillation modes. The primary driven mode is around the z-axis and is excited with electrostatic forces using the inner spoke electrodes of the inner wheel. Attached to the inner wheel, by torsional springs, is a rectangular structure, which, in response to rotation about the sensitive axis (x-axis), will exhibit a secondary rotary oscillation about the y-axis. Owing to the high stiffness of the suspension beam in this direction, the oscillation of the inner wheel is suppressed and only the rectangular structure can move due to a Coriolis force. With this approach the primary and secondary modes are mechanically decoupled, which suppresses mechanical cross-coupling effects such as quadrature error. The oscillation of the secondary mode is detected capacitively by electrodes on the substrate. The sensor reported a dynamic range of 200°/sec, a scale factor of 10 mV/(°/sec), and a rms noise of 0.05°/sec in a 50-Hz bandwidth, which makes it suitable for most automotive applications.

Another popular implementation of a micromachined gyroscope, based on a single oscillating structure with two vibrating modes, is shown in Figure 8.26.

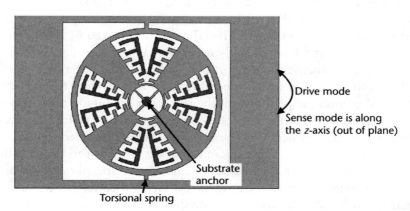

Figure 8.25 Surface-micromachined gyroscope with decoupled drive and sense mode. The drive mode is excited by an electrostatic comb drive and is rotational about the z-axis (out-of-plane). The sense oscillation causes the outer frame to oscillate along the z-axis. (*After:* [59].)

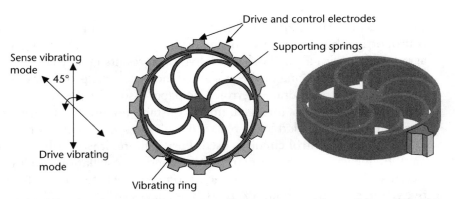

Figure 8.26 Vibrating ring structure gyroscope. Electrodes around the periphery of the ring are used to electrostatically excite the ring into a resonance mode. A secondary mode at 45° is a measure of the angular rate and is sensed capacitively. (*After:* [63].)

It is based on a ring supported by a number of semicircular springs and anchored in the middle. The ring is excited to vibrate electrostatically in-plane, the vibration having an elliptic shape. Any rotation about the axis normal to the ring structure transfers energy to a secondary mode, which is 45° apart from the primary mode. In other words, the antinodes of the primary flexural mode are located at the nodes of the secondary flexural mode. Electrodes placed at these positions are used to capacitively measure the amplitude of the secondary mode, which is proportional to the angular rate to be measured. An obvious advantage of this design is the high degree of symmetry of the sensing element. An early version was presented by Putty and Najafi in 1994 [63]. It relied on a nickel electroplated ring structure, which was fabricated on a wafer containing standard CMOS circuitry for the control and interface electronics. Subsequently, this group presented more advanced versions of this approach. Another electroplated ring gyroscope was presented by Sparks et al. [64], which mainly improved the signal and interface circuitry. More recently, improved designs have been reported based on a high aspect ratio ring made from polysilicon [65, 66]. The fabrication relies on the deep reactive dry etching of 50- to 100-m-deep trenches with near vertical sidewalls into a low-resistive silicon substrate. The trenches are subsequently refilled with highly doped polysilicon over a sacrificial silicon dioxide layer. After various patterning and etching steps of the oxide and the structural polysilicon, the sacrificial oxide is removed by a HF etch step to free the ring structure and form the air gaps between the electrodes and the ring. The ring is 1.1 mm in diameter, the support post in the middle has a diameter of 120 μm, and the width of the ring and support springs is 4 μm. Sixteen fixed electrodes are evenly located around the periphery of the ring; they are 60 μm tall, 150 μm long, and are separated from the ring by a 1.4-μm air gap. The fabrication technology has the advantage that the height of the ring structure and the electrodes can be made in the order of a hundred or more microns and the air gaps can be made in the submicron range. This results in high values of capacitance for vibration measurements; thus, the sensitivity is increased considerably. The fabrication process also allows large air gaps, which can be used to excite the structure in the primary mode with high amplitude, again resulting in higher sensitivity. There is, however, a trade-off between the higher voltages required to electrostatically drive the ring using larger air gaps. Test results were reported for a structure 80 μm tall, operated at a low pressure (1 mTorr), which resulted in a quality factor for the oscillation of 1,000 to 2,000. This was lower than expected and was attributed to anchor losses and voids inside the polysilicon beams. Improved designs were expected to have a quality factor of up to 20,000. Similar to other micromachined gyroscopes, the resonant frequencies of drive and sense mode were designed to be equal in order to amplify the sense mode amplitude by the quality factor. Both resonant frequencies had a nominal value of 28.3 kHz. Any mismatch due to fabrication tolerances can be electrostatically tuned by applying suitable voltages to the electrodes around the periphery of the ring. A 63-Hz mismatch was observed between the sense and drive modes, which required a tuning voltage of only 0.9V. Other prototypes had a higher mismatch of up to 1 kHz for which a tuning voltage of 6V was required to match sense and drive mode resonant frequencies. The resolution of the device was measured to be less than 1°/sec for a 1-Hz bandwidth; however, with some changes in the interface circuitry this should be reduced to 0.01°/sec, which is then limited by the Brownian noise floor of the structure.

8.3.2.2 Dual-Axis Gyroscopes

It is also possible to design micromachined gyroscopes that are capable of sensing angular motion about two axes simultaneously. These devices are based on a rotor-like structure that is driven into a rotary oscillation by electrostatic comb-drives. Angular motion about the x-axis causes a Coriolis acceleration about the y-axis, which, in turn, results in a tilting oscillation of the rotor. Similarly, any rotation of the sensor about the x-axis causes the rotor to tilt about the x-axis. Conceptually, this is shown in Figure 8.27.

An implementation of such a dual-axis gyroscope was reported by Junneau et al. [67]. It was manufactured in a surface-micromachining process with a 2-m-thick proof mass. The interface and control electronics were integrated on the same chip. Underlying pie-shaped electrodes capacitively detect the tilting motion. To

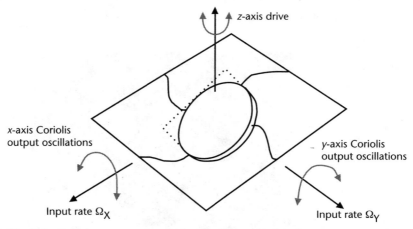

Figure 8.27 A dual-axis gyroscope. A rotor is driven into rotational resonance; angular motion about the x- and y-axes causes the rotor to tilt, which can be measured capacitively by electrodes below it. (*After:* [66].)

distinguish the two different output modes, a different voltage modulation frequency (200 and 300 kHz) is used for each sense electrode pair. The reported performance was 1°/sec in a 25 Hz bandwidth. The natural driving frequency of the rotor is about 25 kHz. Similar to single-axis devices a high quality factor can be used to amplify the output motion. In a 60-mTorr vacuum Junneau et al. report a quality factor of about 1,000. Electrostatic tuning of the different resonant frequencies can be used. Cross-coupling between the two output modes is a major problem and was measured to be as high as 15%. This implies that for a commercially viable version more research has to be done for such a dual-axis gyroscope.

A conceptually similar implementation was reported by An et al. [68]. The authors reported a higher resolution, of 0.1°/sec, which was mainly due to a thicker proof mass (7 μm).

8.3.3 Commercial Micromachined Gyroscopes

Silicon Sensing Systems (a joint venture between BAE SYSTEMS and Sumitomo Precision Products [69]) is producing a very successful commercial gyroscope based upon a ring-type sensing element. It uses magnetic actuation and detection, which may prove to be problematic for further device size reduction. The ring has diameter of 6 mm and is connected by eight radially compliant spokes to a support frame with the dimensions of 10 × 10 mm. It is fabricated by deep reactive ion etching of a 100-m-thick silicon wafer. Current-carrying conductor loops are deposited on the surface of the ring structure. These loops, together with the magnetic field, set up by the permanent magnet provide the signal pick-off and primary oscillation mode drive. This gyroscope has a resolution of 0.005/sec, a bandwidth of 70 Hz, and a noise floor of 0.1/sec in a 20-Hz bandwidth. A picture of the sensor is shown in Figure 8.28. Currently, they are developing a capacitive sensor without a permanent magnet, thereby allowing for further size reduction [70].

Analog Devices has recently released the ADXRS family of integrated angular rate-sensing gyroscopes, which contains the ADXRS300 (with dynamic range of ±300 m°/sec) and the ADXRS150 (with dynamic range of ±150°/sec). It is the first fully integrated commercial gyroscope. A picture of the chip is shown in Figure 8.29(a). It operates from 5V supply over the industrial temperature range of −40°C

Figure 8.28   Commercial micromachined gyroscope from Silicon Sensing Systems. (*From:* [69]. © 1997 BAE Systems. Permission obtained from BAE Systems, who are a part of SSS.)

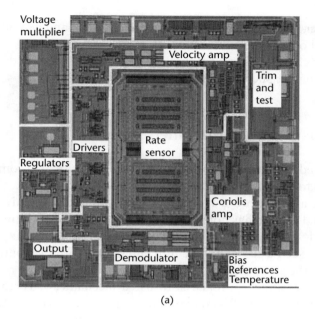

(a)

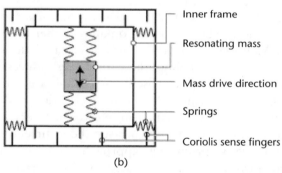

(b)

Figure 8.29   (a) Die photo of the surface-micromachined gyroscope from Analog Devices with the interface and control electronics integrated on the same chip. It contains two identical mechanical structures to achieve differential sensing. (b) Schematic drawing of one of the two identical gyroscope elements. (Courtesy Analog Devices, Inc. Picture taken from ADI Web site, http://www.analog.com.)

to +85°C and is available in a space-saving 32-pin Ball Grid Array surface-mount package measuring 7 × 7 × 3 mm. Both are priced at approximately $30 per unit in thousand-piece quantities. Because the internal resonators require 14V to 16V for proper operation, ADI includes on-chip charge pumps to boost an applied TTL-level voltage. Both the ADXRS150 and ADXRS300 are essentially z-axis gyroscopes based on the principle of resonant-tuning-fork gyroscopes. In these systems, two polysilicon sensing structures each contain a so-called dither frame that is driven electrostatically to resonance. Interestingly, the gyroscope includes two identical structures to enable differential sensing in order to reject environmental shock and vibration. Figure 8.29(b) shows one structure schematically.

A rotation about the z-axis, normal to the plane of the chip, produces a Coriolis force that displaces the inner frame perpendicular to the vibratory motion. This Coriolis motion is detected by a series of capacitive pick-off structures on the edges of the inner frame. The resulting signal is amplified and demodulated to produce the rate signal output.

Other commercial gyroscopes are in the final stages of their commercialization from companies such as Samsung and Sensonor.

8.4 Future Inertial Micromachined Sensors

It is believed that in future years, the major innovation will come from multiaxis sensors, both for linear and angular motion. As described above, three-axis accelerometers using a single proof mass have been presented already as prototypes, but a commercial version has not yet been implemented. As an ultimate goal, a single sensor capable of measuring linear and angular motion for six degrees of freedom is envisaged. Such a sensor can be fully integrated with the control and interface electronics on the same chip.

One interesting approach is to use a mechanical structure similar to the one shown in Figure 8.16. Watanabe et al. [71] report a five-axis capacitive motion sensor. Linear acceleration is sensed in the same way as described in the paper by Mineta et al. [44]: Out-of-plane acceleration causes the proof mass to move along the z-axis, and in-plane acceleration along either the x- or y-axes makes the proof mass tilt. Additionally, the proof mass is vibrated along the z-axis with electrostatic forces. Angular motion about the x- or y-axes induces a Coriolis-based tilting oscillation of the proof mass. The oscillatory signals are of much higher frequency (about 2 kHz) as the signals caused by linear acceleration, and hence, they can be separated easily in the frequency domain using electronic filters. In this way linear acceleration and angular rate signals can be measured concurrently.

Another very promising approach towards such a sensor is to use a micromachined disk that is levitated by electrostatic or magnetic forces and spun about its main axes. This is similar to macroscopic flywheel type gyroscopes; however, the lack of a good bearing in the microworld has excluded this approach so far for micromachined gyroscopes. Using a levitated object alleviates this problem. Any angular motion perpendicular to the spin axis of the disk will cause it to recess, and this can be detected by a capacitive position measurement to provide a measure of the angular velocity. Using a levitated object for inertial sensing has several advantages. First, since there is no mechanical connection from the substrate to the disk, the effective spring constant is solely dependent on the electrostatic forces set up by voltages or currents applied to surrounding electrodes; hence, the characteristics of the sensor, such as bandwidth and sensitivity, can be adjusted on-line, according to the application requirements. Second, when used as a gyroscope, quadrature error is inherently ruled out. The comparable effect, due to the imbalance of the mass, will manifest itself at the rotation frequency, whereas the Coriolis force will cause the disk to recess at the rotational speed of the body of interest. These two frequencies are several orders of magnitude apart and are easy to separate. Furthermore, there is no need to tune the drive and sense resonant frequencies since the scale factor does not depend on the matching of different modal frequencies. Linear acceleration along the three axes can be measured simultaneously by measuring the displacement of the disk.

Levitation using magnetic forces has been investigated by Shearwood et al. [72, 73], who successfully demonstrated a gyroscope based on this approach. The electromagnetic forces are produced by currents up to 1A, which precludes the use of standard integrated electronics, which is a severe disadvantage of this approach. A more promising approach is to use electrostatic forces to levitate and spin a disk. Fukatsu et al. [74] have developed a prototype of such a device and have demonstrated the feasibility of using it for simultaneously detecting linear and angular motion. Houlihan et al. [75] present the design and simulation of a similar device for three-axis acceleration measurement, which is also suitable to detect angular motion about two axes if rotated. Here, the micromachined disk is incorporated in a multipath sigma-delta modulator control system. A system-level diagram of the sensor is shown in Figure 8.30.

It should be emphasized here that these devices are promising and interesting approaches to future inertial sensors. It will take considerable effort and time, however, to develop them into commercial products.

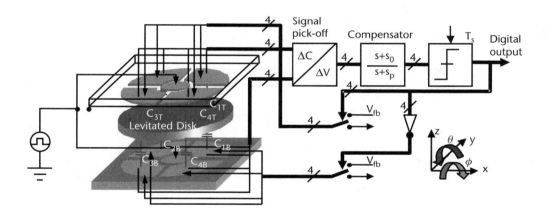

Figure 8.30   An electrostatically levitated disk, which is spun about the z-axis, can be used to measure three-axis linear acceleration and angular velocity about two axes (x and y) and control system. (*After:* [75].)

References

[1]   Roylance, L. M., and J. B. Angell, "A Batch-Fabricated Silicon Accelerometer," *IEEE Trans. Electron Devices*, ED-26, 1979, pp. 1911–1917.

[2]   Greiff, P., et al., "Silicon Monolithic Micro-Mechanical Gyroscope," *Proc. Transducers '91*, 1998, pp. 966–968.

[3]   Andrews, M., I. Harris, and G. Turner, "A Comparison of Squeeze-Film Theory with Measurements on a Microstructure," *Sensors and Actuators*, Vol. A36, 1993, pp. 79–87.

[4]   Starr, J. B., "Squeeze-Film Damping in Solid-State Accelerometers," *IEEE Solid-State Sensor and Actuator Workshop*, Hilton Head, SC, 1990, pp. 44–47.

[5]   Veijola, T., and T. Ryhaenen, "Equivalent Circuit Model of the Squeezed Gas Film in a Silicon Accelerometers," *Sensors and Actuators*, Vol. A48, 1995, pp. 239–248.

[6]   Zhang, L., et al., "Squeeze-Film Damping in Micromechanical Systems," *ASME, Micromechanical Systems*, DSC-Vol. 40, 1992, pp. 149–160.

[7]   Henrion, W., et al., "Wide Dynamic Range Direct Digital Accelerometer," *IEEE Solid-State Sensor and Actuator Workshop*, Hilton Head, SC, 1990, pp. 153–157.

[8]   De Coulon, Y., et al., "Design and Test of a Precision Servoaccelerometer with Digital Output," *7th Intl. Conf. Solid-State Sensors and Actuators (Transducer '93)*, Yokohama, Japan, 1993, pp. 832–835.

[9]   Smith, T., et al., "Electro-Mechanical Sigma-Delta Converter for Acceleration Measurements," *IEEE International Solid-State Circuits Conference*, San Francisco, CA, 1994, pp. 160–161.

[10] Lu, C., M. Lemkin, and B. Boser, "A Monolithic Surface-Micromachined Accelerometer with Digital Output," *IEEE J. Solid-State Circuits*, Vol. 30, No. 12, 1995, pp. 1367–1373.

[11] Allen, H. V., S. C. Terry, and W. Knutti, "Understanding Silicon Accelerometers," *Sensors*, 1989, pp. 1–6.

[12] Barth, P. W., et al., "A Monolithic Silicon Accelerometer with Integral Air Damping and Overrange Protection," *Tech. Dig. Solid-State Sensors and Actuators Workshop*, Hilton Head, SC, 1988, pp. 35–28.

[13] Pourahmadi, F., L. Christel, and K. Petersen, "Silicon Accelerometer with New Thermal Self-Test Mechanism," *Tech. Dig. Solid-State Sensors and Actuators Workshop*, 1992, Hilton Head, SC, pp. 122–125.

[14] Allen, H. V., S. C. Terry, and D. W. DeBruin, "Accelerometer Systems with Self-Testable Features," *Sensors and Actuators*, Vol. 20, 1989, pp. 153–161.

[15] Seidel, H., et al., "A Piezoresistive Accelerometer with Monolithically Integrated CMOS-Circuitry," *Proc. Eurosensors IX and Transducers '91*, Stockholm, Sweden, 1995, pp. 597–600.

[16] Reithmuller, W., et al., "A Smart Accelerometer with On-Chop Electronics Fabricated by a Commercial CMOS Process," *Sensors and Actuators*, Vol. A31, 1992, pp. 121–124.

[17] Kraft, M., C. P. Lewis, and T. G. Hesketh, "Closed Loop Silicon Accelerometers," *IEE Proceedings—Circuits, Devices and Systems*, Vol. 145, No. 5, 1998, pp. 325–331.

[18] Rudolf, F., et al., "Precision Accelerometers with µg Resolution," *Sensors and Actuators*, Vol. A21-23, 1990, pp. 297–302.

[19] Warren, K., "Electrostatically Force-Balanced Silicon Accelerometer," *J. of the Inst. of Nav.*, Vol. 38, No. 1, 1991, pp. 91–99.

[20] Gerlach-Meyer, U. E., "Micromachined Capacitive Accelerometer," *Sensors and Actuators*, Vol. A25-27, 1991, pp. 555–558.

[21] Seidel, H., et al., "Capacitive Silicon Accelerometer with Highly Symmetrical Design," *Sensors and Actuators*, Vol. A21–23, 1990, pp. 312–315.

[22] McDonald, G. A., "A Review of Low Cost Accelerometers for Vehicle Dynamics," *Sensors and Actuators*, Vol. A21–23, 1990, pp. 303–307.

[23] Analog Devices ADXL50-Monolithic Accelerometer with Signal Conditioning. Datasheet, Norwood, MA, 1993.

[24] Analog Devices ADXL05-Monolithic Accelerometer with Signal Conditioning. Datasheet, Norwood, MA, 1993.

[25] Goodenough, F., "Airbags Boom When IC Accelerometer Sees 50G," *Electronic Design*, Vol. 8, 1991, pp. 45–56.

[26] Mukherjee, A., Y. Zhou, and G. K. Fedder, "Automated Optimal Synthesis of Microaccelerometers," *Tech. Dig. 12th IEEE Intl. Conf. Micro Electro Mechanical Systems (MEMS'99)*, Orlando, FL, 1999, pp. 326–331.

[27] Hierold, C., et al., "A Pure CMOS Surface-Micromachined Integrated Accelerometer," *Sensors and Actuators*, Vol. A57, 1996, pp. 111–116.

[28]   Ward, M. C. L., D. O. King, and A. M. Hodge, "Performance Limitations of Surface-Machined Accelerometers Fabricated in Polysilicon Gate Material," *Sensors and Actuators*, Vol. A46–47, 1995, pp. 205–209.

[29]   Kuehnel, W., and S. Sherman, "A Surface Micromachined Silicon Accelerometer with On-Chip Detection Circuitry," *Sensors and Actuators*, Vol. A45, 1994, pp. 7–16.

[30]   Chau, K., et al., "An Integrated Force Balanced Capacitive Accelerometer for Low-g Applications," *Sensors and Actuators*, Vol. A54, 1996, pp. 472–476.

[31]   Lu, C., M. Lemkin, and B. Boser, "A Monolithic Surface Micromachined Accelerometer with Digital Output," *IEEE J. Solid-State Circuits*, Vol. 30, No. 12, 1995, pp. 1367–1373.

[32]   Boser, B. E., and R. T. Howe, "Surface Micromachined Accelerometers," *IEEE J. of Solid-State Circuits*, Vol. 31, No. 3, 1996, pp. 336–375.

[33]   Yazdi, N., and K. Najafi, "An All-Silicon Single-Wafer Micro-g Accelerometer with a Combined Surface and Bulk Micromachining Process," *J. of Micromech. Microeng.*, Vol. 9, No.4, 2000, pp. 544–550.

[34]   Chen, P. L., R. S. Muller, and A. P. Andrews, "Integrated Silicon Pi-FET Accelerometers with Proof Mass," *Sensors and Actuators*, Vol. 5, No.2, 1983, pp. 119–126.

[35]   Nemirovsky, Y., et al., "Design of a Novel Thin-Film Piezoelectric Accelerometer," *Sensors and Actuators*, Vol. A56, 1996, pp. 239–249.

[36]   Beeby, S. P., N. J. Grabham, and N. M. White, "Microprocessor Implemented Self-Validation of Thick-Film PZT/Silicon Accelerometer," *Sensors and Actuators*, A92, 2001, pp. 168–174.

[37]   Rockstad, H. K., et al., "A Miniature, High-Sensitivity, Electron Tunnelling Accelerometer," *Sensors and Actuators*, Vol. A53, 1996, pp. 227–231.

[38]   Burns, D. W., et al., "Sealed-Cavity Resonant Microbeam Accelerometer," *Sensors and Actuators*, Vol. A53, 1996, pp. 249–255.

[39]   Roessig, T. A., et al., "Surface-Micromachined Resonant Accelerometer," *9th Int. Conf. Solid-State Sensors and Actuators (Transducer '97)*, Chicago, IL, 1997, pp. 859–862.

[40]   Roszhart, T. V., et al., "An Inertial Grade, Micromachined Vibrating Beam Accelerometer," *8th Intl. Conf. Solid-State Sensors and Actuators (Transducer '95)*, Stockholm, Sweden, 1995, pp. 659–662.

[41]   Burrer, C., J. Esteve, and E. Lora-Tomayo, "Resonant Silicon Accelerometer in Bulk-Micromachining Technology," *J. of Micromech. Microeng.*, Vol. 5, No. 2, 1996, pp. 122–130.

[42]   Lemkin, M. A., et al., "A Three-Axis Force Balanced Accelerometer Using a Single Proof Mass," *9th Intl. Conf. Solid-State Sensors and Actuators (Transducer '97)*, Chicago, IL, 1997, pp. 1185–1188.

[43]   Lemkin, M. A., and B. Boser, "A Three-Axis Micromachined Accelerometer with a CMOS Position-Sense Interface and Digital Offset-Trim Electronics," *IEEE J. of Solid-State Circuits*, Vol. 34, No. 4, 1999, pp. 456–468.

[44]   Mineta, T., et al., "Three-Axis Capacitive Accelerometer with Uniform Axial Sensitivities," *J. of Micromech. Microeng.*, Vol. 6, 1996, pp. 431–435.

[45]   Takao, H., Y. Matsumoto, and M. Ishida, "A Monolithically Integrated Three-Axis Accelerometer Using CMOS Compatible Stress-Sensitive Differential Amplifiers," *IEEE Trans. on Electron Devices*, Vol. 46, No. 1, 1999, pp. 109–116.

[46]   Takao, H., Y. Matsumoto, and M. Ishida, "Stress-Sensitive Differential Amplifiers Using Piezoresistive Effects of MOSFETs and Their Application to Three-Axial Accelerometers," *Sensors and Actuators*, Vol. A65, 1998, pp. 61–68.

[47]   Schröpfer, G., et al., "Lateral Optical Accelerometer Micromachined in (100) Silicon with Remote Readout Based on Coherence Modulation," *Sensors and Actuators*, Vol. A68, 1998, pp. 344–349.

[48]   Huang, R. S., E. Abbaspour-Sani, and C. Y. Kwok, "A Novel Accelerometer Using Silicon Micromachined Cantilever Supported Optical Grid and PIN Photodetector," *8th Intl.*

Conf. Solid-State Sensors and Actuators (Transducer '95), Stockholm, Sweden, 1995, pp. 663–665.

[49] Leung, A. M., et al., "Micromachined Accelerometer Based on Convection Heat Transfer," *Proc. IEEE Micro Electro Mechanical Systems Workshop (MEMS'98)*, Heidelberg, Germany, 1998, pp. 627–630.

[50] Abbaspour-Sani, E., R. S. Huang, and C. Y. Kwok, "A Linear Electromagnetic Accelerometer," *Sensors and Actuators*, Vol. A44, 1994, pp. 103–109.

[51] Clark, W. A., R. T. Howe, and R. Horowitz, "Surface Micromachined Z-Axis Vibratory Rate Gyroscope," *Digest of Solid-State Sensors and Actuator Workshop*, 1996, pp. 283–287.

[52] Oh, Y., et al., "A Surface-Micromachined Tunable Vibratory Gyroscope," *Proc. IEEE Micro Electro Mechanical Systems Workshop (MEMS'98)*, Nagoya, Japan, 1999, pp. 272–277.

[53] Hashimoto, M., et al., "Silicon Angular Rate Sensor Using Electromagnetic Excitation and Capacitive Detection," *Journal of Microelectromechanical Systems,* Vol. 5, 1995, pp. 219–225.

[54] Choi, J., K. Minami, and M. Esashi, "Application of Deep Reactive Ion Etching for Silicon Angular Rate Sensor," *Microsystem Technologies*, Vol. 2, 1996, pp. 186–190.

[55] Lutz, M., et al., "A Precision Yaw Rate Sensor in Silicon Micromachining," *Proc. 9th Int. Conf. Solid-State Sensors and Actuators (Transducer '97)*, Vol. 2, Chicago, IL, 1997, pp. 847–850.

[56] Paoletti, F., M. A. Gretillat, and N. F. de Rooij, "A Silicon Vibrating Micromachined Gyroscope with Piezoresistive Detection and Electromagnetic Excitation," *Proc. IEEE Micro Electro Mechanical Systems Workshop (MEMS'96)*, San Diego, CA, 1996, pp. 162–167.

[57] Voss, R., et al., "Silicon Angular Rate Sensor for Automotive Applications with Piezoelectric Drive and Piezoresistive Readout," *Proc. 9th Intl. Conf. Solid-State Sensors and Actuators (Transducer '97)*, Vol. 2, Chicago, IL, 1997, pp. 879–882.

[58] Kubena, R. L, et al., "A New Tunnelling-Based Sensor for Inertial Rotation Rate Measurements," *Journal of Microelectromechanical Systems,* Vol. 8, 1999, pp. 439–447.

[59] Degani, O., et al., "Optimal Design and Noise Consideration of Micromachined Vibrating Rate Gyroscope with Modulated Integrative Differential Optical Sensing," *Journal of Microelectromechanical Systems,* Vol. 7, No. 3, 1998, pp. 329–338.

[60] Geiger, W., et al., "A New Silicon Rate Gyroscope," *Sensors and Actuators*, Vol. A73, 1999, pp. 45–51.

[61] Geiger, W., et al., "The Silicon Angular Rate Sensor System DAVED," *Sensors and Actuators*, Vol. A84, 2000, pp. 280–284.

[62] http://www.europractice.bosch.com.

[63] Putty, M. W., and K. Najafi, "A Micromachined Vibrating Ring Gyroscope," *Digest of Solid-State Sensors and Actuators Workshop*, Hilton Head, SC, 1994, pp. 213–220.

[64] Sparks, D. R., et al., "A CMOS Integrated Surface Micromachined Angular Rate Sensor: Its Automotive Applications," *Proc. 9th Intl. Conf. Solid-State Sensors and Actuators (Transducer '97)*, Vol. 2, Chicago, IL, 1997, pp. 851–854.

[65] Ayazi, F., et al., "A High Aspect-Ratio Polysilicon Vibrating Ring Gyroscope," *Digest of Solid-State Sensors and Actuators Workshop*, Hilton Head, SC, 2000, pp. 289–292.

[66] Ayazi, F., and K. Najafi, "A HARPSS Polysilicon Vibrating Ring Gyroscope," *Journal of Microelectromechanical Systems*, Vol. 10, No. 2, 2001, pp. 169–179.

[67] Junneau, T., A. P. Pisano, and J. H. Smith, "Dual Axis Operation of a Micromachined Rate Gyroscope," *9th Int. Conf. Solid-State Sensors and Actuators (Transducers '97)*, Vol. 2, Chicago, IL, 1997, pp. 883–886.

[68] An, S., et al., "Dual-Axis Microgyroscope with Closed Loop Detection," *Sensors and Actuators*, Vol. A73, 1999, pp. 1–6.

[69]   Hopkin, I., "Performance and Design of a Silicon Micromachined Gyro," *Proc. Symp. Gyro Technology*, Stuttgart, Germany, 1997, pp. 1.0–1.11.

[70]   Fell, C., I. Hopkin, and K. Townsend, "A Second Generation Silicon Ring Gyroscope," *Symposium Gyro Technology*, Stuttgart, Germany, September 1999.

[71]   Watanabe, Y., et al., "Five-Axis Motion Sensor with Electrostatic Drive and Capacitive Detection Fabricated by Silicon Bulk Micromachining," *Sensors and Actuators*, Vol. A97–98, 2002, pp. 109–115.

[72]   Shearwood, C., et al., "Levitation of a Micromachined Rotor for Application in a Rotating Gyroscope," *Electronic Letters*, Vol. 31, No 21, 1995, pp. 1845–1846.

[73]   Shearwood, C., et al., "Development of a Levitated Micromotor for Application as a Gyroscope," *Sensors and Actuators*, Vol. A83, 2000, pp. 85–92.

[74]   Fukatsu, K., T. Murakoshi, and M. Esashi, "Electrostatically Levitated Micro Motor for Inertia Measurement System," *Proc. 10th Intl. Conf. Solid-State Sensors and Actuators (Transducer '99)*, Vol. 2, Sendai, Japan, 1999, pp. 1558–1561.

[75]   Houlihan, R., and M. Kraft, "Modelling of an Accelerometer Based on a Levitated Proof Mass," *J. Micromech. Microeng.*, Vol. 12, No. 4, 2002, pp. 495–503.

CHAPTER 9

Flow Sensors

Christian G. J. Schabmueller

Where fluids flow, the question of quantity arises. A fluid flow can be either a gas flow or a liquid flow. Measurands can be either the amount of mass moved (weight per second), the distance moved (meters per second), or the volume moved (volume per second). A variety of conventional flow sensors exist, but they are often of little use in the micro domain. Limited sensitivity, large size, high dead volume, and difficulties in interfacing with microfluidic devices restrict their use. Microfabrication, however, offers the benefits of high spatial resolution, fast time response, integrated signal processing, and potentially low costs. Micromachined flow sensors are able to measure a broad range of fluid flows from liters per minute down to a few droplets an hour. They have matured from the research stage to commercial applications and are now real competitors for conventional sensors and not limited to microfluidic applications, as the examples below will show.

The first micromachined flow sensors were presented by van Putten et al. [1] and van Riet et al. [2] about 30 years ago. They used the thermal domain as the measurement principle. Since then the performance of flow sensors has been improved and several other flow measuring principles were transferred from the macro into the micro world.

The intention of this chapter is to give an overview of the various flow-measuring principles. References to papers published on numerical analysis or analytical models are given at the appropriate places in the text. The necessary parameters of fluids and other materials (e.g. the dynamic viscosity, density, specific heat capacity, thermal conductivity) can be found in [3].

The chapter starts with an introduction to microfluidics, which is relevant for flow sensors. The microfluidic phenomena, the formulas from the fluid mechanics or other relevant aspects are only mentioned briefly, without full explanation, as a detailed description of that matter would exceed the scope of this chapter. Rather, the reader is made aware of these matters and is directed to references where detailed information is available. In the same section, various applications for micro flow sensors are given. Thereafter follows the description of the flow-sensing principles using MEMS fabrication. The section dealing with flow sensors operating in the thermal domain is the most elaborate, as it is one of the most important areas.

9.1   Introduction to Microfluidics and Applications for Micro Flow Sensors

Micromachining has numerous applications in fluidics, and its use in this area has become even more important as people strive to create complete fluidic systems in miniaturized formats. A broad range of devices and systems can be found in the books *Microfluidic Technology and Applications* [4] and *Micromachined Transducers Sourcebook* [5], as well as in various review articles published [6–10]. A brief introduction to microfluidics relevant for flow sensors and applications for micro flow sensors is given in this chapter. The first and most obvious microfluidic devices to integrate with a flow sensor were micropumps and/or valves, to form dosing systems or mass flow controllers [11–17]. Schematics of two typical dosing systems are depicted in Figure 9.1.

Further integration took place including several pumps, valves, flow sensors, and micromixers to form microanalysis systems (μTAS) [18–20]. As an example, a microfluidic system using two pumps, two flow sensors, and a mixer is shown in Figure 9.2 [21]. A microsystem for measurement of flow rate, pressure, temperature, conductivity, UV-absorption, and fluorescence on a single quartz glass chip was presented by Norlin et al. [22]. Another multisensor chip designed for catheter applications has been presented by Goosen et al. [23] and Tanase et al. [24]. It includes blood flow, pressure, and oxygen saturation level sensing.

The automotive industry has been, and is still one of the major driving forces for MEMS-based sensors. For example, in engine control applications, the number of sensors used will increase from approximately 10 in 1995, to more than 30 in 2010 [25]. The micromachined flow sensor has already made the jump into the automobile industry [25–27]. Electronic fuel injection systems need to know the mass flow rate of air sucked into the cylinders to meter the correct amount of fuel. Other areas of application are in pneumatics, bioanalysis [20], metrology (wind velocity and direction [28, 29]), civil engineering (wind forces on building), the transport and process industry (fluidic transport of media, combustion, vehicle performance), environmental sciences (dispersion of pollution), medical technology (respiration and blood flow, surgical tools [30]), indoor climate control (ventilation and air conditioning [31]), and home appliances (vacuum cleaners, air dryers, fan heaters). Flow sensors have even been used in space applications. The microinstrument for life science research, developed at the University of Neuchatel, Switzerland, included

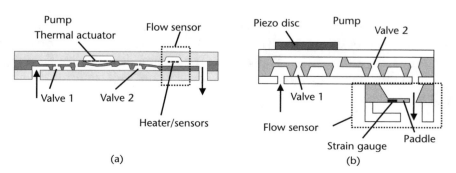

(a)

(b)

Figure 9.1   Schematics: (a) Monolithically assembled dosing system. (*After:* [12].) (b) Hybrid dosing system. (*After:* [11].)

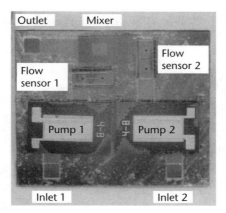

Figure 9.2 Microchemical reaction system realized on a microfluidic circuit board. Dimensions of the system are $3 \times 3.5 \times 0.3$ cm³.

a micromachined differential pressure flow sensor and took measurements aboard a Spacelab [20].

Flow sensors are often used in connection with, or built inside microchannels, which affects the performance of the sensor. The pressure drop within the channel is an important criterion that influences the measurement range and the usability of the flow-sensing device with other devices (e.g., a micropump, which can only pump against a certain backpressure). The pressure drop in a microchannel is given by Gravesen et al. [10]. Koo et al. [32] compare experimental observations with computational analyses of liquid flow. They argue that the entrance effect becomes more important for short channels with high aspect ratios and high Reynolds number conditions. For polymeric liquids and particle suspension flows, the non-Newtonian fluid effects become important. Wall slip effects are negligible for liquid flows in microconduits, and the surface roughness effects are a function of the Darcy number, the Reynolds number, and cross-sectional configurations. For Reynolds numbers above 1,000, turbulence effects become an important part. And finally, viscous dissipation effects on the friction factor are nonnegligible in a microconduit, especially for hydraulic diameters $D_h < 100\,\mu\mathrm{m}\ \{D_h = (4 \cdot \mathrm{area})/\mathrm{circumference}\}$. The Reynolds number is an important parameter in microfluidics and is a measure for the transition from the laminar to the turbulent flow regime. A laminar flow means that the different fluid layers glide over one another smoothly and do not mix. Smooth and connected streamlines are formed around an obstacle [Figure 9.3(a)]. Turbulent flow

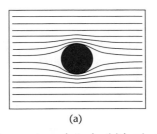
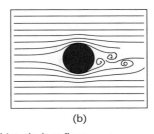

(a)   (b)

Figure 9.3   Flow past an obstacle: (a) laminar flow; (b) turbulent flow.

means that the fluid layers mix. The streamlines are curled [Figure 9.3(b)]. The reader is referred to the book by Koch et al. [4] for the theory of microfluidic flow. General information on fluid mechanics can be found in [33, 34]. It also should be noted that there are two essentially different flow profiles of laminar flow within channels. The pressure-driven flow has a parabolic shaped flow profile with the fastest velocity in the middle of the channel and decreasing velocity towards the channel walls [Figure 9.4(a)]. With an electroosmotically pumped fluid flow, the flow profile is almost flat [Figure 9.4(b)]. For open flow (pressure driven), large flow velocity gradients occur close to the wall [Figure 9.4(c)].

Recently, researchers investigated the slip of liquids in microchannels. In the paper by Tabeling [35], experiments showed a slip of liquids on an atomically smooth solid surface (polished silicon wafer). It is suggested that as a hydrodynamic consequence of this effect the relation of flow rate and pressure drop of laminar Poiseuilles flows between parallel plates must be replaced by a more generalized law, where the slip comes into play as an additional parameter. Experiments using a channel ($1.4 \times 100\,\mu m^2$ cross-section) etched into glass and covered by polished silicon with hexadecane as fluid showed that the pressure required to drive the fluid through the channel is approximately one-third lower than the one given by Poiseuilles law. This pressure reduction, using atomically flat walls, may facilitate the use of nanodevices, making it possible to measure extremely small flow rates. Carbon nanotubes [36], which are mentioned briefly in the conclusion of this chapter, may be used as the sensing element in such devices. Analytical studies to the matter of slippage in circular microchannels can be found in [37]. The study suggests that the efficiency of mechanical and electro-osmotic pumping devices can be greatly improved through hydrophobic surface modification.

Unlike in a whirlpool, bubbles are often a great disturbance within flow sensor channels and hence not very relaxing for the user. In the paper by Matsumoto et al. [38], a theory for the movement of gas bubbles in a capillary is given. It includes formulas for the pressure difference across a gas bubble and the pressure needed to transport such a bubble. For example, the removal of a gas bubble from the exit of a capillary of 1-μm side length, needs a pressure of about 140 kPa (i.e., more than atmospheric pressure) for water as test fluid [10]. To avoid the introduction of gas bubbles during the priming procedure, carbon dioxide can be flushed through the sensor chip prior to filling with the test liquid. The solubility coefficient of CO_2 is three times that of air (O_2/N_2) in water [39]. Other methods for priming involve liquids with low surface tension and wetting angle to silicon like ethanol or

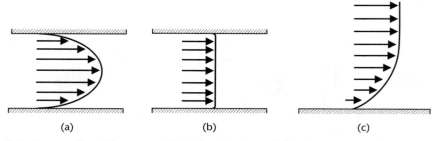

Figure 9.4 Flow profiles: (a) pressure driven flow in channel; (b) electroosmotically pumped fluid flow in channel; and (c) open flow (pressure driven).

isopropanol [39]. After priming, the system has to be flushed for a long time with the working liquid in order to remove the alcohol completely. Prior degassing of the liquids [39] or the use of high pressure for a short time to wash out the bubbles [4] may be successful.

As one can see, flow sensing is very complex. Fluid flow is already a science by itself, and furthermore, various principles can be used for flow sensing.

9.2   Thermal Flow Sensors

The overwhelming majority of micro flow sensors described so far work in the thermal domain. It is also thermal flow sensors that are produced commercially million fold. They are placed in car air intake systems used for motor efficiency control and in air conditioning systems. The commercial production of flow sensors began only about 8 years ago with the replacement of conventional flow sensors in cars [40]. In this section, mostly recent publications have been cited, but there are numerous other publications from the last 20 years that deal with thermal flow sensors. Thermal flow sensors have been classified into three basic categories (see Figure 9.5 [41]):

• Anemometers;
• Calorimetric flow sensors;
• Time of flight sensors.

For most materials, the electrical resistivity changes with temperature. Therefore, this parameter has been chosen for the thermal flow measurements. Various materials have been used to form resistors. The higher the TCR, the better the sensitivity to temperature changes and thus to flow rate. Platinum [17, 29, 42], gold [43], polysilicon [44, 45], Ni-ZrO$_2$ cermet films [46], amorphous germanium [47, 48],

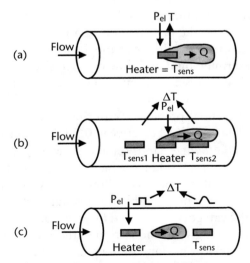

Figure 9.5   Schematic of the working principles of thermal flow sensors: (a) anemometer (heat loss), (b) calorimetric flow sensors (thermotransfer), and (c) time of flight sensors. (*After:* [41].)

and silicon-carbide [49] have been used. Also, thermistors made of germanium (*thermistor*: an electrical resistor making use of a semiconductor whose resistance varies sharply in a known manner with the temperature) were employed [47, 50, 51]. Thermocouples for temperature detection have been made out of aluminum/polysilicon [52], platinum/high boron doped silicon [53], n-polysilicon/p-polysilicon [54], gold/polysilicon [41], and aluminum/p$^+$-doped silicon [28]. The thermocouple uses a self-generating effect due to temperature to measure the flow rate. When there is a temperature difference between two contacts of two materials, a voltage proportional to the temperature difference is generated. This effect is known as the Seebeck effect. The effect is expressed as $\Delta V = \alpha \cdot \Delta T$, where α is the Seebeck coefficient. A thermopile is realized by connecting several thermocouples together.

As a general rule, the lower the mass of the sensing element (resistor, thermistor, thermocouple, and their support structure) and the higher the thermal isolation from the carrier chip, the faster is the sensor in responding to changes in fluid flow and the higher is the sensitivity [49]. Therefore, the sensing elements, including the heater, are suspended on a cantilever to stand free into the flow [55], are placed on very thin membranes [41, 42, 50, 51, 53, 54, 56], or on bridges crossing the flow path [43, 48, 49]. Often a thin-film of silicon nitride is used as membrane or bridge material. An excellent paper on how to obtain low-stress LPCVD silicon nitride was published by Gardeniers et al. [57]. PECVD mixed frequency silicon nitride or oxi-nitride is also an option. It is important that the supporting material has small thermal conductivity or that a thermal barrier is implemented [55]. Using too thin a support for the resistors means that the sensor becomes less robust and is prone to damage.

For the design of a thermal flow sensor, the hydrodynamic boundary layer and the thermal boundary layer need to be taken into account. For pressure-driven flows, large flow velocity gradients occur close to walls. For a detailed explanation and for calculating the thickness of the boundary layers, see [58]. The thickness of the boundary layer is dependent on the thermal conductivity and on the viscosity of the fluid [41]. An analytical model for a calorimetric flow sensor consisting of a heater plus an up- and downstream temperature sensor is given by Lammerink et al. [43]. A similar structure was simulated in SPICE by Rasmussen et al. [59]. The model can be used for electrical, thermal, and fluidic simulations. Ashauer et al. [41] presented a numerical simulation describing the propagation of a heat pulse. Damean et al. [60] modeled the heat transfer in a microfluidic channel with one resistive line across it. The model was used to determine fluid and flow characteristics.

Some thermal flow sensors can also be used as a pressure difference sensor. The differential pressure is indirectly measured with the mass flow, which is generated through the differential pressure. With the sensor from HSG-IMIT (Germany) the sensitivity can be chosen to be between 0.5 mbar up to 5 mbar [61]. For the sensor from Sensirion AG (Switzerland) the measurement range is ±100 Pa with a lowest detectable pressure of ±0.002 Pa, which corresponds to a force of 0.00002 g/cm^2 or a geographic height difference of 0.16 mm [62]. With this setup, a pressure equalization occurs and so it is not suitable for absolute pressure measurement.

Each specific category of thermal flow sensors is discussed below, and examples of MEMS devices are given. The section of thermal flow sensors is spilt into research and commercial devices. So far, commercial devices are using only the thermal measurement principle.

9.2.1   Research Devices

9.2.1.1   Anemometers (Heat Loss)

Anemometers consist generally of a single element, which is heated, and the influence of the fluid flow on that very element is measured [Figure 9.5(a)]. Hot wire or hot film anemometers have very fast response times due to their small thermal mass, but they are not bidirectional. They are operated generally in:

- *Constant power mode:* In the constant power mode, heat is dissipated from the resistor element into the fluid flow, and the resulting temperature of the resistor is a measure for that flow. With increasing fluid flow, the temperature of the element decreases.

- *Constant temperature mode:* The temperature of the heater is directly measured and kept constant above ambient temperature. The electrical power needed to maintain a constant temperature is a measure of the flow. In this mode, the flow sensor is very fast, but an additional control system is necessary.

- *Temperature balance mode:* (Recently proposed by Lammerink et al. [63].) In this concept, the temperature difference between two anemometers (up- and downstream) is kept constant at zero. This is done by a controlled distribution of a constant total heating power. The ratio between the up- and downstream heating power is a measure of the fluid flow. The absolute temperature will not be constant. At constant total power, the average temperate of the up- and downstream sensors will decrease with increasing flow velocity. However, the concept allows nonlinear temperature sensor transfer function as long as it is symmetrical for the two sensors. As it is a balance measurement, the temperature sensor pair should only indicate if the temperature difference is smaller than, equal to, or larger than zero. An advantage of this operating principle is that the system output does not depend on the sensitivity of the sensor. Hence, highly sensitive metal/semiconductor thermopiles, which are strongly nonlinear but with good symmetry, can be used.

Hot wire anemometers have a limited lower range of measurement due to the convection caused by the heat out of the wire. They are sensitive to contamination and therefore need calibration at certain intervals, or they can be damaged by particles. They are kept very thin to achieve fast response time, but at the same time they become fragile. It is important to have a temperature reference resistor in order to make compensation for fluctuations in fluid temperature.

Stemme [55] reported a gas flow sensor where the sensing area was thermally isolated from the silicon body via a polyimide trench [Figure 9.6(a)]. A different anemometer setup is used by Wu et al. [44, 45]. The sensor uses a boron-doped polysilicon thin-film heater that is embedded in the silicon nitride wall of a microchannel, which is formed by surface micromachining [Figure 9.6(b)]. Three sensor designs have been studied to obtain the best sensitivity: (1) the polysilicon heater boron doped at a concentration of 2×10^{19} cm^{-3}; (2) 2×10^{18} cm^{-3}, to increase the temperature coefficient of resistance; and (3) the channel suspended to improve the thermal isolation. As a result, the relative sensitivities for (1), (2), and (3) are 8, 40, and 180 ppm/(nl/min), respectively. This shows that the less doped (higher TCR), suspended sensor setup has a far better sensitivity. This sensor chip has a very high pressure drop due to the small channel size. A three-dimensional anemometer was presented by Ebefors et al. [64] and is described later within the turbulent flow measurement section. Using the same fabrication technology as for the drag force flow sensor described below, Chen et al. [65] presented an out-of-plane hot wire anemometer made of chrome/nickel [Figure 9.6(c)]. In a later publication Chen et al. [66] sandwiches the nickel between platinum to reduce the oxidation of nickel while in operation. Although the sensor is very fast, it is doubtful that it will find a commercial application as the thin wire is prone to be damaged. Researchers from the Forschungszentrum Karlsruhe, Germany, produced a flow sensor made of polymer, combining surface micromachining, molding, and diaphragm transfer technology [Figure 9.6(d)] [67].

A typical measurement curve of an anemometer type micromachined flow sensor operated in constant power mode is shown in Figure 9.7, and data for various sensors are given in Table 9.1.

9.2.1.2 Calorimetric Flow Sensors (Thermotransfer)

For calorimetric flow sensors, at least two elements are required. Most of the sensors presented in this category use a heating element with temperature sensing elements up- and downstream rendering the sensor bidirectional. The upstream sensor is cooled by the flow and the downstream sensor is heated due to the heat transport from the heater in the flow direction [Figure 9.5(b)]. Thus, the amount of heat measured is proportional to the flow rate. The sensors need to be calibrated for each fluid

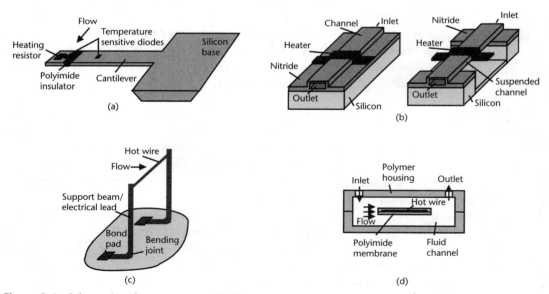

Figure 9.6 Schematics of anemometers: (a) The sensing part is a 400 × 300-μm^2 area suspended at the end of a 30-μm-thick and 1.6-mm-long silicon beam, thermally isolated by a polysilicon trench. (*After:* [55].) (b) The channel dimensions are 2 × 20 × 2,000 μm^3. At the right, the channel is suspended for better thermal isolation. (*After:* [44, 45].) (c) The hot wire is made of 100-nm-thick and 50-μm-long chrome/nickel, suspended above the wafer plane by two 0.4-mm-long beams. (*After:* [65].) (d) A gold or platinum thin-film is enclosed in a 2.4-μm-thick polyimide membrane. (*After:* [67].)

Table 9.1 Data for Anemometer Type Flow Sensors

Author; Year	Flow Range	Sensitivity	Response Time	Fluid	Chip Size
Stemme et al. [55]; 1986	0.8–30 m/s	0.01–0.5 (mW/m/s)/(mW)	50 ms	Air	—
Ebefors et al. [64]; 1998	0–60 l/min	—	120–330 μs	Air	$3.5 \times 3 \times 0.5$ mm^3
Wu et al. [44, 45]; 2000, 2001	<20 nl/min; resolution: 0.4 nl/min	8, 40, and 180 ppm/(nl/min)	—	Water	—
Chen et al. [65]; 2002	—	—	50 μs	—	—
Dittmann et al. [67]; 2001	0.1–500 sccm[1]; 1 μl/min to 2.5 ml/min	—	—	Nitrogen Water	$5.5 \times 4.5 \times 1.2$ mm^3

[1] 1 sccm = 1 ml/min

as the transported heat is connected to the fluid parameter (e.g., the specific heat or the thermal conductivity). For various flow measurement ranges, the distance between the sensors can be adjusted symmetrically up- and downstream of the heater. The output signal is the difference in temperature between the up- and downstream sensors. The prominent measurement circuit is the Wheatstone bridge. Calorimetric flow sensors are able to operate at very low flow rates. A few examples of calorimetric flow sensors are presented below. Table 9.2 gives the reader an idea about the measured flow ranges, sensitivities, and sensor dimensions.

The sensor by Glaninger et al. [50] has thin-film germanium thermistors used as heater and temperature sensors. The flow sensor chip from Oda et al. [53] is composed of one heater and four thermopiles, consisting of 9 or 23 thermocouples each, and has a dynamic range of 1:1000 for air flow measurements. A sensor fabricated only by CMOS compatible technology was presented by Häberli et al. [54]. Lyons et al. [49] use silicon-carbide heater and sensing elements due to the excellent mechanical stability (better than silicon by a factor of 2 to 4) and thermal stability (melting point of silicon-carbide is 2,800°C). The devices are able to sustain harsh environmental and operating conditions. Porous silicon, as thermal isolation, was used by Kaltsas et al. [52]. The very small sensor chip has a polysilicon heater and

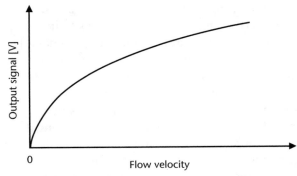

Figure 9.7 Typical measurement curve of an anemometer type micromachined flow sensor operated in constant power mode.

Table 9.2 Data for Calorimetric Type Flow Sensors

Author; Year	Flow Range	Sensitivity	Response Time	Fluid	Chip Size
Häberli et al. [54]; 1997	0–40 m/s	—	—	Air	—
Lyons et al. [49]; 1998	0–3.5 m/s	—	2.5 ms	—	—
Kaltsas et al. [52]; 1999	0.41–40 cm/s	6.0 (mV/W)/(m/s)	—	Nitrogen	1.1 × 1.5 mm²
de Bree et al. [68]; 1999	100 μm/s to 1 m/s	—	A few milliseconds	Air	—
Ashauer et al. [41]; 1999	0.1–150 mm/s	—	2 ms	Liquids and gases	—
Glaninger et al. [50]; 2000	0.01–200 m/s; 0.6 ml/h to 150 l/h	—	20 ms	Air	2 × 4 × 0.3 mm³
Oda et al. [53]; 2002	<12,000 l/h	—	—	Air	2 × 2 mm²
Ernst et al. [47]; 2002	>100 nl/h	—	—	Water	—
Makinwa et al. [28]; 2002	2–18 m/s	—	—	Air	4 × 4 mm²
Park et al. [29]; 2003	5–10 m/s	—	—	Air	6.2 × 6.2 mm²

aluminum/polysilicon thermopiles up- and downstream. A sensor for very small flow rate detection down to 100 nl/h in water was developed by Ernst et al. [47] for biomedical applications like micro dialysis systems or drug infusion systems.

Some sensors are able to measure both the flow velocity and the direction of the flow over 360°. A wind sensor has been realized in a standard CMOS process, consisting of four heaters (polysilicon resistors) and four thermopiles (aluminum/p$^+$-doped silicon) [28]. The sensor electronics are integrated in the silicon chip. Wind speed and direction were measured in a wind tunnel with an accuracy of ±4% and ±2° over a range of 2 to 18 m/s. Earlier work from that group is described in [69–71]. Another flow sensor for direction-sensitive measurements was presented by Park et al. [29]. This sensor is circular with one platinum heater and four platinum detectors arranged in a circle around the heater over a silicon membrane. The sensor was tested between a flow rate of 5 to 10 m/s with an accuracy of 5°. Flow direction and flow velocity were not yet measured at the same time with this sensor. Schematics of both sensors are shown in Figure 9.8.

The calorimetric sensor has a higher sensitivity compared to the anemometer, but at larger flow velocities the anemometer becomes advantageous. Hence, de Bree et al. [68] developed a flow sensor operated by using both principles. The flow sensor, comprised of two resistors, has a very large dynamic range. It measures air flow rates from 100 μm/s to 1 m/s. Also a combination of two operating principles is published by Ashauer et al. [41]. The considerable increase in the measuring range, from 0.1 to 150 mm/s, was done by combining the calorimetric sensing mode and the time of flight mode (described later). Twenty thermocouples are placed in a row on each side of the heater. Measurements were taken for gases and liquids. Another sensor designed to use this same operating principle was proposed by Rodrigues et al. [56].

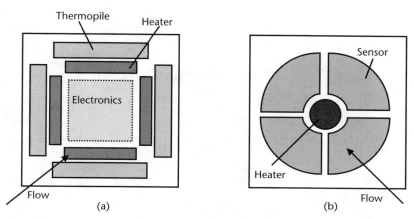

Figure 9.8   (a, b) Schematics of velocity and direction-sensitive flow sensors. (a) (*After:* [28]); and (b) (*After:* [29]).

Figure 9.9 shows a typical micromachined calorimetric flow sensor. Gold resistors sit on a low-stress silicon nitride bridge spanning a fluidic channel (Figure 9.10). A typical measurement curve of a calorimetric type micromachined flow sensor operated in constant power mode is given in Figure 9.11, and simulated sensor temperatures as a function of the volume flow in given in Figure 9.12.

9.2.1.3   Time of Flight Sensors

In this category of thermal sensors, the heater is continually pulsed with a certain amount of electrical energy. This heat pulse is carried away from the heater by the flowing fluid, and the temperature sensor is used to measure the time delay between

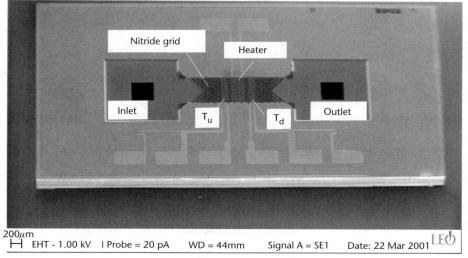

Figure 9.9   SEM photograph of a silicon micromachined calorimetric flow sensor. The chip is 4 × 7.5 mm², shown without the Pyrex cover. The fluidic channel is 580 μm wide. (Courtesy of Southampton University, Microelectronics Center, England.)

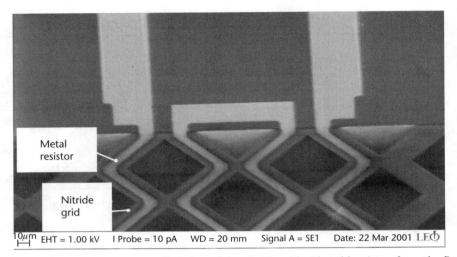

Figure 9.10 The picture is a magnified view of the nitride grid with gold resistors from the figure above. The nitride is 160 nm thick. The resistor lines are about 5 μm wide. (Courtesy of the University Southampton Microelectronics Center, England.)

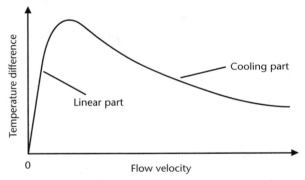

Figure 9.11 Typical measurement curve of a calorimetric type micromachined flow sensor operated in constant power mode. The curve shows the temperature difference between up- and downstream sensor elements. Measurements can be taken at the linear part.

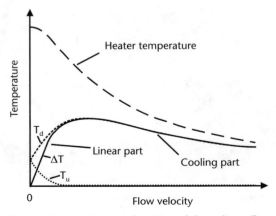

Figure 9.12 Simulated sensor temperatures as function of the volume flow. T_u and T_d are the upstream and downstream sensor temperatures. ΔT is the temperature difference between T_u and T_d. (*After:* [43].)

heat source and heat detection [Figure 9.5(c)]. The time of flight mode is the least sensitive to changes in ambient temperature as the time of arrival of the heat pulse maxima are measured. A minimum of two wires is needed. A third wire renders the sensor bidirectional. The time of flight mode works best in a regime of large flow velocities. In this case, the shape of the heat pulse is not seriously deformed by diffusion, which leads to sharp signals [12]. The measurement range can be set by the distance between the heat source and the heat detection. For lower flow rates, the distance needs to be short, and for large flow rates the distance should be large. However, the fluid flow will broaden the signal and if the detector is too far away from the source, the signal pulse is broadened so much that a peak cannot be distinguished.

This category of thermal flow sensors has not been used as often as anemometers or calorimeters. A silicon micromachined time of flight flow sensor in combination with an anemometer was presented by Ashauer et al. [41] and was described above. Figure 9.13 shows a typical measurement curve for a time of flight flow sensor, giving the signal of the sensor downstream of the heater. It can be seen that for fast-flowing fluids the pulse arrives quickly at the temperature sensor, and for smaller flow rates, the heat pulse broadens, is less intense, and arrives later at the sensor. Analytical and numerical investigations have been done by Durst et al. [72]. Sensors using a nonthermal time of flight measurement principle are described in Section 9.5.

9.2.2 Commercial Devices

As with accelerometers and gyroscopes, the incentive for developing MEMS flow sensors to the commercialization stage came from the car industry. In previous automotive air mass flow sensors, hot wire anemometers were used, which were dynamically fast due to the small thermal mass, but they could not detect reverse flow rates and were prone to damage. Other sensors were made of thin-film platinum resistors on a glass or ceramic film, which were unable to follow fast changing flow (high thermal mass and hence longer heating/cooling times). Depending on the number of revolutions per minute of the engine and the geometry of the suction pipe, the air flow can change from simple pulsation to an oscillating flow with large amplitudes [50]. Considering these aspects, micromachined sensors are of major advantage. They have the dynamic speed of a hot wire, the robustness of a

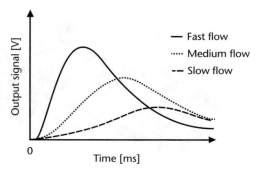

Figure 9.13 Typical measurement curve of a thermal time of flight flow sensor. A heat pulse from the heater is given at $t = 0$.

conventional hot film design, and reverse flow detection; thus, they are able to measure the net air intake. Nitride membranes of 150-nm thickness can withstand more than 1 bar and are therefore stable enough to be used as sensor membranes [41]. Particles, even after air filters, reach the surface of the sensor chip. This can result in changes of the calibrated sensor signal as the particles slowly remove the surface protection layer above the heater/sensor resistors creating shortcuts or even damaging the resistors themselves. Therefore, the sensor can be placed within an aerodynamic bypass as the one developed by Robert Bosch GmbH [73] (Figure 9.14).

A silicon-based bidirectional, thermal air flow sensor is produced by Robert Bosch GmbH, Germany, for the automotive industry and is used by most car manufacturers worldwide within the air intake module (Figure 9.15). Under operating conditions of the car engine, strong oscillations lead to temporal reverse flow. The bypass mentioned above not only prevents particle damage, but leads to a reduction of the pulsation amplification near the sensor element and a correction of the mean value of the flow passing the sensor element. The production of the micromachined sensor started in 1996 and more than 20 million sensors have been sold so far.

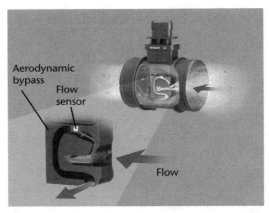

Figure 9.14 Photograph of an aerodynamic bypass for automotive applications. (Courtesy Robert Bosch GmbH, Germany.)

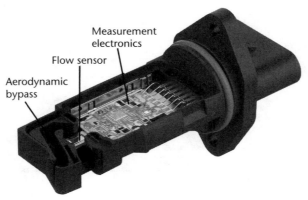

Figure 9.15 Photograph of a mass air flow meter for automotive applications developed by Konzelmann et al. [74]. (Courtesy Robert Bosch GmbH, Germany.)

Other micromachined flow sensors developed for the car industry, but not limited to this application are the sensors by HL Planartechnik GmbH, Germany [69], and by the Fraunhofer Institute for Silicon Technology, Germany [75]. The sensor by HL Planartechnik is a bidirectional mass airflow sensor. The sensor membrane is 1 μm thick with a nickel heater/sensors. No data about the minimum or maximum flow rates is available. The sensor from the Fraunhofer Institute can be manufactured at extremely low cost, as the processing is CMOS compatible. Also, the small chip dimensions enable several hundred sensors to be fabricated on a single wafer. The sensor can measure bidirectional air mass flow velocities. Photographs of the sensor chip can be seen in Figures 9.16 and 9.17. Both sensors, HL Planartechnik and Fraunhofer Institute, are available as unpackaged original equipment manufacturer (OEM) solutions excluding the measurement electronics. (An OEM is one that produces complex equipment from components usually bought from other manufacturers.)

The Hahn-Schickard-Gesellschaft–Institut für Mikro- und Informationstechnik (HSG-IMIT), Germany [61], produces a thermal flow sensor consisting of a heating element and temperature sensing elements made of doped polysilicon sitting on a membrane (Figure 9.18). The measurement elements are covered by silicon oxide

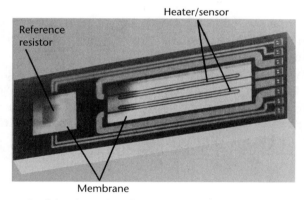

Figure 9.16 Photograph of the thermal air flow sensor by Fraunhofer Institute for Silicon Technology, Germany. The chip size is 2.6 × 7.7 mm². (Courtesy Fraunhofer Institute for Silicon Technology.)

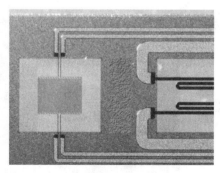

Figure 9.17 Photograph of the thermal air flow sensor by Fraunhofer Institute for Silicon Technology, Germany. The pictures show a magnified view of the reference resistor and heater/sensors. The membrane consists of a stack of nitride/oxide/nitride. The resistors are made of titanium and are covered by a nitride passivation layer. (Courtesy Fraunhofer Institute for Silicon Technology.)

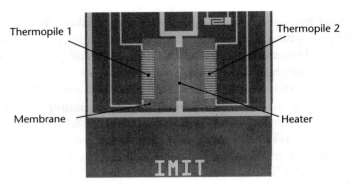

Thermopile 1   Thermopile 2

Membrane   Heater

Figure 9.18 Photograph of the flow sensor developed by Ashauer et al. [41]. The picture shows the 5-μm-wide polysilicon heating element and 20 polysilicon temperature sensing elements (thermopiles) in series on either side. The 100-nm-thick silicon-nitride membrane is $600 \times 600 \ \mu m^2$.

and silicon nitride to render them inert to various gases and liquids. There are several ways to package the sensor to be used for measurements in a nanoliter dosing system, in small tubes (Figure 9.19), for flow in large diameter pipes and in open flow. It is stated that the sensor has been tested with high viscous fluids like glue, paste, and oil [41, 61]. This flow sensor is used in air-conditioning systems. The sensor production started in 2003 with a quantity of 30,000 sensors.

An interesting application developed by HSG-IMIT is the thermodynamic inclination and acceleration sensor [76]. The sensor is built similar to a calorimetric flow sensor having a heater and temperature sensors, but the elements are inside a gas-filled, closed chamber. The heating element produces convection along the gravity field. Any movement of the sensor due to inclination or acceleration causes an imbalance between the temperature sensors, which is detected. The sensor is distributed by Vogt Electronic AG, Germany.

Figure 9.19 Photograph of a packaged sensor chip. (Courtesy HSG-IMIT.)

Sensirion AG, Switzerland [62], produces a flow sensor including the CMOS measurement electronics on the same chip. The flow meters are based on Sensirions's CMOSens technology, and use the calorimetric flow measurement principle. They combine the thermal sensor element with the amplification and A/D converter circuit on one single CMOS chip (Figure 9.20). This renders them very resistant to electromagnetic disturbances. The measurement data is fully calibrated and temperature compensated by means of an internal microcontroller. Chemical resistance and biocompatibility are achieved by measuring heat transfer through the tubing material of a capillary made of PEEK or fused silica (Figure 9.21). Therefore, the media is not in direct contact with the sensor chip. Flow sensors in CMOSens technology have been sold since 1999. The sales for gas flow sensors are significantly higher than for liquid flow sensors, indicating that the market for liquid flow sensing is not yet significant. The sensors can be bought as plug-and-play units for laboratory use, or as an OEM solution.

Micromachined gas flow sensors are also available from Leister Process Technologies, Switzerland [77], and SLS Micro Technology GmbH, Germany [78]. A liquid flow sensor is available from GeSiM mbH, Germany [79]. A wind sensor has been commercialized by Mierij Meteo [80]. Data for the various commercial flow sensors are listed in Table 9.3.

9.3   Pressure Difference Flow Sensors

This flow sensing principle relies on the measurement of the differential pressure p in a flowing fluid. Pressure sensors can be used to measure flow by sampling the

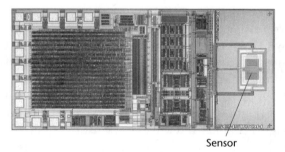

Figure 9.20   Photograph of a CMOSens chip. On the right is the flow sensing element, and on the left the CMOS electronics. (Courtesy Sensirion AG.)

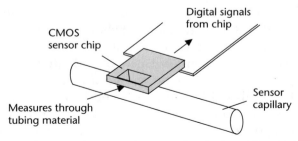

Figure 9.21   Measurement setup of Sensirion's sensor chip for liquid flow. (*After:* [62].)

Table 9.3 Data for Commercial Flow Sensors

Company	Flow Range	Sensitivity/ Resolution	Response Time	Fluid; Operating Temperature	Maximum Overpressure
Robert Bosch GmbH [73]	<1,000 kg/h	—	—	Air; –40°C to +120°C	—
HL Planartech- nik GmbH [74]	—	—	—	Air; –40°C to +120°C	—
Fraunhofer Insti- tute for Silicon Technology [75]	2–700 g/s	—	2 ms	Air	—
HSG-IMIT [61]	10μl/h to 5 l/h	4 mV/K	5 ms	Liquid	—
	0.01–50 slpm[1]	4 mV/K	5 ms	Gas	—
Sensirion AG [62]	150 nl/min to ±1,500 μl/min	50 nl/min	20 ms	Water; +10°C to +50°C	5 bar
	1 nl/min up to 50 μl/min		50 ms	Water	100 bar
	0.01–400 sccm[2]	0.01 sccm[2]		Nitrogen	2 bar
	bypass: <100 l/min			Nitrogen; 0°C to +70°C	—
Leister [77]	0.01–200 sccm[2]	—	2 ms	Gas; –10°C to +70°C	10 bar
SLS Micro Tech- nology [78]	0.01–1,000 sccm[2] (with bypass)	0.3 mV/μl	230 μs	Gas; –20°C to +120°C	3.5 bar
GeSiM [79]	1–70 μl/min	100 μV/(μl/min)	—	Water	40 bar
Mierij Meteo [80]	0.2–25 m/s	0 to 360°	1 sec	Air; –25°C to +70°C	—

[1] slpm = standard liter per minute.
[2] 1,000 sccm = 1 l/min.

pressure drop along a flow channel with known fluidic resistance, R_f, and calculating the flow Q from the fluidic equivalent to Ohm's law: $Q = \Delta p/R_f$. It is comparable to measuring the current (Q) in an electric circuit by sensing the voltage drop (Δp) over a fixed resistance (R_f).

The sensor presented by Cho et al. [81] uses a silicon-glass structure with capacitive read-out [Figure 9.22(a)]. Fluid enters the chip through the inlet at pressure p_1, flows through a channel and leaves the sensor with pressure p_2. If the flow channel is small enough to create a resistance to the flow, a pressure drop Δp appears across the channel. The pressure above the membrane and the pressure at the inlet are kept equal. The pressure difference is measured by a capacitive pressure sensor, which is switched at 100 kHz.

Capacitive pressure sensing principles are also used in the devices described by Oosterbroek [82, 83]. In addition, a hybrid piezoresistive readout was fabricated. Two separate capacitive pressure sensors were used for the sensor shown in Figure 9.22(b). This enables the measurement of both pressure and volume flow rate. For example, a 340-μm-wide channel has a resistance for ethanol of 1.7×10^{-12} Ns/m^5. The paper [83] also gives a detailed model to predict the sensor's behavior. An advantage of this sensor design is that the capacitor electrodes are not in contact with the fluid, thereby avoiding any short circuit and degradation due to aggressive fluids. Also, the sensor has a robust design using a glass/silicon/glass sandwich.

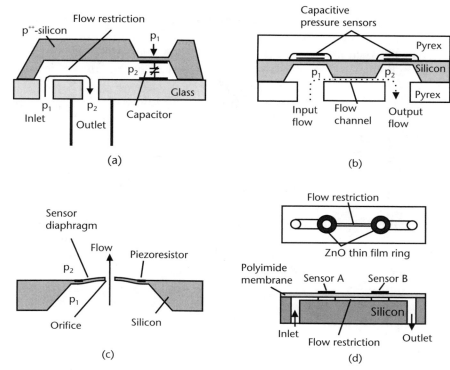

Figure 9.22 (a, b) Schematic drawings of pressure difference flow sensors: (a) (*After:* [81].)
(b) (*After:* [82, 83].) The silicon membranes are 25 μm thick, 1.5 mm long, and 1.5 mm wide. The
flow restriction channel is between 200 and 570 μm wide, 2.9 mm long, and 21 μm deep. (c) The
orifice, acting as flow restriction, has a diameter of 100 to 400 μm in the middle of the membrane,
which is 20 μm thick. (*After:* [84].) (d) The membrane has a diameter of 1 mm, and a thickness of
25 μm. The thin-film sputtered ZnO is 1 μm thick. (*After:* [85].)

Richter et al. [84] uses a commercially available pressure sensor, drills a hole in
the middle, and uses it as a differential pressure flow meter [Figure 9.22(c)]. A
similar principle has been presented by Nishimoto [85] using a self-made pressure
sensor.

A polyimide membrane with thin-film sputtered ZnO piezoelectric sensors for
measuring liquid flow has been presented by Kuoni et al. [86]. Two round piezoelectric sensors are placed before and after a flow restriction [Figure 9.22(d)]. The
restrictor has a hydraulic resistance of 60 mbar/(ml/h) with a channel length of 10
mm. The sensor has been tested in connection with a piezoelectric micropump, and
stroke volumes of 1 to 10 nl could be measured.

A flow velocity sensor based on the classical Prandtl tube was presented by Berberig et al. [87]. It realizes flow velocity detection by measuring the pressure difference between the stagnant fluid pressure in front of the sensor chip and the static
pressure in the flow around the sensor chip. The pressure difference deflects a silicon diaphragm, which is the counter electrode of an integrated capacitor (see
Figure 9.23). Two fluid passages, which are on the side the sensor faces the flow,
connect the cavity with the ambient fluid. The purpose of the fluid passage is the
transmission of the stagnation pressure p_{tot} into the sensor cavity, and in the case a
liquid is used, the multiple passage allows for cavity priming. The outer side of the

membrane is loaded with the flow's static pressure p_{stat}. The pressure difference between p_{tot} and p_{stat} causes a deflection of the membrane, which changes the capacitance between the electrodes (Figure 9.23). A reference capacitor is located around the perimeter of the membrane to compensate for the dielectric coefficient of the fluid between the capacitor electrodes.

The advantage of the differential pressure flow measuring principle is that the heating of the fluid is negligible. This can be important when using temperature-sensitive fluids or during chemical reactions.

A disadvantage of differential pressure flow sensors is that they are affected by particles because of the necessary flow restrictions. Also, the total pressure loss might be a problem if, for example, a micropump is used that can only pump against a certain backpressure. Temperature changes can have strong influences on the sensing signal due to the change in density and viscosity. Therefore, the temperature must also be monitored. The differential pressure sensing principle is better suited for liquids as the compressibility of gases distorts the measurement results. Data for pressure difference type flow sensors are listed in Table 9.4.

9.4 Force Transfer Flow Sensors

9.4.1 Drag Force

This type of flow sensor consists of a cantilever beam, or paddle, with an integrated strain gauge resistor. When the cantilever is immersed in a flowing fluid, a drag force is exerted resulting in a deflection of the cantilever, which can be detected by the piezoresistive elements incorporated in the beam. The figures in the following sections show schematics of devices using this measurement principle.

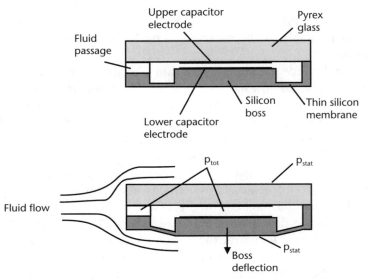

Figure 9.23 Schematic of a micromachined flow sensor based on the Prandtl tube. The fluid passage is 250 μm wide. The gap between the capacitor electrodes is 8 μm and the membrane thickness is 14 μm. (*After:* [87].)

Table 9.4 Data for Pressure Difference Type Flow Sensors

Author; Year	Flow Range	Sensitivity	Response Time	Fluid	Chip Size
Cho et al. [81]; 1991	0.001–4 Torr	200 ppm/mTorr	—	Nitrogen	9.7 × 3 mm^2
Nishimoto et al. [86]; 1994	0–800 μl/min	0.5 (μV/V)/(μl/min)	—	Water	—
Oosterbroek et al. [82, 83]; 1997, 1999	0–4.5 l/s	—	—	Water	10 × 5 mm^2
Berbering et al. [87]; 1998	0–23 m/s	—	—	Air	8 × 5 × 1.4 mm^3
Richter et al. [84]; 1999	2–32 ml/min	—	1 ms	Water	—
Kuoni et al. [85]; 2003	30–300 μl/h	—	—	Water	—

In-Plane Drag Force Flow Sensors   Gass et al. [88], Nishimoto et al. [85], and Zhang et al. [89] presented in-plane paddle flow sensors (Figure 9.24). Zhang proposed that their sensor can have two working modes: drag force and pressure difference. Simulation showed that drag force mode is more suitable for small flow rates (e.g., below 10 μl/min for water) and pressure difference is more suitable for high flow rates (e.g., above 100 μl/min for water) [85]. The pressure difference mode is feasible due to the pressure drop through the small gap around the paddle at high flow rates (Figure 9.24), since the pressure drop increases with increasing flow rate. However, the high pressure drop is a disadvantage if the sensor is to be used with other devices as mentioned above. Other disadvantages of this type of flow sensor setup are the disturbance of the flow profile, the sensitivity to particles, and the fragility of the paddle suspension.

Out-of-Plane Drag Force Flow Sensors   Su et al. [90], Ozaki et al. [91], Fan et al. [92, 93], and Chen et al. [66] discuss out-of-plane drag force flow sensors, thereby avoiding the high pressure drop. The sensor described by Su et al. employs a paddle suspended on two beams [Figure 9.25(a)]. The beams and the paddle are only 2.5 μm thick, and therefore, a high sensitivity is achieved. The air flow sensor by Ozaki et al. is modeled on wind receptor hair of insects. Structures are designed as one-dimensional [Figure 9.26(a)] and two-dimensional sensors [Figure 9.26(b)]. The angle of attack could be sensed with the two-dimensional arrangement. In this case, a thin long wire (dimensions

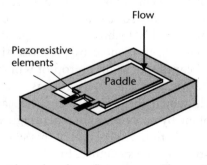

Figure 9.24 Schematic of in-plane drag force flow sensors. Zhang et al. [89] use a 10-μm-thick cantilever beam (100 × 124 μm^2) attached to a square paddle (500 × 500 μm^2). A narrow gap (200 mm) around the cantilever paddle forms a flow channel. The size of the cantilever beam for the sensor by Gass et al. [88] was 1 × 3 μm^2 with a thickness of 10 μm.

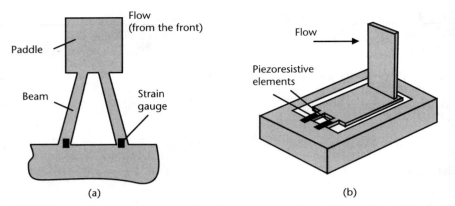

Figure 9.25 Schematics of wind receptor hair flow sensor structures: (a) one-dimensional structure: sensory hairs are 400 to 800 μm long, 230 μm wide, and 10 μm thick; and (b) two-dimensional structure: beams crossing at the center are 3 mm long, 250 μm wide, and 8 μm thick. (*After:* [91].)

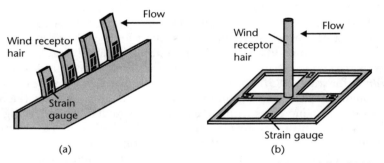

Figure 9.26 Schematics for out-of-plane drag force flow sensors. (a) A paddle of 100 × 100 μm^2 or 250 × 250 μm^2 is suspended on two 200- to 550-μm-long beams. (*After:* [90].) (b) The cantilever beam has a size of 1,100 × 180 × 17 μm^3. The vertical beam is 820 × 100 × 10 μm^3. (*After:* [66, 92, 93].)

and material were not given in the paper) was manually glued to the center of the beams. The manual assembly has a negative influence on the reproducibility of the measurement and ultimate mass production.

Also, a look to the natural world produced a sensor that tries to imitate the lateral line sensor of fish, which consists of a large number of fine hairs attached to nerve cells. Fan et al. realized a vertical beam, representing a single hair, using a three-dimensional assembly technique called plastic deformation magnetic assembly. The nerve cells are represented by piezoresistive elements. The sensor is based on a conventional cantilever beam on top of which another beam with a sacrificial layer between is fabricated. The top beam has electroplated magnetic material (permalloy) attached, which, after removing the sacrificial layer (copper), can be brought out-of-plane by an external magnet [Figure 9.25(b)]. The hinge is made out of a 600-nm-thick gold film. A problem of this sensor fabrication is the reproducibility and the robustness of the structure. In a later design [66] parylene is deposited to increase the stiffness and to avoid electrolysis and shorting. However, the thicker the parylene, the less sensitive the sensor. The overall sensor system may use an array of those sensors with varying positions, height, and orientation.

A general disadvantage of the drag force flow sensors is the possible damage through high-speed particles, which can destroy the petit paddle suspension, or low-speed particles, which clog the fluid pathway and block the paddle in case of in-plane sensor arrangement. There is a trade-off between robustness and sensitivity of the sensor. It is difficult to imagine this sensor being applied in harsh environments like car engines. Sensors do not induce heat to the fluid, which is advantageous in some applications, as mentioned in the last section, and the chip size is generally smaller than the pressure difference flow sensors. Data for drag force type flow sensors is shown in Table 9.5.

9.4.2   Lift Force

Another type of flow-force sensor has been presented by Svedin et al. [94, 95]. The silicon chip to measure bidirectional gas flow rates consists of a pair of bulk-micromachined torsional airfoil plates connected to a center support beam as shown in Figure 9.27. Each plate is suspended from the center support beam by two flexible, stress-concentrating beams containing polysilicon piezoresistor on either side to detect the deflection of the plates. The strain gauges are connected in a Wheatstone bridge. The output of the Wheatstone bridge measuring the differential deflection is proportional to the square of the flow velocity. The center beam is connected to two side supports, which are used to fix the sensor in the flow stream. The sensor is mounted at an optimum angle of 22° in a flow channel of 16×16 mm^2. If the mounting angle becomes too large, the viscous drag force dominates with the result that the deflection of both airfoil plates becomes symmetric. The lift force principle is based on fundamental airfoil theory, and the generated force acts perpendicular to the flow. Due to the nonuniform lift force distribution, the airfoil plates are deflected in the same direction, but with different magnitudes. Measurements have shown that the upstream plate was deflected about five times more than the downstream plate (Figure 9.28). Owing to the symmetric design, the devices are insensitive to acceleration forces. Data for the lift force type flow sensor are given in Table 9.6.

Table 9.5   Data for Drag Force Type Flow Sensors

Author; Year	Flow Range	Sensitivity	Response Time	Fluid	Chip Size
Nishimoto et al. [86]; 1994	0–140 μl/min	1.5 $(\mu$V/V)/(μl/min)	—	Water	—
Gass et al. [88]; 1993	5–500 μl/min	4.3 $(\mu$V/V)/(μl/min)	—	Water	—
Su et al. [90]; 1996	—	$(\Delta R/R)/y(0)$ 0.23–2.91 $\times 10^{-6}$ nm^{-1}	—	Air	—
Zhang et al. [89]; 1997	10–200 ml/min for 200-μm gap; 3–35 ml/min for 50-μm gap	—	—	Air	3.5×3.5 mm^2
Ozaki et al. [91]; 2000	A few centimeters per second to 2 m/s	—	—	Air	—
Fan et al. [92, 93]; 2002. Chen et al. [66]; 2003	0.2–0.9 m/s	—	—	Water	—

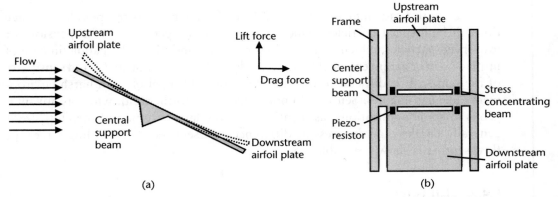

Figure 9.27 Schematic of the lift force sensor: (a) side view, and (b) top view. The airfoil plates are 15 μm thick and have an area of 5 × 5 mm^2. (*After:* [95].)

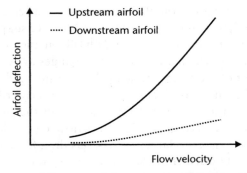

Figure 9.28 Measurement curves of the up- and downstream airfoil plate deflection. (*After:* [95].)

Table 9.6 Data for Lift Force Type Flow Sensors

Author; Year	Flow Range	Sensitivity	Response Time	Fluid	Chip Size
Svedin et al. [95]; 1998	0–6 m/s	7.4 $(\mu V/V)/(m/s)^2$	—	Gas	—

9.4.3 Coriolis Force

A silicon resonant sensor structure for Coriolis mass-flow measurement was developed by Enoksson et al. [96]. The Coriolis force is usually exploited for MEMS gyroscopes as described in Chapter 8. The sensor consists of a double-loop tube resonator structure, which is excited electrostatically into a resonance bending or torsion vibration mode. An excitation voltage of 100V amplitude was applied between the electrode and the sensor structure (Figure 9.29). A liquid mass flow passing through the tube induces a Coriolis force F_c, resulting in a twisting angular motion θ_C, phase-shifted and perpendicular to the excitation θ_{exc}. The excitation and Coriolis-induced angular motion are detected optically by focusing a laser beam on the loop structure and detecting the deflected beam using a two-dimensional

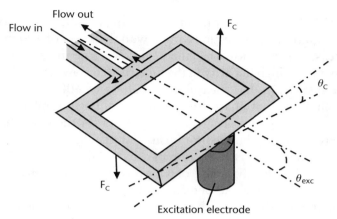

Figure 9.29   Coriolis force loop twisting due to mass flow. (*After:* [96].)

high-linearity position photodetector. The amplitude of the induced angular motion is linearly proportional to the mass flow and therefore a measure of the flow. A single-loop configuration is possible for Coriolis mass-flow sensing, but the balanced double-loop configuration gives a higher Q value and relatively large amplitudes and hence easier detection [96].

The sensor is fabricated by anisotropic etching and silicon fusion bonding. Two 500-μm-thick silicon wafers are masked with silicon dioxide and etched in KOH-solution to a depth of 400 μm as shown in Figure 9.30(a). Then the oxide is removed and the wafers bonded together by silicon fusion bonding. A second silicon oxide layer is grown and patterned [Figure 9.30(b)]. Next, the wafer is etched in KOH to

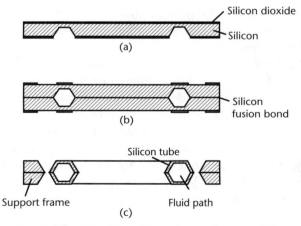

Figure 9.30   Cross-sectional view of the fabrication sequence based on micromachining of (100) single-crystal silicon: (a) KOH wet etching of a silicon wafer using silicon dioxide as masking material; (b) silicon fusion bonding of two wafers after the patterning of the silicon dioxide mask; and (c) after KOH wet etching of the bonded silicon wafers and removal of the silicon dioxide mask. The resulting tube wall thickness is about 100 μm and the double wafer thickness is 1 mm. The chip has a size of 9 × 18 × 1 mm^3. (*After:* [96].)

full wafer thickness resulting in a free-hanging silicon tube system with six-edged 1-mm-high tube cross-sections and a wall thickness of 100 m [Figure 9.30(c)].

Measurements show that the device is a true mass-flow sensor with direction sensitivity and high linearity in the investigated flow range. The micromachined silicon tube structure has measured Q factors of 600 to 1,500, depending on their vibration mode (antiphase and in-phase bending, antiphase and in-phase torsion), with water filling and operation in air. Data for the sensor is shown in Table 9.7. The sensor can also be used for measuring the fluid density since the resonance frequency of the sensor is a function of the fluid density.

The major disadvantage of Coriolis mass-flow sensors is that they require rather complex drive and detection electronics. It is quite difficult to measure the very small Coriolis force when the twisting amplitude is in the nanometer range. These amplitudes, however, are sufficient for capacitive detection and make it possible to produce a more compact sensor structure, for instance, by anodic bonding of glass lids with integrated electrodes for electrostatic excitation and capacitive detection [96].

A sensor using a U-shaped resonant silicon microtube measuring fluid flow also with the Coriolis force is proposed by Sparks et al. [97]. So far, the resonant microtube is used to sense chemical concentration, but experimental results for flow measuring are proposed for an upcoming publication.

9.4.4 Static Turbine Flow Meter

A silicon micromachined torque sensor is used to measure the volume flow converted by a static turbine wheel (the wheel does not rotate) [98]. The flow sensor has been developed for monitoring respiratory flow of ventilated patients. The application requires a bidirectional flow sensor with a low pressure drop, resistance to humidity, and temperature variations of the respiratory gas. The sensor setup consists of a wheel, which is fixed to the torque sensor and, in turn, is connected to the pipe wall. A schematic is shown in Figure 9.31. The flow is deflected as it passes the turbine wheel blades, providing a change in momentum [Figure 9.31(a, b)], which excerpts forces on the blade generating a torque, which is measured by the torque sensor. The torque depends on the flow velocity, the fluid density, the length of the blade, and the blade angle. The flow passing the wheel is distributed over the circumference of the wheel, thus levelling out effects of nonuniform flow profiles and leading to a profile-independent volumetric flow measurement. The torque-sensing element has been DRIE etched to form three different parts: the mounting part, the supporting part, and two stiffness reduction beams, as shown in Figure 9.31(c). The wheel is fixed to the mounting part just above the stiffness reduction beams. On each side of the stiffness reduction beams are boron doped polysilicon resistors connected to a Wheatstone bridge. When a flow passes the turbine wheel, the strain gauges (polysilicon resistors) on one side are tensed and on the other side compressed,

Table 9.7 Data for Coriolis Force Type Flow Sensor

Author; Year	Flow Range	Sensitivity	Q-Factors	Fluid	Chip Size
Enoksson et al. [96]; 1997	0–0.5 g/s	2.95 (mV/V)/(g/s)	600–1,500	Water	12 × 21 × 1 mm^3

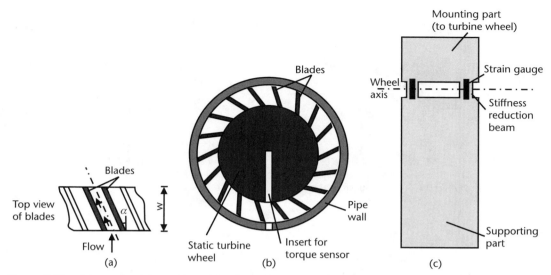

Figure 9.31   Schematic of the static turbine flow meter setup. (a) Top view of the static turbine wheel. When the flow passes between the blades it changes direction and the momentum change transfer gives rise to a force on the wheel, which is detected by the torque sensor. (b) Side view of the static turbine wheel of 15.8-mm diameter in a channel. (c) Torque sensor; the two sides of the sensor are identical. The torque-sensing element is a 300-μm-thick, 2-mm-wide, and 16-mm-long silicon cantilever. The stiffness reduction beams are 20 μm wide and 100 μm long. (*After:* [98].)

resulting in a measurement of the bending moment from the turbine wheel. The most efficient wheel in the published analysis had a blade length of 2.7 mm and a blade angle of 30°. Data for the flow sensor can be found in Table 9.8.

9.5   Nonthermal Time of Flight Flow Sensors

9.5.1   Electrohydrodynamic

This method is based on the measurement of the ion transit time between two grids [99]. The principle of such a sensor is based on the injection of charge at one electrode grid and the subsequent detection of a charge pulse at a second grid. The charge is carried along by ionic species. The transit time will increase or decrease depending on the flow rate and is therefore a direct measure of the fluid flow rate. The charge density is influenced by the electrochemistry of the pumping fluid, the electrode material, the electrode shape, and the applied voltage. The sensor is fabricated using two silicon wafers structured with KOH and bonded by an intermediated, 4-μm-thick, sputtered Pyrex layer. The metallization is made out of NiCr/Ni/Au. A schematic of the sensor is depicted in Figure 9.32(a). A voltage of

Table 9.8   Data for Flow Sensor Using a Static Wheel and Torque Sensor

Author; Year	Flow Range	Sensitivity	Response Time	Fluid	Chip Size
Svedin et al. [98]; 2001	80 l/min	4.0 (μV/V)/(l/min)	—	Air	—

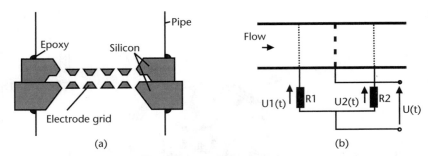

Figure 9.32 Schematics of the electrohydrodynamic flow sensor. (a) The grid size is 2.5 × 2.5 mm². The orifices in the 35-μm-thick grid structure are 100 × 100 µm². The grid distance is 10 to 60 μm. (*After:* [99].) (b) Double injector. (*After:* [94].)

300V with a repetition rate of 1 Hz was applied between the grids. Data for the electrohydrodynamic type flow sensor is given in Table 9.9.

Since this measurement principle is based on the electrohydrodynamic micropump [100], where the injected charges are used to pump the liquid, the sensor itself produces pressure, which can influence the flow rate. To reduce such disturbances, a symmetrical double injector was proposed [Figure 9.32(b)]. It produces two equal opposite pressure components compensating each other. At the same time, the flow sensor can now be used for bidirectional flow measurement. The high voltages needed are a serious disadvantage of this type of sensor.

9.5.2 Electrochemical

A somewhat similar principle is used by the electrochemical flow sensor published by Wu et al. [101]. The sensor uses an in-situ electrochemically produced molecular tracer. An upstream electrochemical cell functions as an oxygen producer, and the downstream cell as an amperometric oxygen sensor. Since the geometry of the flow channel is known, the flow rate is derived from the time difference between the signals. Unlike the time of flight flow sensors using thermal tracers, there is no need for delicate microstructures to avoid heat conduction to the wall of the channel in this sensor since the diffusion of oxygen into the wall of the channel is negligible. However, this sensor is restricted to aqueous solutions. A schematic of the sensor is given in Figure 9.33.

Two electrochemical cells are integrated in the flow channel consisting of a platinum working electrode, a platinum counter electrode, and a reference electrode made of silver. The silver electrode can be set up as a pseudo Ag/AgCl reference electrode as the concentration of Cl^- in phosphate buffered saline (PBS) is constant. PBS (in this publication: 0.04 M phosphate, 4.5% NaCl) is one of the most common solutions in bioanalysis. The cells are connected to potentiostats. Upon application

Table 9.9 Data for Electrohydrodynamic Flow Sensor

Author; Year	Flow Range	Sensitivity	Response Time	Fluid	Chip Size
Richter et al. [99]; 1991	8–50 μl/min	—	—	Ethanol	4 × 4 mm²
	8–1,700 μl/min	—	—	Deionized water	

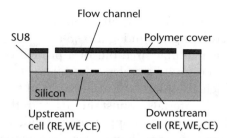

Figure 9.33 Schematic cross-section of an electrochemical time of flight flow sensor (RE: reference electrode, WE: working electrode, CE: counter electrode). The electrodes are 1 mm wide and 100 nm thick, and the two cells have a distance of 1.5 cm from each other. The height of the channel is 100 μm. (*After:* [101].)

of an electrical pulse (2V, 100 ms), water is dissociated at the working electrode and a pulse of oxygen is released [see (9.1)] and transported downstream by the fluid flow. There, the electrochemical cell functions as an amperometric oxygen sensor. At a potential of −600 mV, the dissolved oxygen in the solution is reduced [see (9.1)]. The current of the oxygen sensor is determined by the oxygen concentration in the solution. Data for the electrochemical type flow sensor is given in Table 9.10.

$$2H_2O \Leftrightarrow 4H^+ + O_2 + 4e^- \tag{9.1}$$

The diffusion coefficient of oxygen, D, varies with temperature. Normally, D changes by 2% per degree Celsius in aqueous solutions. Convection, however, is usually very fast compared to diffusion (D is $\sim 10^{-5}$ cm^2/s for oxygen at 25°C in aqueous solutions) [101]. Therefore, it is possible to omit the diffusion in the flow direction as long as the flow rate is not extremely slow, which implies that the measurement of the flow sensor is temperature independent.

The flow sensor can also be set up for impedance measurement. Then the platinum working electrode and platinum counter electrode are connected to an impedance meter. The produced oxygen forms microbubbles in the solution, which increases the impedance of the solution. This setup was used for sensing the flow in tap water. The applied potential to the oxygen producer was 4V.

9.6 Flow Sensor Based on the Faraday Principle

This technique uses the physical principle that an electric potential is developed when a fluid of low conductivity passes perpendicularly through a magnetic filed [102]. By using an ac magnetic field (here 65 kHz), it is possible to extract the electric potential across capacitors formed through plates on both channel sides and an isolating layer. This eliminates the need for electrodes in contact with the fluid. The

Table 9.10 Data for Electrochemical Flow Sensor

Author; Year	Flow Range	Sensitivity	Response Time	Fluid	Chip Size
Wu et al. [101]; 2002	1–15 μl/min	—	—	Water	—

sensor, however, is limited to applications in conducting fluids. A schematic of the sensor cross-section is given in Figure 9.34.

The sensor is manufactured out of a printed circuit board (PCB) with copper structures and is covered by a glass plate. The PCB is glued to the glass plate with epoxy resin, which also forms the inside walls of the fluidic channels. The resin film protects the board materials against the effects of fluids (e.g., swelling of the board), reduces the influence of toxic copper on the fluid, and isolates the copper structures preventing electrical current flowing between copper and fluid. A potential difference between left and right copper tracks of 5V was applied. The magnetic field strength in the fluidic channel was approximately 1,200 A/m. The sensor output signal is linear. Although the sensor is not strictly a MEMS flow sensor, the principle is interesting and can be transferred to MEMS. Data for the flow sensor can be found in Table 9.11.

9.7 Flow Sensor Based on the Periodic Flapping Motion

Lee et al. [103] fabricated a micromachined flow sensor using the periodic flapping motion of a planar jet impinging on a V-shaped plate. The sensor detects the oscillating frequency of the periodically flapping jet either optically with the help of a colored fluid inserted into the middle of the flow stream or by a pair of resistors in front of the V-shaped plate, which has opening angles between 70° and 110°. The resistors were connected within a Wheatstone bridge and the output voltage was measured by an ADC. A schematic drawing of the micro flow sensor with a convergent nozzle and a V-shaped plate downstream is shown in Figure 9.35(a). Experimental data shows that the flow velocity is linear proportional to the frequency of the jet

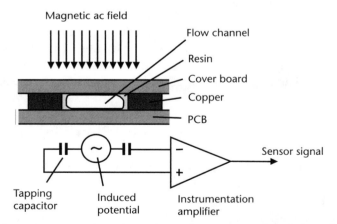

Figure 9.34 Schematic cross-section of capacitance flow sensor and equivalent circuit diagram. The copper tracks are 35 μm high and have a width of 100 μm. They are separated by 200 μm and the resin thickness above and below the channel is 5 μm. (*After:* [102].)

Table 9.11 Data for Flow Sensor Based on Faraday Principle

Author; Year	Flow Range	Sensitivity	Response Time	Fluid	Chip Size
Merkel et al. [102]; 2000	2–15 μl/min	—	—	Salty water	—

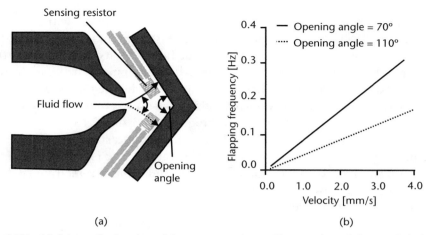

(a)   (b)

Figure 9.35 (a) Schematic drawing of the sensor structure. The opening of the nozzle is 360 μm, the height of the structure is 48 μm, and the distance from nozzle to plate is around 2.8 mm. (b) Graph showing the flapping frequency versus the flow velocity. (*After:* [103].)

flapping motion. The flapping frequency for flow velocities of up to 4 mm/s was below 0.2 Hz with water as the test fluid [Figure 9.35(b)].

As an explanation for the flapping motion, Lee et al. [103] say that a larger pressure field may be developed with the presence of the V-shaped plate, and therefore, the impinging jet column interacts with the pressure wave propagating upstream, resulting in a periodic flapping motion. The authors propose the required work to investigate why the flapping motion occurs at those low Reynolds numbers (0.2 to 5.4 for this device). Data for the sensor is given in Table 9.12.

Another interesting aspect is that the sensor was manufactured in commercially available quartz photomask plates with unpatterned chromium and resist, normally used for electron-beam mask writing. The resist was patterned by standard photolithography using a film generated from a high-resolution laser plotter (10,000 dot/inch). Subsequently the chromium and quartz were etched. The quartz plate is bonded to another quartz plate (on which the resistors were patterned in the chromium) at 50°C for 8 hours with an intermediate layer of sodium silicate solution (SiO_2:NaOH). However, this is only a cheap fabrication option if there is no metal evaporator available, as the photomask plates are rather expensive compared to quartz wafers.

9.8 Flow Imaging

Various methods for flow imaging have been proposed and are described in this section. The flow imaging can be used for measuring the fluid flow velocity or to take snapshots of the fluid flow to visualize flow profiles or eddies. With some

Table 9.12 Data for Flow Sensor Based on Periodic Flapping Motion

Author; Year	Flow Range	Sensitivity	Response Time	Fluid	Chip Size
Lee et al. [103]; 2002	>0.15 mm/s	—	—	Water	3 × 6 mm^2

techniques it can be used to follow the motion of fluids within a silicon chip, for example, to show the droplet formation within an inkjet printer nozzle [104] or the spinning of a microrotor [105]. For true flow velocity measurement, the flow imaging technique is rather expensive due to high equipment costs and the requirement of extensive computation. Therefore, it may only be used for specialized applications, where not only the flow velocity but also the flow profile is of interest. For measured flow range data of the various techniques, see Table 9.13.

The technique published by Leu et al. [104] measures steady-state flow in micropipes of various shapes by way of illustration. The experimental setup includes a wide bandwidth X-ray monochromator and a high frame rate CCD camera (160 frames/sec). Flow image sequences were collected for micropipes of 100- to 400-μm diameter. A flow recovery algorithm derived from fluid mechanics was applied to recover velocity flow profiles. The X-ray stroboscopic technique can be used to image periodic motion up to kilohertz rates. Dynamic behavior of flow fields inside MEMS structures can be measured by an animation of a sequence of phase corresponding images.

Another system to allow observation of motion inside silicon-based microdevices uses infrared diagnostics [105]. A Nd:YAG laser is tuned to a wavelength of 1,200 nm, where silicon reaches a high level of transparency. The pulse is coupled to a fiber optic delivery system and directed to the target, which is flood illuminated. The scattered light is collected by a near-infrared microscope objective and imaged using an Indigo System Indium Gallium Arsenide Near-Infrared camera. The camera has a 320 × 256 pixel array. For flow measurements, the fluid needs to be seeded with particles. Flow rates in water seeded with 0.06% by volume with 1-μm polystyrene particles were investigated. The resolution is 360 nm.

A simpler and cheaper imaging system was presented by Chetelat et al. [106]. This system can only be used for devices with an optical window. The particle image velocimetry system has a field of view of 6 × 5 mm^2 and can measure 50 velocity vectors for liquid flows slower than 1 m/s. Twelve super-bright LEDs are used as a strobe light in forward-scatter configuration. The signal is detected using a one-chip-only CMOS camera with digital output. A computer is needed to calculate the velocity field. Experiments were performed in water with hollow glass particles (10 μm), in air with water spray droplets (50 μm), and with water fog (20 μm).

Characterization of microfluidic flow profiles from slow laminar flow to fast near-turbulent flow was presented by Shelby et al. [107]. Using a photo-activated fluorophore (fluorescein), nanosecond duration photolysis pulses from a nitrogen laser, and high-sensitivity single-molecule detection with argon laser excitation, flow speeds up to 47 m/s in a 33-μm-wide straight channel and the mapping of flow

Table 9.13   Data for Flow Sensors Using Imaging Techniques

Author; Year	Flow Range	Fluid
Leu et al. [104]; 1997	4–8 nl/s	—
Han et al. [105]; 2002	250 μm/s to 62 mm/s	Water
Chetelat et al. [106]; 2002	1 m/s	Water
	10 m/s	Air
Shelby et al. [107]; 2003	47 m/s	—

profiles in a 55-μm-wide microchamber were measured. This technique permits the high-resolution three-dimensional mapping and analysis of a wide range of velocity profiles in confined spaces that measure a few micrometers in dimension. The particle trajectories are mapped and it is assumed that the particles trace out the flow lines.

9.9   Optical Flow Measurement

Although almost all optical flow sensors are not strictly MEMS-based, they are, however, included in this chapter as they can be used in areas, which are important, but for which MEMS cannot yet cater for. Fiber optic sensors have a number of advantages over their electrical counterparts. They are safe around volatile chemicals, are free from electromagnetic interference, and provide electrical isolation. In some applications, fiber sensors show higher durability at elevated temperatures, and they are corrosion resistant. For example, Eckert et al. [108] developed a mechanooptical sensor to measure flow in metallic melts of about 350°C. Flow rates between 1 and 14 cm/s in eutectic InGaSn melt could be measured. Borosilicate glass can be used up to temperatures of 350°C and quartz glass up to 1,000°C [108]. The major disadvantage of optical measurement systems is their size. Lasers, optical power meters, lenses, couplers, and mirrors are needed, making the system setup rather expensive and not suitable for portable systems or for use in small, confined spaces. Optical devices are not suitable for operation in unclean conditions for long periods of time (e.g., on the engine block of a car) because dirt and condensation lead to problems.

9.9.1   Fluid Velocity Measurement

A flow sensor using a silicon cantilever with a wave guide on its surface is described by Chun et al. [109]. It uses a similar principle to the sensors based on drag force, but here, the sensing is not detected by an implanted piezoresistor but rather optically. Light is transmitted across the wave guide and is used to detect the movement of the cantilever. The intensity of the optical beam changes with the deformation of the silicon cantilever due to fluid flow [Figure 9.36(a)]. An optical fiber is used for the light input to the wave guide, and a second optical fiber is used to detect the light intensity. The optical fibers are fixed to the silicon chip by V-grooves. Unfortunately, neither minimum or maximum flow rate nor the sensitivity of the sensor is given in the paper.

   An optical fiber drag force flow sensor to measure the speed and direction of fluid flow was published by Philip-Chandy et al. [110]. The flow sensor comprises a fiber optic strain gauge, a cantilever element made of rubber, and a spherical drag element. The fiber optic strain gauge was produced by inserting six grooves into a multimode optical fiber of 1-mm diameter. As the fiber bends, the variation in the angle of the grooves causes an intensity modulation of the light transmitted through the fiber [Figure 9.36(b)]. The flow sensor has a repeatability of 0.3% and measures wind velocity up to 30 m/s with a resolution of 1.4 m/s.

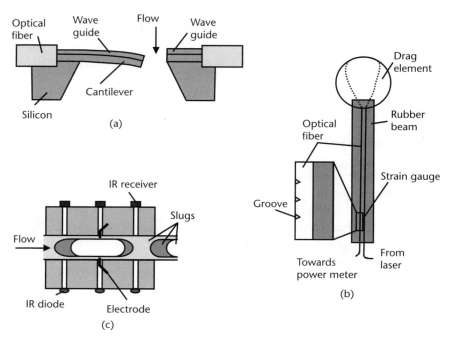

Figure 9.36 Schematic diagrams of optical flow sensor: (a) The sensor employs a cantilever beam and wave guides made of N^-/N^{++}. V-grooves in the silicon are used to place the optical fibers. (*After:* [109].) (b) Optical drag force flow sensor. (*After:* [110].) (c) Flow sensor for detection of multiphase flow (light gray: water, dark gray: decane, white: air). (*After:* [110].)

9.9.2 Particle Detection and Counting

Other optical setups are used to detect particles or cells out of the fluid flow. A micromachined version in soda-lime glass using two embedded optical fibers was presented by Lin et al. [111] and using integrated optical waveguides by Lee et al. [112]. The particles/cells (5- to 20-µm polystyrene beads and diluted whole blood) are squeezed hydrodynamically into a narrow stream by two neighboring sheath flows so that they flow individually through the detection region. The resulting scattered light is then detected optically. The flow rates for the sheath flows and the sample flow were 0.2 and 0.05 μl/min, respectively. Various micromachined devices for particle counting, separation, deflection, sorting, or transporting within fluids are described in the book by Koch et al. [4].

9.9.3 Multiphase Flow Detection

The flow of fluids in pipes, where more than one immiscible phase is present, are of major importance in several industries, like the power-generation industry (steam generators and some types of nuclear reactors) and the petroleum extraction industry [113]. Multiphase flow could be, for example, steam/water, water/kerosene, and crude oil/water/natural gas. Fordham et al. [113–115] use standard silica fibers and internal reflection to distinguish drops, bubbles, or other regions of fluid in multiphase flows on the basis of refractive-index contrast. They tested various geometries for the tip of the optical fiber and used a self-assembled monolayer of alkyl functional groups to the silica surface to render the surface strongly hydrophobic. The latter is necessary as the refractive index needs to be controlled when the fiber is immersed in liquids. A dual optical probe for local volume fraction, drop velocity, and drop size measurement in a kerosene-water two-phase flow was published by Hamad et al. [116]. A method for nonintrusive measurement of velocity and slug length in multiphase flow in glass capillaries of 1- or 2-mm inner diameter was presented by Wolffenbuttel et al. [117]. (Slug here means an amount of fluid and not the animals that eat, uninvited, the salad leaves of hobby gardeners.) A combination of an impedance meter and two infrared sensors is used to distinguish between air, water, and decane [Figure 9.36(c)].

9.10 Turbulent Flow Studies

An area where MEMS sensors have considerably broadened the field of study is fluid dynamics. A typical MEMS sensor is at least one order of magnitude smaller than conventional sensors used to measure instantaneous flow quantities such as pressure and velocity [118]. The micromachined sensors are able to resolve all relevant scales, even in high Reynolds number turbulent flows. Due to their small size, the inertial mass and the thermal capacity are reduced. Thus, they can be used for the study of turbulent flows, where a high-frequency response and a fine spatial resolution are essential. The smallest scales of eddies in turbulent flow are in the order of 100 μm [64]. Arrays of microsensors could make it possible to achieve complete information on the effective small-scale coherent structures in turbulent wall-bounded flows. Applications of turbulent flow study include the optimization of wing sections of aircraft, the minimization of noise generation of vehicles, or mixing enhancement for fluids.

The goal of measuring turbulent flows is to resolve both the largest and smallest eddies that occur in the flow. In order to obtain meaningful results, both wall pressure and wall shear stress need to be measured [118]. The wall shear stress is the friction force that a flow exerts on the surface of an object.

The wall pressure can be measured with the sensors described in Chapter 6. Löfdahl et al. [118] recommends that the pressure sensor needs to have a membrane size between $100 \times 100\,\mu m^2$ and $300 \times 300\,\mu m^2$, it needs to have a high sensitivity of ± 10 Pa, and the frequency characteristic should be in the range of 10 Hz to 10 kHz. The wall shear stress is a parameter of small magnitude. For example, a submarine cruising at 30 km/h has an estimated value of the shear stress of 40 Pa; an aircraft flying at 420 km/h, 2 Pa; and a car moving at 100 km/h, 1 Pa [118]. Therefore, the sensitivity of shear stress sensors needs to be very high. For wall shear stress sensors, there are direct and indirect measurement methods. For MEMS devices, the direct measurement method is the floating element method. Here, the sensor needs to have an element movable in the plane of the wall, which is laterally displaced by the tangential viscous force. The movement can be measured using resistive, capacitive, or optical detection principles. It is an important requirement that the sensor is mounted flush to the wall. Misalignment and gaps around the sensing element, needed to allow small movements, are sources of error. For conventional "macro" sensors, effects that could cause measurement errors are pressure gradients, heat transfer, suction/blowing, gravity, and acceleration. With micromachined devices, these errors are less severe. Early floating element shear stress sensors were published by [119–122]. A schematic drawing of a floating element shear stress sensor is given in Figure 9.37(a).

The indirect measurement is the thermal element method [64, 123]. Here, a time-dependent, convective heat transfer to the fluid is measured. An example of such a sensor is the three-dimensional silicon triple-hot-wire for turbulent gas flow measurement by Ebefors et al. [64]. To achieve good spatial resolution, the hot-wire needs a length-to-diameter ratio larger than 100. Time constants in the microsecond range were obtained. A schematic of the sensor is shown in Figure 9.38. Two wires are located in the wafer plane and a third wire is rotated out of plane using the thermal shrinkage of polyimide in V-grooves.

Recently, von Papen et al. [124] presented a surface fence sensor for wall shear stress measurement. The sensor consists of a silicon fence mounted flush to a channel wall [Figure 9.37(b)]. A pressure difference between both sides of the fence occurs in a fluid flow and deflects the fence structure. Four piezoresistors connected to a Wheatstone bridge detect the deflection. This shear stress measurement technique is also indirect.

For a detailed summary and critical evaluation of MEMS-based sensors for turbulent flow measurement, the reader is refereed to the paper by Löfdahl et al. [118].

Dao et al. [125] proposed a sensor not to measure the turbulent flow itself, but the force and moment acting on boundary particles in a turbulent liquid flow. The micro multiaxis force-moment sensor is mounted inside a sphere. The sensor ($3 \times 3 \times 0.4$ mm^3) is designed to independently detect three components of force and three components of momentum in three orthogonal directions. Detection is done by 18 piezoresistors spread along two cross beams with a center plate at their intersection. No measurement results are presented in the paper.

9.11 Conclusion

The large variety of different flow-sensing devices with applications in various areas clearly shows that micromachined flow sensors have attracted a lot of interest, not

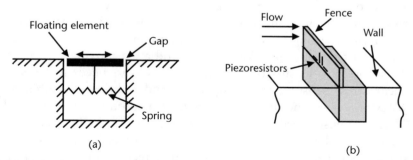

(a)   (b)

Figure 9.37 Schematics of a floating element shear stress sensor: (a) Working principle of a floating element shear stress sensor. The element is free to displace laterally due to the shear force act- ing on the plate. (*After:* [119].) (b) Drawing of the surface fence sensor for wall shear stress measurements (5-mm-long, 100- to 300-μm-high, and 7- to 10-μm-thick silicon fence). (*After:* [117].)

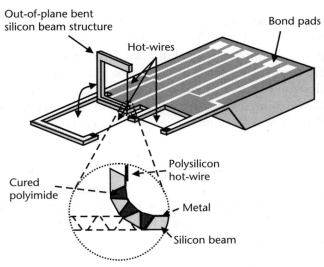

Figure 9.38 Schematic drawing of a triple-hot-wire anemometer. The polysilicon hot-wires are $500 \times 5 \times 2 \, \mu m^3$. (*After:* [64].)

only in the research field but also from industry, which has already commercialized millions of MEMS flow sensors. Examining these in detail, it is noted that, to date, gas flow sensing is more popular than liquid flow sensing. Devices are used for car air intake modules or air-conditioning systems. The BioMEMS field is a promising candidate for further commercializing of microfluidic devices and systems, including flow sensors working with liquids.

CMOS fabrication compatibility is an enormous advantage for a micromachined sensor. The fabrication of MEMS devices can run in parallel with other processes in microelectronics fabrication cleanrooms, thereby reducing the costs enormously. This means that metals like platinum or gold, or KOH etching to form thermal isolation structures cannot be used. Specialized MEMS-only cleanrooms would not be economical, as the selling numbers of MEMS sensor chips are still far below microelectronic devices (apart from ink jet printer nozzles and hard disk drive heads). An overview of micromachined thermally based CMOS sensors was presented by Baltes et al. [126].

Packaging of a flow sensor is not an easy task. This has a great influence on the sensor's performance, as described in Chapter 4. For example, the diameter of the channel in which a sensor sits has an impact on the minimum or maximum flow rate. The packaging can protect the sensor from damage by particles, as seen in the aerodynamic bypass developed by Bosch GmbH. Sensors for wall shear stress measurement need to be mounted flush to the wall. Up to now, each sensor needs to have an individual packaging solution, depending on the measurement principle adopted, the required flow range, and the measurement environment.

Ultrasonic macroflow measurement systems are commonly used [127]. They are based on drift, Doppler, and attenuation or diffraction effects. Ultrasound is normally generated by piezoelectric transducers. A miniaturized ultrasonic wave velocity and attenuation sensor for liquids was developed in 1993 by Hashimoto et al. [128]. The device is made of silicon and glass with sputtered ZnO to create the ultrasonic waves and it has the potential to measure liquid flow rates. However, the idea of this novel MEMS flow sensing principle has not been picked up by future researchers, and thus, it was not discussed above. Ultrasonic wave sensors to measure liquid density and viscosity have also been described [129, 130]. Ultrasound enables instruments to be noninvasive because the acoustic wave can often penetrate the walls of channels. Other sensing principles that have gained little attention so far are the fluid flow detection via a pyroelectric element [131] and resonant flow sensing mechanism [132, 133].

Turbulent flow studies, with considerable impact from micromachining, may open a new area. In the future an aircraft wing could be covered with wall shear stress sensors and actuators to actively influence the flow profile. A first step in this direction is the micromachined flexible shear stress sensor skin applied to an unmanned aerial vehicle presented recently by Xu et al. [134]. Here, an array of 36 shear stress sensors was mounted over the 180° surface of the leading edge of a wing, and data during flight was collected for an aerodynamic study.

Flow sensors have made the jump from the MEMS into the NEMS world (nanoelectro mechanical systems). Ghosh et al. [36] measured flow rates for various liquids using carbon nanotubes. They reported that the flow of a liquid on single walled carbon nanotube bundles induces a voltage in the sample along the direction of the flow. The magnitude of the voltage depends on the ionic conductivity and on the polar nature of the liquid. Nanotube bundles with an average tube diameter of 1.5 nm were densely packed between two metal electrodes. The dimensions of the sensor were $1 \times 0.2 \times 2$ mm^3. Flow rates between micrometers per second and millimeters per second were measured. This approach using carbon nanotubes may have the potential to measure extremely low flow rates.

Lerch et al. [135] writes, "The research is often technology driven and does not necessarily fit industrial or market requirements. Beyond scientific and technical interest, the market finally decides if the developed devices are of practical significance." As for all MEMS sensors, this is also true for micro flow sensors. Without the basic research, however, there would be a lack of variety of principles. Not all applications are in the automotive field.

References

[1]   van Putten, A. F. P., and S. Middelhoek, "Integrated Silicon Anemometer," *IEE Electronics Letters*, Vol. 10, 1974, pp. 425–426.

[2]   van Riet, R. W. M., and J. H. Huysing, "Integrated Direction-Sensitive Flow-Meter," *IEE Electronics Letters*, Vol. 12, 1976, pp. 647–648.

[3]   Weast, R. C., (ed.), *Handbook of Chemistry and Physics*, 69th ed., Boca Raton, FL: CRC Press, 1989.

[4]   Koch, M., A. G. R. Evans, and A. Brunnschweiler, *Microfluidic Technology and Applications*, Baldock, England: Research Studies Press Ltd., 2000.

[5]   Kovacs, G. T. A., *Micromachined Transducers—Sourcebook*, New York: McGraw-Hill, 1998.

[6]   Dario, P., et al., "Micro-Systems in Biomedical Applications," *Journal of Micromechanics and Microengineering*, Vol. 10, 2000, pp. 235–244.

[7]   Harrison, D. J., et al., "From Micro-Motors to Micro-Fluidics: The Blossoming of Micromachining Technologies in Chemistry, Biochemistry and Biology," *Proc. Transducers*, No. 1A2.2, Sendai, Japan, 1999.

[8]   Abramowitz, S., "DNA Analysis in Microfabricated Formats," *Journal of Biomedical Microdevices*, Vol. 1, No. 2, 1999, pp. 107–112.

[9]   Shoji, S., and M. Esashi, "Microflow Devices and Systems," *Journal of Micromechanics and Microengineering*, Vol. 4, 1994, pp. 157–171.

[10]   Gravesen, P., J. Branebjerg, and O. S. Jensen, "Microfluidics—A Review," *Journal of Micromechanics and Microengineering*, Vol. 3, 1993, pp. 168–182.

[11]   Gass, V., et al., "Integrated Flow-Regulated Silicon Micropump," *Proc. Transducers*, Yokohama, Japan, 1993, pp. 1048–1051.

[12]   Elwenspoek, M., et al., "Towards Integrated Microliquid Handling Systems," *Journal of Micromechanics and Microengineering*, Vol. 4, 1994, pp. 227–245.

[13]   Nguyen, N. T., et al., "Hybrid-Assembled Micro Dosing System Using Silicon-Based Micropump/Valve and Mass Flow Sensor," *Sensors and Actuators*, Vol. A69, 1998, pp. 85–91.

[14]   Nguyen, N.-T., et al., "Integrated Flow Sensor for In Situ Measurement and Control of Acoustic Streaming in Flexural Plate Wave Micropumps," *Sensors and Actuators*, Vol. A79, 2000, 115–121.

[15]   Lo, L.-H., et al., "A Silicon Mass Flow Control Micro-System," *Mecanique and Industries*, Vol. 2, 2001, pp. 363–369.

[16]   Hirata, K., and M. Esashi, "Stainless Steel-Based Integrated Mass-Flow Controller for Reactive and Corrosive Gases," *Sensors and Actuators*, Vol. A97–98, 2002, pp. 33–38.

[17]   Liu, Y., et al., "A Modular Integrated Microfluidic Controller," *Proc. Actuator*, Bremen, Germany, 2002, pp. 231–234.

[18]   Lammerink, T. S. J., et al., "Modular Concept for Fluid Handling System," *Proc. IEEE Micro Electro Mechanical Systems*, San Diego, CA, 1996, pp. 389–394.

[19]   Richter, M., et al., "A Chemical Microanalysis System as a Microfluid System Demonstrator," *Proc. Transducers*, Chicago, IL, 1997, pp. 303–306.

[20]   van der Schoot, B., M. Boillat, and N. de Rooij, "Micro-Instruments for Life Science Research," *IEEE Trans. on Instrumentation and Measurement*, Vol. 50, No. 6, 2001, pp. 1538–1542.

[21]   Schabmueller, C. G. J., et al., "Micromachined Chemical Reaction System Realized on a Microfluidic Circuitboard," *Proc. Eurosensors XII*, Southampton, England, Vol. 1, Institute of Physic Publishing, Bristol, England, 1998, pp. 571–574.

[22]   Norlin, P., et al., "A Chemical Micro Analysis System for the Measurement of Pressure, Flow Rate, Temperature, Conductivity, UV-Absorption and Fluorescence," *Sensors and Actuators*, Vol. B49, pp. 34–39, 1998.

[23]   Goosen, J. F. L., P. J. French, and P. M. Sarro, "Pressure, Flow and Oxygen Saturation Sensors on One Chip for Use in Catheters," *Proc. International Conference on Micro Electro Mechanical Systems (MEMS)*, Miyazaki, Japan, 2000, pp. 537–540.

[24]   Tanase, D., et al., "Multi-Parameter Sensor System with Intravascular Navigation for Catheter/Guide Wire Application," *Sensors and Actuators*, Vol. A97–98, 2002, pp. 116–124.

[25]   Fleming, W. J., "Overview of Automotive Sensors," *IEEE Sensors Journal*, Vol. 1, No. 4, 2001, pp. 296–308.

[26]   Marek, J., and M. Illing, "Microsystems for the Automotive Industry," *Proc. International Electron Devices Meeting*, San Francisco, CA, 2000, pp. 3–8.

[27]   Brasseur, G., "Robust Automotive Sensors," *IEEE Instrumentation and Measurement Technology Conference*, Ottawa, Canada, 1997, pp. 1278–1283.

[28]   Makinwa, K. A. A., and J. H. Huijsing, "A Smart Wind Sensor Using Thermal Sigma-Delta Modulation Techniques," *Sensors and Actuators*, Vol. A97–98, 2002, pp. 15–20.

[29]   Park, S., et al., "A Flow Direction Sensor Fabricated Using MEMS Technology and Its Simple Interface Circuit," *Sensors and Actuators*, Vol. B91, 2003, pp. 347–352.

[30] Chen, X., and A. Lal, "Integrated Pressure and Flow Sensor in Silicon-Based Ultrasonic Surgical Actuator," *Proc. IEEE Ultrasonics Symposium*, 2001, pp. 1373–1376.

[31] Kang, J., et al., "Comfort Sensing System for Indoor Environment," *Proc. Transducers*, Chicago, IL, 1997, pp. 311–314.

[32] Koo, J., and C. Kleinstreuer, "Liquid Flow in Microchannels: Experimental Observations and Computational Analyses of Microfluidics Effects," *Journal of Micromechanics and Microengineering*, Vol. 13, 2003, pp. 568–579.

[33] White, F. M., *Fluid Mechanics*, New York: McGraw-Hill, 1979.

[34] Shames, I. H., *Mechanics of Fluids*, New York: McGraw-Hill, 1992.

[35] Tabeling, P., "Some Basic Problems of Microfluidics," *Proc. 14th Australasian Fluid Mechanics Conference*, Adelaide, Australia, 2001.

[36] Ghosh, S., A. K. Sood, and N. Kumar, "Carbon Nanotube Flow Sensor," *Science Express Reports*, Vol. 299, 2003, pp. 1042–1044.

[37] Yang, J., and D. Y. Kwok, "Analytical Treatment of Flow in Infinitely Extended Circular Microchannels and the Effect of Slippage to Increase Flow Efficiency," *Journal of Micromechanics and Microengineering*, Vol. 13, 2003, pp. 115–123.

[38] Matsumoto, H., and J. E. Colgate, "Preliminary Investigation of Micropumping Based on Electrical Control of Interfacial Tension," *Proc. Micro Electro Mechanical Systems (MEMS)*, Napa Valley, CA, 1993, pp. 105–110.

[39] Zengerle, R., et al., "Carbon Dioxide Priming of Micro Liquid Systems," *Proc. Micro Electro Mechanical Systems (MEMS)*, Amsterdam, the Netherlands, 1995, pp. 340–343.

[40] Konzelmann, U., H. Hecht, and M. Lembke, "Breakthrough in Reverse Flow Detection—A New Mass Air Flow Meter Using Micro Silicon Technology," SAE Technical Paper Series 950433, International Congress and Exposition, Detroit, MI, 1995, pp. 105–111.

[41] Ashauer, M., et al., "Thermal Flow Sensor for Liquids and Gases Based on Combinations of Two Principles," *Sensors and Actuators*, Vol. A73, 1999, pp. 7–13.

[42] Bedö, G., H. Fannasch, and R. Müller, "A Silicon Flow Sensor for Gases and Liquids Using AC Measurements," *Sensors and Actuators*, Vol. A85, 2000, pp. 124–132.

[43] Lammerink, T. S. J., et al., "Micro-Liquid Flow Sensor," *Sensors and Actuators*, Vol. A37–38, 1993, pp. 45–50.

[44] Wu, S., et al., "MEMS Flow Sensors for Nano-Fluid Applications," *Proc. International Conference on Micro Electro Mechanical Systems (MEMS)*, Miyazaki, Japan, 2000, pp. 745–750.

[45] Wu, S., et al., "MEMS Flow Sensors for Nano-Fluid Applications," *Sensors and Actuators*, Vol. A89, 2001, pp. 152–158.

[46] Sundeen, J. E., and R. C. Buchanan, "Thermal Sensor Properties of Cermet Resistor Films on Silicon Substrates," *Sensors and Actuators*, Vol. 90, 2001, pp. 118–124.

[47] Ernst, H., A. Jachimowicz, and G. A. Urban, "High Resolution Flow Characterization in Bio-MEMS," *Sensors and Actuators*, Vol. A100, 2002, pp. 54–62.

[48] Kohl, F., et al., "A Micromachined Flow Sensor for Liquid and Gaseous Fluids," *Sensors and Actuators*, Vol. A41–42, 1994, pp. 293–299.

[49] Lyons, C., et al., "A High-Speed Mass Flow Sensor with Heated Silicon Carbide Bridges," *Proc. International Conference on Micro Electro Mechanical Systems (MEMS)*, Heidelberg, Germany, 1998, pp. 356–360.

[50] Glaninger A., et al., "Wide Range Semiconductor Flow Sensors," *Sensors and Actuators*, Vol. A85, 2000, pp. 139–146.

[51] Bracio, B. R., et al., "A Smart Thin-Film Flow Sensor for Biomedical Application," *Proc. Annual EMBS International Conference*, Chicago, IL, 2000, pp. 2800–2801.

[52] Kaltsas, G., and A. G. Nassiopoulou, "Novel C-MOS Compatible Monolithic Silicon Gas Flow Sensor with Porous Silicon Thermal Isolation," *Sensors and Actuators*, Vol. A76, 1999, pp. 133–138.

[53] Oda, S., et al., "A Silicon Micromachined Flow Sensor Using Thermopiles for Heat Transfer Measurement," *Proc. IEEE Instrumentation and Measurement Technology Conference*, Anchorage, AK, 2002, pp. 1285–1289.

[54] Häberli, A., et al., "IC Microsensors—Between System and Technology," *Proc. IEEE-CAS Region 8 Workshop on Analog and Mixed IC Design*, Baveno, Italy, 1997, pp. 36–40.

[55] Stemme, G., "A Monolithic Gas Flow Sensor with Polyimide as Thermal Insulator," *IEEE Trans. on Electron Devices*, Vol. 33, No. 10, 1986, pp. 1470–1474.

[56] Rodrigues, R. J., and R. Furlan, "Design of Microsensor for Gases and Liquids Flow Measurement," *Microelectronics Journal*, Vol. 34, 2003, pp. 709–711.

[57] Gardeniers, J. G. E., H. A. C. Tilmans, and C. C. G. Visser, "LPCVD Silicon-Rich Silicon Nitride Films for Applications in Micromechanics, Studied with Statistical Experimental Design," *Journal of Vacuum Science Technology*, Vol. A14, No. 5, 1996, pp. 2879–2892.

[58] Elwenspoek, M., and R. Wiegerink, *Mechanical Microsensors*, Berlin: Springer-Verlag, 2001.

[59] Rasmussen, A., et al., "Simulation and Optimization of a Microfluidic Flow Sensor," *Sensors and Actuators*, Vol. A88, 2001, pp. 121–132.

[60] Damean, N., P. P. L. Regtien, and M. Elwenspoek, "Heat Transfer in a MEMS for Microfluidics," *Sensors and Actuators*, Vol. 105, 2003, pp. 137–149.

[61] http://www.hsg-imit.de.

[62] http://www.sensirion.com.

[63] Lammerink, T. S. J., et al., "A New Class of Thermal Flow Sensors Using $\Delta T=0$ as a Control Signal," *Proc. International Conference on Micro Electro Mechanical Systems (MEMS)*, Miyazaki, Japan, 2000, pp. 525–530.

[64] Ebefors, T., E. Kälvesten, and G. Stemme, "Three Dimensional Silicon Triple-Hot-Wire Anemometer Based on Polyimide Joints," *Proc. International Conference on Micro Electro Mechanical Systems (MEMS)*, Heidelberg, Germany, 1998, pp. 93–98.

[65] Chen, J., J. Zou, and C. Liu, "A Surface Micromachined, Out-of-Plane Anemometer," *Proc. International Conference on Micro Electro Mechanical Systems (MEMS)*, Las Vegas, NV, 2002, pp. 332–335.

[66] Chen, J., et al., "Two-Dimensional Micromachined Flow Sensor Array for Fluid Mechanics Studies," *Journal of Aerospace Engineering*, Vol. 16, No. 2, 2003, pp. 85–97.

[67] Dittmann, D., et al., "Low-Cost Flow Transducer Fabricated with the AMANDA-Process," *Proc. Transducers*, Munich, Germany, 2001, pp. 1472–1475.

[68] de Bree, H.-E., et al., "Bi-Directional Fast Flow Sensor with a Large Dynamic Range," *Journal of Micromechanics and Microengineering*, Vol. 9, 1999, pp. 186–189.

[69] van Oudheusden, B. W., and J. H. Huijsing, "An Electronic Wind Meter Based on a Silicon Flow Sensor," *Sensors and Actuators*, Vol. A22, 1990, pp. 420–424.

[70] van Oudheusden, B. W., and A. W. van Herwaarden, "High-Sensivity 2-D Flow Sensor with an Etched Thermal Isolation Structure," *Sensors and Actuators*, Vol. 22, 1990, pp. 425–430.

[71] van Oudheusden, B. W., "Silicon Thermal Flow Sensor with a Two-Dimensional Direction Sensitivity," *Measurement Science and Technology*, Vol. 1, 1990, pp. 565–575.

[72] Durst, F., A. Al-Salaymeh, and J. Jovanovic, "Theoretical and Experimental Investigations of a Wide-Range Thermal Velocity Sensor," *Measurement Science and Technology*, Vol. 12, 2001, pp. 223–237.

[73] http://www.bosch.com.

[74] http://www.hl-planar.com.

[75] http://www.isit.fraunhofer.com.

[76] Billat, S., et al., "Micromachined Inclinometer with High Sensitivity and Very Good Stability," *Proc. Transducers*, Munich, Germany, Vol. 2, 2001, pp. 1488–1491.

[77] http://www.leister.com.

[78] http://www.sls-micro-technology.com.

[79] http://www.gesim.de.

[80] http://www.mierijmeteo.demon.nl.

[81] Cho, S. T., and K. D. Wise, "A High-Performance Microflowmeter with Built-In Self Test," *Proc. Transducers*, San Francisco, CA, 1991, pp. 400–403.

[82] Oosterbroek, R. E., et al., "Designing, Realization and Characterization of a Novel Capacitive Pressure/Flow Sensor," *Proc. Transducers*, Chicago, IL, 1997, pp. 151–154.

[83] Oosterbroek, R. E., et al., "A Micromachined Pressure/Flow Sensor," *Sensors and Actuators*, Vol. A77, 1999, pp. 167–177.

[84] Richter, M., et al., "A Novel Flow Sensor with High Time Resolution Based on Differential Pressure Principle," *Proc. International Conference on Micro Electro Mechanical Systems (MEMS)*, Orlando, FL, 1999, pp. 118–123.

[85] Nishimoto, T., S. Shoji, and M. Esashi, "Buried Piezoresistive Sensor by Means of MeV Ion Implantation," *Sensors and Actuators*, Vol. A43, 1994, pp. 249–253.

[86] Kuoni, A., et al., "Polyimide Membrane with ZnO Piezoelectric Thin Film Pressure Transducers as a Differential Pressure Liquid Flow Sensor," *Journal of Micromechanics and Microengineering*, Vol. 13, 2003, pp. S103–107.

[87] Berberig, O., et al., "The Prantl Micro Flow Sensor (PMFS): A Novel Silicon Diaphragm Capacitive Sensor for Flow-Velocity Measurement," *Sensors and Actuators*, Vol. 66, 1998, pp. 93–98.

[88] Gass, V., B. H. van der Schoot, and N. F. de Rooij, "Nanofluid Handling by Microflow Sensor Based on Drag Force Measurement," *Proc. IEEE Micro Electro Mechanical Systems*, Fort Lauderdale, FL, 1993, pp. 167–172.

[89] Zhang, L., et al., "A Micromachined Single-Crystal Silicon Flow Sensor with a Cantilever Paddle," *Proc. IEEE International Symposium on Micromechatronics and Human Science*, Nagoya, Japan, 1997, pp. 225–229.

[90] Su, Y., A. G. R. Evans, and A. Brunnschweiler, "Micromachined Silicon Cantilever Paddles with Piezoresistive Read-out for Flow Sensing," *Journal of Micromechanics and Microengineering*, Vol. 6, 1996, pp. 69–72.

[91] Ozaki, Y., et al., "An Air Flow Sensor Modeled on Wind Receptor Hairs of Insects," *Proc. of the 13th IEEE International Conference on Micro Electro Mechanical Systems (MEMS)*, Miyazaki, Japan, 2000, pp. 531–536.

[92] Fan, Z., et al., "Design and Fabrication of Artificial Lateral Line Flow Sensors," *Journal of Micromechanics and Microengineering*, Vol. 12, 2002, pp. 655–661.

[93] Fan, Z., et al., "Development of Artificial Lateral-Line Flow Sensors," *Proc. Solid-State Sensor, Actuator and Microsystems Workshop*, Hilton Head, SC, 2002, pp. 169–172.

[94] Svedin, N., et al., "A Lift-Force Flow Sensor Designed for Acceleration Insensitivity," *Sensors and Actuators*, Vol. A68, 1998, pp. 263–268.

[95] Svedin, N., et al., "A New Silicon Gas-Flow Sensor Based on Lift Force," *Journal of Microelectromechanical Systems*, Vol. 7, No. 3, 1998, pp. 303–308.

[96] Enoksson, P., G. Stemme, and E. Stemme, "A Silicon Resonant Sensor Structure for Coriolis Mass-Flow Measurements," *Journal of Microelectromechanical Systems*, Vol. 6, No. 2, 1997, pp. 119–125.

[97] Sparks, D., et al., "In-Line Chemical Concentration Sensor," *Proc. Sensor Expo and Conference*, Chicago, IL, 2003.

[98] Svedin, N., E. Stemme, G. Stemme, "A Static Turbine Flow Meter with a Micromachined Silicon Torque Sensor," *Proc. International Conference on Micro Electro Mechanical Systems (MEMS)*, Interlaken, Switzerland, 2001, pp. 208–211.

[99] Richter, A., et al., "The Electrohydrodynamic Micro Flow Meter," *Proc. Transducers*, San Francisco, CA, 1991, pp. 935–938.

[100] Richter, A., et al., "Electrohydrodynamic Pumping and Flow Measurement," *Proc. Transducers*, San Francisco, CA, 1991, pp. 271–276.

[101] Wu, J., and W. Sansen, "Electrochemical Time of Light Flow Sensor," *Sensors and Actuators*, Vol. A97–98, 2002, pp. 68–74.

[102] Merkel, T., L. Pagel, and H.-W. Glock, "Electric Fields in Fluidic Channels and Sensor Applications with Capacitance," *Sensors and Actuators*, Vol. A80, 2000, pp. 1–7.

[103] Lee, G.-B., T.-Y. Kuo, and W.-Y. Wu, "A Novel Micromachined Flow Sensor Using Periodic Flapping Motion of a Planar Jet Impinging on a V-Shaped Plate," *Experimental Thermal and Fluid Science*, Vol. 26, 2002, pp. 435–444.

[104] Leu, T.-S., et al., "Analysis of Fluidic and Mechanical Motion in MEMS by Using High Speed X-Ray Micro-Imaging Techniques," *Proc. Transducers*, Chicago, IL, 1997, pp. 149–150.

[105] Han. G., et al., "Infrared Diagnostics for Measuring Fluid and Solid Motion Inside MEMS," *Proc. Solid-State Sensor, Actuator and Microsystems Workshop*, Hilton Head, SC, 2002, pp. 185–188.

[106] Chetelat, O., and K. C. Kim, "Miniature Particle Image Velocimetry System with LED In-Line Illumination," *Measurement Science and Technology*, Vol. 13, 2002, pp. 1006–1013.

[107] Shelby, J. P., and D. T. Chiu, "Mapping Fast Flows over Micrometer-Length Scales Using Flow-Tagging Velocimetry and Single-Molecule Detection," *Analytical Chemistry*, Vol. 75, 2003, pp. 1387–1392.

[108] Eckert, S., W. Witke, and G. Gerbeth, "A New Mechano-Optical Technique to Measure Local Velocities in Opaque Fluids," *Flow Measurement and Instrumentation*, Vol. 11, 2000, pp. 71–78.

[109] Chun, D., et al., "Discussion on the Optical-Coupling in Silicon Optical-Type Micromechanical Sensors," *Proc. International Conference on Solid-State and Integrated Circuit Technology*, Beijing, China, 1998, pp. 953–956.

[110] Philip-Chandy, R., P. J. Scully, and R. Morgan, "The Design, Development and Performance Characteristics of a Fiber Optic Drag-Force Flow Sensor," *Measurement Science and Technology*, Vol. 11, 2000, pp. N31–N35.

[111] Lin, C.-H., and G.-B. Lee, "Micromachined Flow Cytometers with Embedded Etched Optic Fibers for Optical Detection," *Journal of Micromechanics and Microengineering*, Vol. 13, 2003, pp. 447–453.

[112] Lee, G.-B., C.-H. Lin, and G.-L. Chang, "Micro Flow Cytometers with Buried SU-8/SOG Optical Waveguides," *Sensors and Actuators*, Vol. A103, 2003, pp. 165–170.

[113] Fordham, E. J., et al., "Multi-Phase-Fluid Discrimination with Local Fiber-Optical Probes: I. Liquid/Liquid Flows," *Measurement Science and Technology*, Vol. 10, 1999, pp. 1329–1337.

[114] Fordham, E. J., et al., "Multi-Phase-Fluid Discrimination with Local Fiber-Optical Probes: II. Gas/Liquid Flows," *Measurement Science and Technology*, Vol. 10, 1999, pp. 1338–1346.

[115] Fordham, E. J., et al., "Multi-Phase-Fluid Discrimination with Local Fiber-Optical Probes: III. Three-Phase Flows," Vol. 10, 1999, pp. 1347–1352.

[116] Hamad, F. A., B. K. Pierscionek, and H. H. Brunn, "A Dual Optical Probe for Volume Fraction, Drop Velocity and Drop Size Measurement in Liquid-Liquid Two-Phase Flow," *Measurement Science and Technology*, Vol. 11, 2000, pp. 1307–1318.

[117] Wolffenbuttel, B. M. A., et al., "Novel Method for Non-Intrusive Measurement of Velocity and Slug Length in Two- and Three-Phase Slug Flow in Capillaries," *Measurement Science and Technology*, Vol. 13, 2002, pp. 1540–1544.

[118] Löfdahl, L., and M. Gad-el-Hak, "MEMS-Based Pressure and Shear Stress Sensors for Turbulent Flows," *Measurement Science and Technology*, Vol. 10, 1999, pp. 665–686.

[119] Schmidt, M., et al., "Design and Calibration of a Microfabricated Floating-Element Shear-Stress Sensor," *IEEE Trans. Electron Devices*, Vol. 35, 1988, pp. 750–757.

[120] Ng, K., K. Shajii, and M. Schmidt, "A Liquid Shear-Stress Sensor Fabricated Using Wafer Bonding Technology," *Proc. Transducers*, San Francisco, CA, 1991, pp. 931–934.

[121] Padmanabhan, A., et al., "A Silicon Micromachined Floating-Element Shear-Stress Sensor with Optical Position Sensing by Photodiodes," *Proc. Transducers*, Stockholm, Sweden, Vol. 2, 1995, pp. 436–439.

[122] Pan, T., et al., "Calibration of Microfabricated Shear Stress Sensors," *Proc. Transducers*, Stockholm, Sweden, Vol. 2, 1995, pp. 443–446.

[123] Oudheusen, B., and J. Huijsing, "Integrated Flow Friction Sensor," *Sensors and Actuators*, Vol. A15, 1988, pp. 135–144.

[124] von Papen, T., et al., "A MEMS Surface Fence Sensor for Wall Shear Stress Measurement in Turbulent Flow Areas," *Proc. Transducers*, Munich, Germany, Vol. 2, 2001, pp. 1476–1479.

[125] Dao, D. V., et al., "Micro Force-Moment Sensor with Six-Degree of Freedom," *Proc. International Symposium on Micromechatronics and Human Science*, Nagoya, Japan, 2001, pp. 93–98.

[126] Baltes, H., O. Paul, and O. Brand, "Micromachined Thermally Based CMOS Microsensors," *Proc. of IEEE*, Vol. 86, No. 8, 1998, pp. 1660–1678.

[127] Hauptmann, P., N. Hoppe, and A. Püttmer, "Application of Ultrasonic Sensors in the Process Industry," Review Article, *Measurement Science and Technology*, Vol. 13, 2002, pp. R73–R83.

[128] Hashimoto, K., T. Ienaka, and M. Yamaguchi, "Miniaturized Ultrasonic-Wave Velocity and Attenuation Sensors for Liquid," *Proc. Transducers*, Yokohama, Japan, 1993, pp. 700–704.

[129] Costello, B. J., S. W. Wenzel, and R. M. White, "Density and Viscosity Sensing with Ultrasonic Flexural Plate Waves," *Proc. Transducers*, Yokohama, Japan, 1993, pp. 704–707.

[130] Rajendran, V., et al., "Mass Density Sensor for Liquids Using ZnO-Film/Al-Foil Lamb Wave Device," *Proc. Transducers*, Yokohama, Japan, 1993, pp. 708–711.

[131] Yu, D., H. Y. Hsieh, and J. N. Zemel, "Microchannel Pyroelectric Anemometer," *Sensors and Actuators*, Vol. A39, 1993, pp. 29–35.

[132] Bouwstra, S., P. Kemna, and R. Legtenberg, "Thermally Excited Resonating Membrane Mass Flow Sensor," *Sensors and Actuators*, Vol. A20, 1989, pp. 213–223.

[133] Joshi, S. G., "Flow Sensors Based on Surface Acoustic Waves," *Sensors and Actuators*, Vol. 44, 1994, pp. 191–197.

[134] Xu, Y., et al., "Flexible Shear-Stress Skin and Its Application to Unmanned Aerial Vehicles," *Sensors and Actuators*, Vol. 105, 2003, pp. 321–329.

[135] Lerch, P., O. Dubochet, and P. Renaud, "From Simple Devices to Promising Microsystems Applications," *Proc. International Conference on Microelectronics*, Nis, Yugoslavia, Vol. 1, 1997, pp. 59–62.

About the Authors

Stephen Beeby is a senior research fellow in the School of Electronics and Computer Science at the University of Southampton, United Kingdom. He holds a B.Eng. (Hons.) in mechanical engineering from the University of Portsmouth, United Kingdom, and a Ph.D. in mechanical engineering from the University of Southampton.

Graham Ensell is a senior research fellow in the School of Electronics and Computer Science at the University of Southampton. He received a B.Sc. in physics from Imperial College at the University of London and a Ph.D. in medical physics from the Royal Free Hospital School of Medicine at the University of London.

Michael Kraft is a lecturer in the School of Electronics and Computer Science at the University of Southampton. He holds a Dipl.-Ing. in electronics from Alexander von Humboldt University in Erlangen, Germany, and a Ph.D. in electronics and control from Coventry University in Coventry, United Kingdom.

Neil White is a professor of intelligent sensor systems in the School of Electronics and Computer Science at the University of Southampton. A Fellow of the Institution of Electrical Engineers (IEE) and the Institute of Physics (IOP), as well as a Senior Member of the IEEE, he earned a Ph.D. in sensors at the University of Southampton and has been a full-time member of its academic staff since 1990. Professor White has researched extensively in the area of sensor technology and materials, and his work has been published in refereed journals and textbooks and has been presented at international conferences.

Contributing Authors

Barry Jones is an experienced teacher, practitioner, and researcher in the fields of measurement, sensors, transducers and actuators, instrumentation, metrology, automatic inspection, condition monitoring and preventative maintenance, and nondestructive testing and evaluation. He is the author and editor of five books and has published more than 280 papers and articles. Since 1986, he has been a professor of manufacturing metrology at Brunel University, in West London, England, and the director of the Brunel Centre for Manufacturing Metrology. He received a D.Sc. from the University of Manchester in 1985 and the Dr. Honoris Causa from the Technical University of Sofia in 2001, and holds fellowships of five professional bodies. He received a 1995 Metrology for World Class Manufacturing Award.

Christian G. J. Schabmueller received his first degree in microsystems technology from the University of Applied Sciences, Regensburg, Germany. During his studies he worked for an extended period of time at Yokogawa Electric Corporation in Tokyo, Japan. In 2001, he was awarded a Ph.D. from the University of Southampton, United Kingdom, for a thesis entitled "Microfluidic Devices for Integrated Bio/Chemical Systems." In 2002, he worked as a postdoctoral research fellow at the University of Washington in Seattle, Washington, within the Center of Applied Microsystems. Currently, he is with the Fraunhofer Institution for Silicon Technology in Itzehoe, Germany, working in the biotechnical microsystems group. In 2001, he was awarded the Ayrton Premium by the Institute of Electrical Engineers (IEE), United Kingdom.

John Tudor obtained a Ph.D. in physics from Surrey University and a B.Sc. (Eng.) in electronic and electrical engineering from University College London. In 1987 he joined Schlumberger Industries, working first at their transducer division in Farnborough and then at their research center in Paris, France. In 1990 he joined Southampton University as a lecturer. In 1994 Dr. Tudor joined ERA Technology as the microsystems program manager. In 2001, he returned to Southampton University as a senior research fellow in the School of Electronics and Computer Science to pursue university-based research in microsystems. Dr. Tudor has 25 publications and seven patents and served on the IEE Microengineering Committee for 4 years. He is both a chartered physicist and an engineer. Dr. Tudor has contributed material to Chapters 5 and 6.

Tinghu Yan received a B.Sc. in 1988 and an M.Sc. in 1991, both in mechanical engineering, and a Ph.D. in instrumentation science and engineering in 1994, all from Southeast University, China. He is currently a research fellow in the Department of Design and Systems Engineering at Brunel University, United Kingdom. His research involves the design of metallic resonant force sensors, load cells, torque transducers, and associated electronics including wireless and batteryless sensing technologies. He was formerly a lecturer and an associate professor within the Department of Mechanical Engineering at Southeast University. His other research interests include acoustic emission and nondestructive testing, condition monitoring and intelligent fault diagnosis, signal processing and pattern recognition, modeling, and optimization. He has more than 40 academic publications.

Index

3-D Builder, 50

A

Absolute pressure sensors, 121
Accelerometers, 68, 175–95
    applications, 174
    capacitive, 178, 182–85
    capped at wafer level, 68
    closed loop, 177–80
    commercial, 192–95
    companies and, 196–97
    dynamic performance, 176
    layout, 45
    lumped parameter model, 175
    mechanical sensing element, 175–76
    multiaxis, 188–91
    open loop, 176–77
    piezoelectric, 185–86
    piezoresistive, 181–82
    principle of operation, 175–80
    research prototype, 180–92
    resonant, 187–88
    tunneling, 186–87
    See also Inertial sensors
Actuators, 3
Additive materials, 11
Adhesive bonding, 31–32, 63
Amorphous silicon, 5
    deposition, 13
    LPCVD, 13
    piezoresistive, 88
Analog force-feedback, 177–79
    capacitive accelerometers in, 178
    illustrated, 180
    See also Closed loop accelerometers
Anemometers, 219–20
    data, 221
    hot wire, 219
    measurement curve, 220, 221
    operation, 219
    schematics, 220

    triple-hot-wire schematic, 249
    See also Thermal flow sensors
Aneroid barometers, 122
AnisE, 50
Anisotropic etching
    dry, 27–28
    wet, 22–27
    See also Isotropic etching
Anodic bonding, 30–31
    defined, 30
    process, 30
    setup, 31
    See also Wafer bonding
ANSYS, 50–54
    example MEMS applications and, 51
    MEMS capability, 50
    Multiphysics software, 50–51
    routine illustration, 52
    simulations, 51
    structural analysis, 54
    See also Simulation tools
Atmospheric pressure, 115
Atmospheric pressure CVD (APCVD), 13
Atomic force microscope (AFM), 164–66
Audience, this book, 5
Automatic gain control (AGC) control loop,
    200
AutoSpring, 46

B

Behavioral modeling simulation tools, 40–43
    Matlab, 40–42
    Saber, 43
    Simulink, 40–42
    Spice, 42–43
    VisSim, 43
    See also Simulation tools
Bonding, 29–32
    adhesive, 31–32
    anodic, 30–31
    eutectic, 31

Bonding (continued)
    silicon fusion, 29–30
    vacuum, 32
Boron etch stop technique, 23
Bossed diaphragms, 128–29
    analysis, 128–29
    defined, 128
    deflection of, 129
    fabrication, 131
    geometry, 128
    resistor placement on, 133
    See also Diaphragm-based pressure sensors;
Diaphragms
Bourdon tubes, 122–23
    defined, 122
    elements, 122
    use of, 123
    See also Pressure sensors
Brazing, 65
Breakdown voltage, 4

C

Calorimetric flow sensors, 220–23
    data for, 222
    measurement curve, 224
    required elements, 220
    schematics, 223
    SEM photograph, 223
    sensitivity, 222
    See also Flow sensors; Thermal flow sensors
Capacitance, pressure vs., 138
Capacitance sensors
    cross-section through SOI, 139
    differential, 93
    displacement, 92
    simplicity, 92
Capacitive accelerometers, 182–85
    in analog force-feedback loop, 178
    defined, 182
    high-performance bulk-micromachined, 183
    in open-loop mode, 183
    surface/bulk-micromachined, 186
    surface-micromachined, 184
    three-axis, 189, 190
    Yazdi/Najafi, 185
    See also Accelerometers
Capacitive microphones, 143–44
Capacitive pressure sensors, 137–39
    acceleration compensated quartz, 137
    defined, 137
    drawbacks, 138
    silicon/Pyrex, 137

    See also Pressure sensors
Capacitive techniques, 92–94
    noise and, 94
    types of, 94
    See also Mechanical transduction
        techniques
Capacitive torque sensors, 160–62
Ceramic packages, 58–59
Chemical vapor deposition (CVD), 13
    atmospheric pressure (APCVD), 13
    low-pressure (LPCVD), 13
    plasma enhanced (PECVD), 13
Closed loop accelerometers, 177–80
    actuation mechanisms, 177
    advantages, 177
    analog force-feedback, 177–79
    defined, 177
    digital feedback, 179–80
    drawback, 177
    See also Accelerometers
Commercial gyroscopes, 204–6
    ADXRS, 204–5
    die photo, 205
    illustrated, 204
    Silicon Sensing Systems, 204
Commercial micromachined accelerometers,
        192–95
    ADXL50, 192–93
    ADXL105, 193
    ADXL150, 193
    ADXL202, 193–94
    MMA1201P, 194–95
    MS7000/MS8000, 195
Commercial thermal flow sensors, 225–29
    bidirectional, 226
    data for, 230
    Fraunhofer Institute, 227
    gas, 229
    HSG-IMIT, 227–28
    Sensirion, 229
    See also Thermal flow sensors
Coriolis force, 195, 197, 198
Coriolis force flow sensors, 236–38
    cross-sectional view, 237
    data, 238
    defined, 236–37
    disadvantages, 238
    fabrication, 237
    See also Force transfer flow sensors
Corrugated diaphragms, 129
CoSolveEM, 45
CoventorWare, 44–47

Analyzer, 44, 45–47
Architect, 44
AutoSpring, 46
bundles, 44
CoSolveEM, 45
defined, 44
Designer, 44–45
Integrator, 44, 47
MemCap, 45
MemDamping, 46
MemETherm, 46
MemHenry, 46
MemMech, 45
MemPackage, 46
MemPZR, 46
MemTrans, 47
SimMan, 46
Solid Model tool, 45
See also Finite element simulation tools
Crystalline silicon, 4–5
diamond structure, 7–8
wafers, 5
See also Silicon (Si)

D

DampingMM, 47
Deposition, 12–17
amorphous silicon, 13
chemical vapor (CVD), 13
epitaxy, 14
FIB, 36
illustrated, 12
metals, 17
polysilicon, 13
silicon dioxide, 15–17
silicon nitride, 14–15
thermal growth, 12–13
Diaphragm-based pressure sensors, 123–30
Diaphragms
bossed, 128–29
corrugated, 129
defined, 123
edge conditions, 136
flat, 136
medium deflection, 127
metal, 124
rigidly clamped, 124
simply-supported, 125
small deflection, 125–27
stepped, 136
transduction mechanisms, 129–30
Die attachment, 63–64

adhesive, 63
eutectic bonding, 64
glass, 64
method comparison, 64
soldering, 63
See also Packaging
Differential pressure sensors, 121
Direct chip attach (DCA). *See* Flip chip
Doped silicon, 17
Dosing systems, 214
Double-axis gyroscopes, 203–4
Double-ended tuning fork (DETF), 101
Double-sided lithography, 18
Drag force flow sensors, 232–35
data, 235
defined, 232
disadvantages, 235
in-plane, 233
out-of-plane, 233
wind receptor hair, 234
See also Force transfer flow sensors
DRIE, 27
Druck resonant pressure sensor, 140
Dry etching, 21–22
anisotropic, 27–28
isotropic, 27
See also Etching
Dual-axis gyroscopes, 203–4
defined, 203
illustrated, 203
implementation, 203–4
See also Gyroscopes
Dynamic pressure, 116
sensing, 120–21
sensor response, 120

E

Electret microphones, 144
Electrical interconnects, 60–63
flip chip, 61–63
tape automated bonding, 61
wafer level packages, 68–70
wire bonding, 60–61
Electric power-assisted steering (EPAS)
systems, 159
Electrochemical etching, 35
Electrochemical etch stop, 35–36
process, 35
setup, 36
See also Fabrication
Electrochemical flow sensors, 240–41
data, 241

Electrochemical flow sensors (continued)
    for impedance measurement, 241
    schematic cross-section, 241
Electroplating, 33–34
    defined, 33
    illustrated, 34
    process, 33–34
Electrostatic bonding. *See* Anodic bonding
Epitaxy, 14
Etching, 21–28
    deep, 22
    dry, 21–22
    dry anisotropic, 27–28
    dry isotropic, 27
    electrochemical, 35
    FIB, 36
    illustrated, 12
    KOH, 22
    types of, 22
    wet, 21
    wet anisotropic, 22–27
    wet isotropic, 22
Eutectic bonding, 31, 64

F
Fabrication, 11–36
    bossed diaphragms, 131
    deposition, 12–17
    electrochemical etch stop, 35–36
    electroplating, 33–34
    etching, 21–28
    FIB technology, 36
    LIGA, 34
    lithography, 17–21
    porous silicon, 35
    surface micromachining, 28–29
    thick-film screen printing, 32–33
    wafer bonding, 29–32
Finite element modeling (FEM), 39–40
Finite element simulation tools, 43–56
    ANSYS, 50–54
    CoventorWare, 44–47
    defined, 43–44
    IntelliSuite, 48–50
    MEMS Pro/MEMS Xplorer, 54–55
    See also Simulation tools
First order packaging, 67
    defined, 67
    pressure sensor, 73
    stage, 73–74
    See also Packaging
Flip chip, 61–63

    advantages, 62
    bonding cross-section, 62
    defined, 61
    disadvantages, 62
    See also Packaging
Flow
    laminar, 215
    measurement principles, 213
    profiles, 216
    turbulent, 215, 247–48
Flow imaging, 243–45
    data for, 244
    use of, 243–44
Flow sensors, 213–50
    anemometers, 219–20
    bubbles and, 216
    drivers, 214
    Faraday principle, 241–42
    force transfer, 232–39
    MEMS to NEMS, 250
    micro, 214–17
    in microchannels, 215
    nonthermal time of flight, 239–41
    optical, 245–47
    packaging, 249
    with periodic flapping motion, 242–43
    pressure difference, 229–32
    thermal, 217–29
    time of flight, 223–25
Focused ion beam (FIB) technology, 36
Force balance, 143
Force sensors
    capacitive, 161
    load cells as, 156, 157
    piezoresistive, 156
    PZT, 164
    surface-machined, 157
    variable gap capacitor, 161
    See also Torque sensors
Force transfer flow sensors, 232–39
    Coriolis force, 236–38
    drag force, 232–35
    lift force, 235–36
    static turbine flow meter, 238–39
    See also Flow sensors
Frequency, 96–97

G
Gauge pressure sensors, 121
Glass die attach, 64
Glasses, 10–11
Grayscale lithography, 18–19

defined, 18
fabrication using, 19
See also Lithography
Gyroscopes, 195–206
    applications, 174
    commercial, 204–6
    double-axis, 203–4
    dual-axis, 203–4
    macroscopic mechanical, 197
    principle of operation, 195–99
    research prototypes, 199–204
    single-axis, 199–203
    surface-machined, 200
    surface-micromachined, 201
    two-gimbal structure, 199
    vibrating ring structure, 202
    vibratory rate, 198
    See also Inertial sensors

H
Hydraulic force multiplication, 116
Hysteresis, 118–19
    defined, 118
    illustrated, 119

I
Inductive coupling, 143
Inertial sensors, 173–207
    defined, 173
    future, 206–7
    introduction, 173–75
    micromachined accelerometers, 175–95
    micromachined gyroscopes, 195–206
    research, 174
InertiaMM, 47
IntelliFab, 49
IntelliSuite, 48–50
    3-D Builder, 50
    AnisE, 50
    defined, 48
    electromechanical solver, 49–50
    electrostatic solver, 49
    IntelliFab, 49
    MEMaterial, 48M   microfluidic analysis
        module, 50
    See also Finite element simulation tools
Intensity, 94–95
Interdigital (IDT) electrodes, 142
Isotropic etching
    dry, 27
    wet, 22
    See also Anisotropic etching

K
KOH etching, 22

L
Levitation, 206–7
Lift force flow sensors, 235–36
    data, 236
    defined, 235
    measurement curves, 236
    schematic, 236
    See also Force transfer flow sensors
Liftoff process, 20
    example, 20
    process flow for, 20
LIGA process, 34
Linearity
    baselines, 118
    independent, 118
    terminal-based, 118
    zero-based, 118
    See also Pressure sensors
Linear variable differential transformer
        (LVDT), 130
Lithography, 17–21
    defined, 17
    double-sided, 18
    grayscale, 18–19
    illustrated, 12
    liftoff process, 20
    photoresists, 19
    resist, 17
    topography, 20–21
    See also Fabrication
Load cells, 156, 157
    defined, 156
    distributed capacitive, 162
    principle based on compression of silicon,
        157
Long-term drift, 119
Low-pressure CVD (LPCVD), 13
    amorphous silicon, 13
    polysilicon, 13
    silicon nitride, 15
    See also Chemical vapor deposition (CVD)

M
Magnetic field resistors (MAGRES), 163
Manometers, 121–22
    defined, 121
    U-tube, 121, 122
Martian pressure sensor, 79

Materials, 7–11
   additive, 11
   gauge factors, 88
   piezoelectric, 91–92
   substrates, 7–11
Matlab, 40–42
Mechanical decoupling, 74–75
   defined, 74
   economics benefits, 75
   V-grooves, 74
Mechanical transducers, 3–4
Mechanical transduction techniques, 85–112
   capacitive, 92–94
   optical, 94–97
   piezoelectricity, 89–92
   piezoresistivity, 85–89
   resonant, 97–104
Medium deflection diaphragms, 127
MEMaterial, 48
Membrane analysis, 127–28
MemCap, 45
MemDamping, 46
MemETherm, 46
MemHenry, 46
MemMech, 45
MemPZR, 46
MEMS
   defined, 2–3
   devices, 3
   fabrication process, 5
   fabrication techniques, 11–36
   market, 3
   materials, 7–11
   simulation/design tools, 39–56
MEMS mechanical sensors, 10
   additive materials, 11
   environment protection from, 71
   mechanical isolation of chips, 71–80
   packaging, 66–80
   protection from environmental effects,
      67–70
   quartz in, 10
   See also specific types of sensors
MEMS Pro, 54–55
   defined, 54
   Suite illustration, 55
   Verification Suite, 54
   See also Finite element simulation tools
MEMS technology pressure sensors, 130–43
   capacitive, 137–39
   micromachined silicon diaphragms, 130–32
   piezoresistive, 132–37

   resonant, 139–42
   techniques, 142–43
MEMS Xplorer, 55–56
   defined, 55
   illustrated, 56
   See also Finite element simulation tools
MemTrans, 47
Metal(s)
   deposition, 17
   foil strain gauge, 85, 88
   packages, 59
Microchannels, 215, 216
Microchemical reaction system, 215
Microelectromechanical systems. *See* MEMS
Micro flow sensors, 214–17
Micromachined accelerometers.
      See Accelerometers
Micromachined gyroscopes. *See* Gyroscopes
Micromachined sensor head, 163
Micromachined silicon diaphragms, 130–32
   anisotropically etched, 131
   damping and, 132
   dynamics, 131
Micromachining
   defined, 22
   surface, 28–29
Microphones, 143–44
   capacitive, 143–44
   condenser, 144
   defined, 143
   electret, 144
   piezoelectric, 144
   piezoresistive, 144
Microsystems technology (MST), 2–3
MOS transistors, 143
Motivation, this book, 1–2
Multiaxial strain transducer, 167
Multiaxis accelerometers, 188–91
   defined, 188–89
   with modified piezoresistive pick-off, 191
   pick-off circuit, 190
   sensing element, 189
   with single proof mass, 191
   three-axis, 189, 190
   three wafers, 190
   See also Accelerometers

N

Near-field scanning optical microscopy
(NSOM), 166
Nonlinear effects, 102–4
   hard, 103

hysteresis, 104
soft, 103
Nonthermal time of flight flow sensors, 239–41
  electrochemical, 240–41
  electrohydrodynamic, 239–40
  See also Flow sensors

O

Open loop accelerometers, 176–77
  defined, 176
  illustrated, 177
  See also Accelerometers
Optical flow measurement, 245–47
  fluid velocity, 245–46
  multiphase flow detection, 246–47
  particle detection and counting, 246
Optical techniques, 94–97
  frequency, 96–97
  intensity, 94–95
  phase, 95
  polarization, 97
  spatial position, 96
  wavelength, 96
  See also Mechanical transduction techniques
Optical torque sensors, 159–60
  modified moiré fringe method, 160
  optoelectronic, 160
  torsion angle measurement, 160

P

Packaging, 57–81
  ceramic, 58
  die attachment methods, 63–64
  electrical interconnects, 60–63
  first order, 67
  introduction, 57
  low-cost approach, 77
  metal, 59
  plastic, 59
  processes, 59–66
  requirements, 66
  sealing techniques, 65–66
  second order, 67
  standard IC, 58–59
  wafer level, 67–68
Periodic flapping motion, 242–43
Phase, 95
Photoresists, 19
Piezoelectric accelerometers, 185–86
  design, 187

macroscopic, 185
PZT, 186
sensing element SEM photograph, 187
See also Accelerometers
Piezoelectric cantilevers, 163
Piezoelectricity, 89–92
  anisotropic nature, 90
  effect illustration, 89
  material properties, 91
  voltage coefficient, 90
Piezoelectric microphones, 144
Piezoresistive accelerometers, 181–82
  cross-sectional view, 181
  defined, 181
  integrated, 192
  See also Accelerometers
Piezoresistive microphones, 144
Piezoresistive pressure sensors, 132–37
  commercial availability, 133
  cross-section, 132
  dual beam configuration, 136
  fusion bonded, 135
  plan view, 132
  temperature cross-sensitivity, 134
  See also Pressure sensors
Piezoresistivity, 85–89
  defined, 85
  in silicon, 88
Pitot tube arrangement, 116
Plasma enhanced CVD (PECVD), 13
  defined, 13
  silicon nitride, 15
  See also Chemical vapor deposition (CVD)
Plastic packages, 59
Poisson's ratio, 86, 87
Polarization, 97
Polysilicon, 5
  deposition, 13
  epipoly, 14
  LPCVD, 13
  piezoresistive, 88
  See also Silicon (Si)
Porous silicon, 35
Pressure difference flow sensors, 229–32
  advantages/disadvantages, 232
  capacitive pressure sensing principles, 230
  defined, 229–30
  schematics, 231
  thermal flow sensors as, 218
  velocity, 231
  See also Flow sensors

Pressure(s)
  atmospheric, 115
  capacitance vs., 138
  direction, 114
  dynamic, 116
  in static fluid, 114
  on submerged block, 115
  units, 115
  volume vs., 117
Pressure sensing die, 72
Pressure sensors, 113–45
  absolute, 121
  aneroid barometers, 122
  Bourdon tubes, 122–23
  capacitive, 137–39
  diaphragm-based, 123–30
  differential, 121
  dynamic, 120–21
  gauge, 121
  hysteresis, 118–19
  introduction, 113–14
  linearity, 118
  long-term drift, 119
  manometers, 121–22
  Martian, 79
  MEMS technology, 130–43
  microphones, 143–44
  mounting of, 79
  optical techniques, 142
  physics of, 114–21
  piezoresistive, 132–37
  resonant, 139–42
  sensitivity, 119
  specifications, 117–19
  temperature effects, 119
  traditional, 121–23
  types of, 121
  vacuum, 123
  zero/offset, 117
Printed circuit boards (PCBs), 242

Q
Q-factor, 99–102
  calculation, 100
  defined, 99
  high, 99–100
  limitation, 100, 102
Quartz, 10
  properties, 10
  for resonant applications, 141

R
Radial stress, 126, 127
Research prototype accelerometers, 180–92
  capacitive, 182–85
  classification, 180
  multiaxis, 188–91
  piezoelectric, 185–86
  piezoresistive, 181–82
  resonant, 187–88
  tunneling, 186–87
  See also Accelerometers
Research prototype gyroscopes, 199–204
  double-axis, 103–4
  single-axis, 199–203
  See also Gyroscopes
Resist
  defined, 17
  negative, 17
  positive, 17
  profiles, 21
Resistors, thick-film, 89
Resonant accelerometers, 187–88
  bulk/surface micromachining, 188
  defined, 187–88
  high resonant frequency, 188
  See also Accelerometers
Resonant pressure sensors, 139–42
  defined, 139
  Druck, 140
  quartz, 141–42
  Yokogawa differential, 141
  See also Pressure sensors
Resonant sensors, 97–98
  block diagram, 98
  performance features, 97
Resonant techniques, 97–104
  resonator design characteristics, 99–104
  vibration excitation and detection
    mechanisms, 98–99
  See also Mechanical transduction
    techniques
Resonators
  coupling, 98
  defined, 97
  design characteristics, 99–104
  metallic, 157
  nonlinear behavior and hysteresis, 102–4
  Q-factor, 99–102
  SAW, 142–43, 159

S

Saber, 43
Scanning force microscope (SFM), 164
Scanning hall probe microscope (SHPM), 164–65
Screen printing, 32–33
    defined, 32
    process, 33
    thick-film, 32–33
Sealing techniques, 65–66
    brazing, 65
    processes, 65
    soldering, 65
    welding, 65
    See also Packaging
Second order packaging, 67
    defined, 67
    displacing sensor from, 75–77
    See also Packaging
Sensitivity, 119
Sensors, 3
Silicon-based torque sensors, 154–57
Silicon dioxide, 15–17
    properties, 16
    use of, 15
Silicon fusion bonding, 29–30
Silicon nitride, 14–15
    LPCVD, 15
    PECVD, 15
    properties, 15
    protective films, 70
Silicon on insulator (SOI), 7
    layers, 9
    See also SOI wafers
Silicon oxide, protective films, 70
Silicon (Si), 1, 7–10
    amorphous, 5
    crystalline, 4–5, 7–8
    dominance, 7
    doped, 17
    epitaxial, 14
    forms, 4–5
    needle, 76
    polycrystalline, 5
    porous, 35
    processing elements, 12
    properties, 5
    reasons for using, 4
    Young's modulus of, 86
SimMan, 46

Simulation tools, 39–56
    ANSYS, 50–54
    approaches, 39–40
    behavioral modeling, 40–43
    CoventorWare, 44–47
    finite element analysis (FEA), 43–56
    finite element modeling (FEM), 39–40
    IntelliSuite, 48–50
    Matlab, 40–42
    MEMS Pro/MEMS Xplorer, 54–56
    Saber, 43
    Simulink, 40–42
    Spice, 42–43
    system level modeling, 39
    VisSim, 43
Simulink, 40–42
    defined, 40
    sensing element model, 40
    sensor system model, 41
    See also Behavioral modeling simulation tools
Single-axis gyroscopes, 199–203
Small deflection diaphragms, 125–27
Soft adhesives, 77–79
    die mount with glass spacers, 78
    mounting of pressure sensor, 79
    RTV silicone, 77
SOI wafers, 8
    bonded and etched (BESOI), 9
    manufacturing processes, 9
    UNIBOND, 9
Soldering
    die attach, 63
    as sealing technique, 65
Solid Model tool, 45
Spatial position, 96
Spice, 42–43
    defined, 42
    illustrated use of, 43
    See also Behavioral modeling simulation tools
SpringMM, 47
Static turbine flow meter, 238–39
    data, 239
    defined, 238
    schematic, 239
    See also Force transfer flow sensors
Strain gauges
    cantilever integrated, 155
    metal foil, 85, 88
    sensitivity, 86

Strain gauges (continued)
    temperature-compensated semiconductor,
        154
Substrates, 7–11
    glasses, 10–11
    quartz, 10
    silicon, 7–10
    types of, 10
Surface acoustic wave sensors (SAWS)
    operation mode use, 91
    resonators, 142–43, 159
    technology, 154
Surface micromachining, 28–29
    drying method, 29
    structure, 28
    See also Fabrication
System level modeling, 39

T

Tactile sensors, 166–68
    integrated tooth-like, 166
    silicon micromachined array, 167
Tape automated bonding, 61
Temperature effects, 119
Thermal expansion coefficients, 78
Thermal flow sensors, 217–29
    anemometers, 219–25
    bidirectional, 226
    calorimetric, 220–23
    commercial devices, 225–29
    defined, 217
    as pressure difference sensors, 218
    research devices, 219–25
    time of flight, 223–25
    working principle schematic, 217
    See also Flow sensors
Thermal growth, 12–13
Thermistors, 218
Thick-film screen printing, 32–33
Through-wafer contacts, 69
Time of flight sensors, 223–25
    defined, 223–25
    measurement curve, 225
    nonthermal, 239–41
    uses, 225
    See also Flow sensors; Thermal flow sensors
Topography, 20–21
Torque sensors, 153–68
    AFM, 164–66
    capacitive devices, 160–62

future devices, 168
introduction, 153–54
magnetic devices, 162–64
micro, 156
optical devices, 159–60
resonant/SAW devices, 157–59
scanning probes, 164–66
SFM, 164
silicon-based devices, 154–57
size, 162
surface acoustic wave (SAW) technology,
    154
tactile, 166–68
Transducers, 3
    mechanical, 3–4
    multiaxial strain, 167
Triple-beam tuning fork (TBTF), 101
Tunneling accelerometers, 186–87
    defined, 186–87
    illustrated, 187
    See also Accelerometers
Turbulent, 215, 247–48
Turbulent flows, 215
    issues, 247–48
    measuring, 247
    See also Flows; Flow sensors

U

Universal Transducer Interface (UTI) chip,
    138–39

V

Vacuum bonding, 32
Vacuum sensors, 123
Vibration excitation/detection mechanisms,
    98–99
VisSim, 43

W

Wafer bonding, 29–32
    adhesive, 31–32
    aligning, 32
    anodic, 30–31
    eutectic, 31
    silicon fusion, 29–30
    vacuum, 32
    See also Fabrication
Wafer level packaging, 67–68
    advantages, 67–68

defined, 67
electrical interconnects, 68–70
protective coatings, 69
sealing, 68
Wavelength, 96
Wet etching, 21
anisotropic, 22–27

isotropic, 22
See also Etching
Wind receptor hair flow sensor, 234
Wire bonding, 60–61

Y

Yokogawa differential resonant pressure
sensor, 141

Recent Titles in the Artech House
Microelectromechanical Systems (MEMS) Series

Fundamentals and Applications of Microfluidics, Nam-Trung Nguyen
and Steven T. Wereley

Introduction to Microelectromechanical (MEM) Microwave Systems,
Héctor J. De Los Santos

An Introduction to Microelectromechanical Systems Engineering, Nadim Maluf

MEMS Mechanical Sensors, Stephen Beeby, Graham Ensell, Michael Kraft,
and Neil White

RF MEMS Circuit Design for Wireless Communications, Héctor J. De Los Santos

For further information on these and other Artech House titles,
including previously considered out-of-print books now available through our
In-Print-Forever® (IPF®) program, contact:

Artech House
685 Canton Street
Norwood, MA 02062
Phone: 781-769-9750
Fax: 781-769-6334
e-mail: artech@artechhouse.com

Artech House
46 Gillingham Street
London SW1V 1AH UK
Phone: +44 (0)20 7596-8750
Fax: +44 (0)20 7630-0166
e-mail: artech-uk@artechhouse.com

Find us on the World Wide Web at:
www.artechhouse.com